MICROWAVE ELECTRONICS

RONALD F. SOOHOO
Professor and Chairman
Department of Electrical Engineering

University of California, Davis

ADDISON-WESLEY PUBLISHING COMPANY
Reading, Massachusetts · Menlo Park, California · London · Don Mills, Ontario

This book is in the
ADDISON-WESLEY SERIES IN ELECTRICAL ENGINEERING

Consulting Editors:
DAVID K. CHENG
LEONARD A. GOULD
FRED K. MANASSE

Copyright © 1971 by Addison-Wesley Publishing Company, Inc.
Philippines copyright 1971 by Addison-Wesley Publishing Company, Inc.

All rights reserved. No part of this publication may be reproduced, stored in a retrieval system, or transmitted, in any form or by any means, electronic, mechanical, photocopying, recording, or otherwise, without the written permission of the publisher. Printed in the United States of America. Published simultaneously in Canada. Library of Congress Catalog Card No. 75-127893.

*To my wife Rosie and our
children, Bonita and Roland*

CONTENTS

Chapter 1 Introduction

    1.1 Scope of microwave electronics 1
    1.2 Interaction of electromagnetic fields with electron beams 2
    1.3 Interaction of electromagnetic fields with solids 4

Chapter 2 Theory of Microwaves

    2.1 Maxwell's equations 6
    2.2 Electromagnetic boundary conditions 11

Chapter 3 Waveguides and Cavities

    3.1 General solution of wave equation 17
    3.2 Relationship between longitudinal and transverse components . . . 21
    3.3 TE and TM modes 22
    3.4 Waveguides 25
    3.5 Impedance, reflection coefficient, and SWR 31
    3.6 Resonant cavities 35
    3.7 Velocity, energy, and power flow 40

Chapter 4 Electromagnetic Fields and Charges

    4.1 Inhomogeneous wave equation 45
    4.2 Motion of charges in E.M. fields 47
    4.3 Energy relations and Poynting's vector 50
    4.4 Energy, stress and momentum in E.M. fields 51

Chapter 5 Space-Charge Effects

    5.1 Interaction between stationary charges 55
    5.2 Transformations of relativity 57
    5.3 Interaction between moving charges 64

Chapter 6 Comparison Between Conventional and Microwave Tubes

    6.1 Deficiency of conventional vacuum tubes at high frequencies 66
    6.2 Lead inductance effects 66
    6.3 Transit time effects 70
    6.4 Gain-bandwidth product limitation 74

Chapter 7 Cavity-Type Microwave Tubes

- 7.1 Velocity and density modulation 76
- 7.2 The Klystron amplifier 78
- 7.3 Electron-field interaction 82
- 7.4 The Klystron oscillator 88
- 7.5 Application of Klystrons 96

Chapter 8 Microwave Devices with Periodic Structures

- 8.1 Periodic structures 100
- 8.2 Active microwave devices 109
- 8.3 Traveling wave tubes 114
- 8.4 Backward wave oscillators 121
- 8.5 Linear accelerators 125
- 8.6 Magnetron oscillators 129

Chapter 9 Beam Formation, Focusing, and Plasma Oscillations

- 9.1 Thermionic emission 134
- 9.2 Current-voltage relation 139
- 9.3 Electron gun design 142
- 9.4 Beam focusing 144
- 9.5 Space-charge effects in electron beams 154
- 9.6 Plasma oscillations 156
- 9.7 Space-charge waves 158

Chapter 10 Parametric Amplifiers

- 10.1 Parametric amplification 163
- 10.2 Analysis of parametric amplifiers 169
- 10.3 Coupling element 174

Chapter 11 Microwave Generation and Amplification in Junction Semiconductors

- 11.1 Quantum-mechanical tunneling 184
- 11.2 Tunnel diodes 191
- 11.3 Tunnel diode characteristics 196
- 11.4 Zener diodes and avalanche breakdown 202
- 11.5 Avalanche-transit-time diodes 206

Chapter 12 Microwave Generation and Amplification in Bulk Semiconductors

- 12.1 Gunn effect 213
- 12.2 Two-valley model 217
- 12.3 High-field domains 222
- 12.4 Microwave generation 224

12.5	Limited space-charge accumulation mode (LSA)	226
12.6	Microwave amplification	226

Chapter 13 Quantum Electronics

13.1	Introduction	229
13.2	Thermodynamic equilibrium and statistical steady state	230
13.3	Population inversion	231
13.4	Inversion methods	234
13.5	The three-level maser	237
13.6	The two-level maser	241
13.7	Maser structures	241
13.8	Laser structures	243
13.9	Laser cavity field	247
13.10	Types of lasers	251
13.11	Laser applications	253

Appendix A Lorentz Transformation 257

Appendix B Maxwell Boltzmann Distribution 260

Appendix C Einstein Coefficients 263

Index 265

PREFACE

The purpose of this book is to introduce the reader to the study of microwave electronics, including the analysis of microwave-solid state as well as the better known electron-beam devices. The material is based in part on notes for a first-year graduate course in microwave electronics given by the author at the California Institute of Technology and the University of California at Davis. It is to a large extent self-contained so that it can be used as a student text as well as for self-study by research workers in the field of microwave electronics. Problems are provided at the end of each chapter to further the reader's understanding and application of the text material.

Historically, microwave generation and amplification were accomplished by means of electron-beam devices such as magnetrons, Klystrons, and traveling wave tubes. In recent years, however, parametric amplifiers and solid state devices such as tunnel diodes, masers, IMPATT diodes, and Gunn effect devices have been used increasingly to perform these functions. Although much research is still in progress in the areas of IMPATT and Gunn diodes, the general theory underlying their behavior is now reasonably well understood. Consequently, treatment of these subjects here is deemed timely and appropriate.

The chapters in this book can be divided into three groups. The first group deals with the theory of microwaves and their interaction with electronic charges. Chapters in the second group analyze the operation of the electron-beam microwave devices, while those in the third group examine the behavior of solid-state microwave oscillators and amplifiers. The treatment throughout emphasizes physical understanding of the fundamental interaction between r.f. energy and electronic charges in gaseous aggregates or solids without sacrificing mathematical rigor. This approach will enable the reader to acquire not only an understanding of the workings of existing devices but also, hopefully, the ability to perceive phenomena and devices yet unborn.

Chapters 1 through 5, the first section of the book, treat the theory of microwaves and their interaction with electronic charges. Chapter 1 introduces the subject of microwave electronics and Chapters 2 and 3 deal respectively with Maxwell's equations, and waveguides and cavities. Chapter 4 examines the motions of charges in electromagnetic fields while Chapter 5 considers interaction between the charges. The latter treatment includes the behavior of particles at relativistic speeds, laying the ground work for the study of microwave particle accelerators in Chapter 8.

Chapters 6 through 9, the second group, deal with electron-beam interaction phenomena and devices. Chapter 6 examines the limitations of conventional vacuum tubes. Chapter 7 shows how a cavity-type microwave tube such as the Klystron can overcome two of these limitations, those due to lead inductance and transit-time effects. Chapter 8 discusses how microwave devices with periodic structures, such as the traveling wave tube (TWT) and backward wave oscillator (BWO), overcome a third limitation of conventional vacuum tubes as a result of the constancy of the gain-bandwidth product. Some topics common to all electron-beam devices, such as beam formation, focusing, and plasma oscillations, are discussed in Chapter 9.

Chapters 10 through 13 examine mainly the generation and amplification of microwaves in solids. Chapter 10 treats parametric amplifiers with electron beams, semiconductors, or ferrimagnetic materials as coupling elements. Chapters 11 and 12 consider microwave generation and amplification in junction and in bulk semiconductors respectively. In these chapters we discuss in some detail such topics as tunnel diodes, the IMPact ionization Avalanche Transit Time (IMPATT) phenomenon, and the Gunn effect. Finally, Chapter 13 examines the field of quantum electronics, particularly masers. By a novel extension of maser theory, we also examine the behavior of lasers. In the choice of topics for this book, emphasis has been placed on the behavior of *active* microwave electronics devices.

In the preparation of an extensive treatise of this type, I am indebted to many scientists and engineers in the field of microwave electronics for their original calculations and experiments. Of course, only part of the material covered here is new, but the style and the manner of presentation are entirely my own. I am also indebted to my wife for her perseverance, understanding, and encouragement during the course of my work on this book.

Davis, California R.F.S.
January 1971

CHAPTER 1

INTRODUCTION

In this chapter we shall discuss briefly the subject matter of microwave electronics, examining in particular the interaction between electromagnetic waves and the electronic charge and spin.

1. SCOPE OF MICROWAVE ELECTRONICS

Broadly stated, microwave electronics deals with the interaction of electromagnetic fields with the charge, and to a lesser extent the spin, of an electron. When viewed in this light, there is an underlying unity in the subject matter of microwave electronics in spite of the fact that there is a rather large number of phenomena and associated devices based on the field-electron interaction principle.

Indeed, we need only a few basic equations to understand microwave electronics, among them Maxwell's equations, the Lorentz force law, Poisson's equation, Schrödinger's equation, and the equations of state for an atomic system. In general, an electron-beam-type microwave device problem involves the simultaneous solution of Maxwell's equations and the Lorentz force law governing the dynamics of electron motion with the application of appropriate boundary conditions. Similarly, a typical problem involving generation or amplification in semiconductors usually requires the solution of Poisson's equation for a given impurity profile and Schrödinger's equation leading to some knowledge of the band structure. In the case of a maser, we must consider the distribution of electrons among the spin states in the presence of microwave fields.

One must not surmise from this underlying simplicity, however, that the solutions of these problems are necessarily simple. In fact, the geometry of most of the electron-beam-type microwave devices is such that exact solutions are far from feasible, except perhaps with the aid of a large-capacity digital computer. In practice, therefore, simplifying approximations must be made so that these problems can be solved analytically. Thus a good engineer must be able to make valid assumptions and approximations based on his physical understanding of the problem.

A large number of conventional microwave electronics devices such as Klystrons, traveling wave tubes, etc., as well as solid state microwave devices such as parametric amplifiers, tunnel diodes, IMPATT diodes, Gunn devices, and masers, will be studied in this book, not as ends in themselves but as illustrations of basic principles. Nonetheless, one must keep the basic engineering objective in mind when performing complicated calculations in microwave electronics lest they become solely exercises in applied mathematics.

2. INTERACTION OF ELECTROMAGNETIC FIELDS WITH ELECTRON BEAMS

In Maxwell's equations, the electromagnetic fields are given in terms of the currents and charges as follows:

$$\nabla \times \mathbf{E} = -\frac{\partial \mathbf{B}}{\partial t}, \tag{1.1}$$

$$\nabla \times \mathbf{H} = \mathbf{i} + \frac{\partial \mathbf{D}}{\partial t}, \tag{1.2}$$

$$\nabla \cdot \mathbf{D} = \rho, \tag{1.3}$$

$$\nabla \cdot \mathbf{B} = 0, \tag{1.4}$$

where $\mathbf{E}$ and $\mathbf{D}$ are the electric field and electric displacement, $\mathbf{H}$ and $\mathbf{B}$ are magnetic field and magnetic flux density, respectively, $\mathbf{i}$ is the area current density, and ρ is the charge density. Equations (1.1) through (1.4) are partial differential equations since $\mathbf{E}$, $\mathbf{H}$, $\mathbf{i}$, ρ, etc. may vary simultaneously with space and time. Before the above equations can be solved, they must be supplemented by another set of equations relating the quantities $\mathbf{D}$ to $\mathbf{E}$ and $\mathbf{B}$ to $\mathbf{H}$,

$$\mathbf{D} = \varepsilon \mathbf{E}, \tag{1.5}$$

$$\mathbf{B} = \mu \mathbf{H}, \tag{1.6}$$

where ε is the dielectric constant and μ is the permeability. (All equations in this book are written in MKS units, consistent with current electrical engineering practice.)

For certain media, such as plasmas in a magnetic field, ε may be a tensor rather than a scalar. In this case,

$$\mathbf{D} = \|\varepsilon\| \cdot \mathbf{E},$$

where $\|\varepsilon\|$ is the dielectric tensor. Similarly, for magnetic materials such as ferrites,

$$\mathbf{B} = \|\mu\| \cdot \mathbf{H},$$

where $\|\mu\|$ is the permeability tensor.

If ε and μ are known by either experimental measurement or theoretical computation, then Eqs. (1.1) through (1.4) can be solved with the aid of Eqs. (1.5) and (1.6).

In microwave electron-beam devices, an electron beam interacts with an electromagnetic field. Thus an equation of motion for the electrons involving field quantities and the electronic charge and mass is required. According to the

Lorentz force law, the force **F** acting on an electric charge due to the presence of the electric field **E** and the magnetic flux **B** is given by the relation

$$\mathbf{F} = q(\mathbf{E} + \mathbf{v} \times \mathbf{B}), \tag{1.7}$$

where q is the charge, equal to $-e$ for the electron, and **v** is its velocity. Now, according to Newton's second law, the force is equal to the rate of change of the angular momentum:

$$\mathbf{F} = \frac{d}{dt}(m\mathbf{v}) \tag{1.8}$$

where m is the mass of the electron.

Combining Eqs. (1.7) and (1.8), we obtain

$$\mathbf{F} = q(\mathbf{E} + \mathbf{v} \times \mathbf{B}) = \frac{d}{dt}(m\mathbf{v}). \tag{1.9}$$

At first glance, the solution of Eq. (1.9) appears straightforward, for if **E** and **B** are given, **v** can be found. Knowing **v**, we can obtain the relationship between **i** and ρ:

$$\mathbf{i} = \rho\mathbf{v}. \tag{1.10}$$

If μ and ε are given, as is usually the case in microwave electronics problems, the field quantities can be found in terms of the current density **i** and charge density ρ. Furthermore, **i** and ρ are also related by the equation of continuity:

$$\nabla \cdot \mathbf{i} + \frac{\partial \rho}{\partial t} = 0. \tag{1.11}$$

We can easily derive Eq. (1.11) by taking the divergence of both sides of Eq. (1.2), noting that $\nabla \cdot (\nabla \times \mathbf{A}) \equiv 0$ where **A** is any vector, and by using Eq. (1.3). However, the situation is not really so simple inasmuch as **i**, **E**, and **H** are related so that **E** and **H** cannot be specified independent of **i**. In other words, the presence and motion of charges give rise to electric and magnetic fields, which in turn affect the motion of the charges. Thus in many microwave electronics problems we must use an iterative process to simultaneously solve Maxwell's equations (1.1) through (1.4) and the Lorentz force law (1.9) with the aid of Eq. (1.10) and the equation of continuity (1.11). In this process, we first assume a current density distribution, then calculate the corresponding electromagnetic fields by solving Maxwell's equations subjected to appropriate boundary conditions. The resulting fields, acting on the electrons, will give rise to a certain current density **i** and charge density ρ via Eqs. (1.9), (1.10), and (1.11). In general, this current density will not be equal to the current density originally assumed for the calculation of the electromagnetic fields. However, a new current density, intermediate between the assumed and calculated ones, say, can be used to calculate a new set of field values. If this process is repeated again and again, eventually the resulting current will be essentially equal to the originally assumed current, ensuring self-consistency.

To summarize, for a given microwave electron-beam device we must solve

Maxwell's equations (1.1) through (1.4) simultaneously with Eq. (1.9), which is a combination of the Lorentz force law and Newton's second law. Inasmuch as motions of charges give rise to fields and fields in turn influence the motion of charges, many microwave electronics problems must be solved by iteration. We should also mention that, whereas the boundary conditions are in principle fixed by the device geometry, the configuration is often such that an exact solution is analytically unobtainable and judicious assumptions and approximations must be made.

The conventional microwave electronics devices discussed in this book include Klystrons, magnetrons, traveling wave tubes, backward wave oscillators, and linear electron accelerators. In the case of Klystrons, lead inductance and transit time limitations of conventional vacuum tubes are overcome by the employment of cavities with small gaps and preacceleration of the electron beam before modulation. In traveling wave tubes, an extended periodic propagating structure is used to overcome the remaining gain-bandwidth product limitation characteristic of resonance structures.

3. INTERACTION OF ELECTROMAGNETIC FIELDS WITH SOLIDS

Let us first consider the case of microwave generation and amplification in junction semiconductors. In problems of this type, we must begin by specifying the impurity profile of the semiconductor. For example, with the profile specified, Poisson's equation relating potential V and change density ρ;

$$\nabla^2 V = -\rho/\varepsilon, \tag{1.12}$$

can be solved, leading to an expression for the depletion-layer capacitance of a normally-doped junction diode. The incremental depletion-layer capacitance of such a reverse-biased diode can be used as the coupling element in a parametric amplifier.

Similarly, for a heavily-doped junction diode, a solution of Eq. (1.12) will show that the depletion layer can be made very thin, on the order of 100 Å. We can further show that for a thin potential barrier, quantum mechanical tunneling through the barrier can occur. We can demonstrate this by solving the Schrödinger equation

$$\nabla^2 \psi + \frac{2m}{\hbar^2}(E - U)\psi = 0, \tag{1.13}$$

where ψ is the wave function, $\hbar$ is Planck's constant divided by 2π, E is the total energy, and U is the coordinate-dependent potential energy. To understand the unusual I-V characteristic of a tunnel diode, including the existence of a negative resistance region responsible for microwave amplification, it is necessary to have some knowledge of the band structure of a heavily-doped semiconductor. In principle at least, the energy-band diagram can be deduced from a solution of Eq. (1.13).

Without going into further detail, we can also understand the behavior of Zener and avalanche diodes, or IMPATT (IMPact ionization Avalanche Transit Time) diodes, based on the diodes' band structure. This can be determined in principle by solving Eqs. (1.12) and (1.13).

In the case of microwave generation in bulk semiconductors, we must again solve Poisson's equation (1.12) with some assumed impuring profile. We must also solve the Schrödinger equation (1.13) to determine the band structure of a given semiconductor crystal. It turns out that for microwave generation to occur in a semiconductor, it must have overlapping valence and conduction bands. In addition, the effective mass of the conduction band must be larger than that of the valence band in order to obtain a negative slope in the semiconductor's I-V characteristic.

At present, Gunn oscillators, which require merely a dc or pulsed voltage applied across a bulk semiconductor, are coming into increasing use as microwave sources. The oscillations in these devices can be attributed to the transit of a high field domain between diode terminals. In the LSA (Limited Space-charge Accumulation) mode of operation, high-field domain formation is prohibited, making it possible to build oscillators with higher frequencies, or with higher power at a given frequency, than can be obtained with a transit time device.

In a maser (microwave amplification by stimulated emission of radiation) material, such as ruby, a dilute paramagnet is dissolved in a diamagnetic host. In the presence of an applied magnetic field, original energy levels in the crystal split into Zeemann levels containing spin-up and spin-down states. Microwave pumping and amplification can then be achieved between such states. To calculate the radiated power of the maser, we must know the relative population of the states between which amplification occurs. This can be found by solving the equations of state for the relevant maser levels. These equations are merely mathematical statements of the population of different levels as a function of time under a given pumping and amplifying condition. By a novel extension of maser theory, we will also briefly study the behavior of lasers (light amplification by stimulated emission of radiation).

PROBLEMS

1.1 Consider the case of an electron in free space acted upon by a passing plane wave. Show that the magnetic force is much smaller than the electric force if the velocity of the electron is much smaller than that of light.

1.2 Discuss the similarities and differences between the ordinary wave equation and the Schrödinger equation (1.13).

CHAPTER 2

THEORY OF MICROWAVES

In this chapter we shall study the solutions of Maxwell's equations and the wave equations along with the derivation and application of the appropriate electromagnetic boundary conditions. We shall also introduce and discuss the concepts of propagation constant, attenuation constant, phase constant, and wave impedance.

1. MAXWELL'S EQUATIONS

A. The Macroscopic Maxwell Equations

In MKS practical units, Maxwell's equations are:

$$\nabla \times \mathbf{E} = -\frac{\partial \mathbf{B}}{\partial t}, \tag{2.1}$$

$$\nabla \times \mathbf{H} = \mathbf{i} + \frac{\partial \mathbf{D}}{\partial t}, \tag{2.2}$$

$$\nabla \cdot \mathbf{D} = \rho, \tag{2.3}$$

$$\nabla \cdot \mathbf{B} = 0. \tag{2.4}$$

We can see from these equations that the electromagnetic field quantities $\mathbf{D}$, $\mathbf{E}$, $\mathbf{B}$, and $\mathbf{H}$ are dependent upon the area current density $\mathbf{i}$ and the volume charge density ρ and vice versa. Furthermore, the relationships between the electric displacement $\mathbf{D}$ and the electric field $\mathbf{E}$ and between the flux density $\mathbf{B}$ and magnetic field $\mathbf{H}$ are given by

$$\mathbf{B} = \mu \mathbf{H}, \tag{2.5}$$

$$\mathbf{D} = \varepsilon \mathbf{E}, \tag{2.6}$$

where μ and ε are usually scalars but may be complex. However, the permeability may be a tensor, usually denoted by $\|\boldsymbol{\mu}\|$, as in the case of ferrites subjected to static and r.f. magnetic fields. Likewise, the dielectric constant is a tensor $\|\boldsymbol{\varepsilon}\|$ for a plasma in a static magnetic field. It is not always easy to calculate μ and ε from first principles, although it is theoretically possible to find the exact expressions for μ and ε by solving the quantum-mechanical many-body problem. In this respect, modern advances in quantum mechanics do not invalidate Maxwell's equations but instead give us a firmer foundation for the calculation of μ and ε. In practice, μ and ε are independently measured so that Eqs. (2.1) through (2.4) can be solved in a straightforward manner with the help of Eqs. (2.5) and (2.6).

B. The Microscopic Maxwell Equations

At this point it would be instructive to inquire into the atomic origin of the field quantities **E**, **H**, etc., especially for our study later of the electric and magnetic properties of solids and gases at microwave frequencies. This discussion is based on the Maxwell-Lorentz equations first proposed by H. A. Lorentz.[1,2] These equations are the microscopic counterparts of the macroscopic Maxwell's equations (2.1) through (2.4).

The beginning of the twentieth century brought to light the electrical origin of matter, unknown to Maxwell when he developed his macroscopic equations in 1861–73. This discovery suggests that by probing into subatomic levels it should be possible to formulate the equations of electrodynamics in terms of charges in vacuum, i.e., without the introduction of the concepts of dielectric and magnetic media. Lorentz proposed a set of microscopic field equations similar in form to the macroscopic Maxwell's equations (2.1) through (2.4):

$$\nabla \times \mathbf{e} = -\mu_0 \frac{\partial \mathbf{h}}{\partial t}, \tag{2.7}$$

$$\nabla \times \mathbf{h} = \rho' \mathbf{v} + \varepsilon_0 \frac{\partial \mathbf{e}}{\partial t}, \tag{2.8}$$

$$\nabla \cdot \mathbf{e} = \frac{\rho'}{\varepsilon_0}, \tag{2.9}$$

$$\nabla \cdot \mathbf{h} = 0. \tag{2.10}$$

Note, of course, that $\mu = \mu_0$ and $\varepsilon = \varepsilon_0$ in vacuum, $i = \rho' \mathbf{v}$ is the convection current density, and ρ' is the charge density with velocity **v**. The microscopic fields **e** and **h** and charge density ρ' are not the same as their macroscopic counterparts **E**, **H**, and ρ. Indeed, whereas Eqs. (2.7) through (2.10) are atomic or molecular in nature, Eqs. (2.1) through (2.4) are essentially statistical. Hence **D**, **E**, **B**, **H**, and ρ must be correlated in some way with the average of microscopic fields and charges over a large number of molecules.

Now, let $\bar{\mathbf{e}}$ and $\bar{\mathbf{h}}$ be the average value of **e** and **k** over a volume which is physically so small that it is inaccessible to ordinary means of measurement but still large enough to contain a great many molecules. It then seems reasonable

---

[1] H. A. Lorentz, *Theory of Electrons and Its Applications to the Phenomena of Light and Radiant Heat*, B. G. Teubner, Dover Publishing Co., New York, 1952.
[2] J. H. VanVleck, *The Theory of Electric and Magnetic Susceptibilities*, Oxford University Press, New York, 1962; p. 2.

(although not obvious) that the macroscopic fields $\mathbf{E}$ and $\mathbf{B}/\mu_0$ are respectively identical to the microscopic fields $\mathbf{e}$ and $\mathbf{h}$ averaged over such a volume element:[3]

$$\mathbf{E} = \bar{\mathbf{e}} = \frac{1}{N_0} \sum_{i=1}^{N_0} \mathbf{e}_i, \tag{2.11}$$

$$\mathbf{B} = \mu_0 \bar{\mathbf{h}} = \frac{\mu_0}{N_0} \sum_{i=1}^{N} \mathbf{h}_i, \tag{2.12}$$

where $\mathbf{e}_i$ and $\mathbf{h}_i$ are the electric and magnetic fields of the ith molecule and N_0 is a large number. Similarly, the macroscopic polarization $\mathbf{P}$ and the macroscopic magnetization $\mathbf{M}$ are related to the average values of the microscopic polarization $\mathbf{p}$ and magnetization $\mathbf{m}$ as follows:

$$\mathbf{P} = N\bar{\mathbf{p}} = \frac{N}{N_0} \sum_i \mathbf{p}_i, \tag{2.13}$$

$$\mathbf{M} = N\bar{\mathbf{m}} = \frac{N}{N_0} \sum_i \mathbf{m}_i, \tag{2.14}$$

where $\bar{\mathbf{p}}$ and $\bar{\mathbf{m}}$ are microscopic fields over a physically small volume as defined above and N is the number of molecules per unit volume. It further follows that the relationships between $\mathbf{D}$, $\bar{\mathbf{e}}$, $\bar{\mathbf{p}}$ and $\mathbf{H}$, $\bar{\mathbf{h}}$, $\bar{\mathbf{m}}$, etc. are given by

$$\mathbf{D} = \varepsilon_0 \bar{\mathbf{e}} + N\bar{\mathbf{p}}, \tag{2.15}$$

$$\mathbf{H} = \bar{\mathbf{h}} - N\bar{\mathbf{m}}. \tag{2.16}$$

Averaged quantities such as $\bar{\mathbf{e}}$, $\bar{\mathbf{h}}$, $\bar{\mathbf{p}}$, and $\bar{\mathbf{m}}$ are taken for a large number of molecules at a given time. Presumably, if these quantities are to have any meaning they must be independent of the time at which the average is taken. Such an average is called an "ensemble average." If all molecules concerned are identical, then the ensemble average is equal to the "temporal average." The latter is the average value of a microscopic quantity such as $\mathbf{e}$ of a given molecule for all times.

C. Wave Equations

Equations (2.1) and (2.2) are first-order coupled equations which may be decoupled to yield second-order differential equations involving $\mathbf{E}$ or $\mathbf{H}$ only. To accomplish this, first take the curl of both sides of Eq. (2.1) to yield

$$\nabla \times \nabla \times \mathbf{E} = -\nabla \times \frac{\partial \mathbf{B}}{\partial t} = -\frac{\partial}{\partial t}(\nabla \times \mathbf{B}). \tag{2.17}$$

(The last step follows because space and time derivatives are independent.) Multiplying both sides of Eq. (2.2) by μ and noting that $\nabla \times \mathbf{B} = \mu \nabla \times \mathbf{H}$ for an

---
[3] Van Vleck, *Electric and Magnetic Susceptibilities*; pp. 3–4.

isotropic medium, we find that Eq. (2.17), with the help of Eq. (2.2), becomes

$$\nabla(\nabla \cdot \mathbf{E}) - \nabla^2 \mathbf{E} = -\mu \frac{\partial \mathbf{i}}{\partial t} - \mu \frac{\partial^2 \mathbf{D}}{\partial t^2}, \qquad (2.18)$$

where we have made use of the vector identity $\nabla \times \nabla \times \mathbf{A} = \nabla(\nabla \cdot \mathbf{A}) - \nabla^2 A$. For regions where $\mathbf{i} = \rho = 0$, as in the case of free space, Eq. (2.18) becomes

$$\nabla^2 \mathbf{E} = \mu\varepsilon \frac{\partial^2 \mathbf{E}}{\partial t^2}, \qquad (2.19)$$

where Eq. (2.6) was used. Consider a wave which has only an x-component of $\mathbf{E}$ and whose amplitude varies solely in the z-direction, i.e., a plane wave. Then the solution of Eq. (2.19) is of the general form

$$E_x = C_1 f_1(z - vt) + C_2 f_2(z + vt). \qquad (2.20)$$

The first term of Eq. (2.20) represents a forward traveling wave and the second a backward traveling wave. The distinction can be seen by noting that as t in the first term increases, z must also increase to keep $z - vt$ constant; whereas to maintain the constancy of $z + vt$ in the second term z must decrease as t increases. Because of the traveling wave character of solution (2.20), Eq. (2.19) is known as a *wave equation*. Here $v = 1/\sqrt{\mu\varepsilon}$ is the velocity of the traveling waves.

In a similar manner, we can obtain an equation for H analogous to Eq. (2.18) for $\mathbf{E}$ by first taking the curl of Eq. (2.2) to give

$$\nabla \times \nabla \times \mathbf{H} = \nabla \times \left(\mathbf{i} + \frac{\partial \mathbf{D}}{\partial t}\right). \qquad (2.21)$$

Noting that $\nabla \times (\partial \mathbf{D}/\partial t) = \varepsilon(\partial/\partial t)(\nabla \times \mathbf{E})$, we can show that Eq. (2.21), with the help of Eq. (2.1), becomes

$$\nabla(\nabla \cdot \mathbf{H}) - \nabla^2 \mathbf{H} = \nabla \times \mathbf{i} - \varepsilon \frac{\partial^2 \mathbf{B}}{\partial t^2}, \qquad (2.22)$$

where we have again used the vector identity $\nabla \times \nabla \times \mathbf{A} = \nabla(\nabla \cdot \mathbf{A}) - \nabla^2 \mathbf{A}$. According to Eq. (2.4), $\nabla(\nabla \cdot \mathbf{H}) = 0$. For regions where $\mathbf{i} = 0$, as in the case of free space, Eq. (2.22) reduces to

$$\nabla^2 \mathbf{H} = \mu\varepsilon \frac{\partial^2 \mathbf{H}}{\partial t^2}, \qquad (2.23)$$

where we have used Eq. (2.5). Equation (2.23) is identical in form to Eq. (2.19) and therefore its solution must also be of the form (2.20).

Let us now return to the general wave equations (2.18) and (2.22) for $\mathbf{E}$ and $\mathbf{H}$ respectively. To obtain the solution for $\mathbf{E}$ and $\mathbf{H}$, $\mathbf{i}$ must be expressed in terms of $\mathbf{E}$. In a conductor, $\mathbf{i}$ and $\mathbf{E}$ are related by Ohm's law:

$$\mathbf{i} = \sigma \mathbf{E}, \qquad (2.24)$$

where σ is the conductivity of the material. With the help of Eq. (2.24), Eqs. (2.18) and (2.22) become

$$\nabla(\nabla \cdot \mathbf{E}) - \nabla^2 \mathbf{E} = -\mu\sigma \frac{\partial \mathbf{E}}{\partial t} - \mu\varepsilon \frac{\partial^2 \mathbf{E}}{\partial t^2}, \tag{2.25}$$

$$\nabla(\nabla \cdot \mathbf{H}) - \nabla^2 \mathbf{H} = -\mu\sigma \frac{\partial \mathbf{H}}{\partial t} - \mu\varepsilon \frac{\partial^2 \mathbf{H}}{\partial t^2}, \tag{2.26}$$

where we have again used Eqs. (2.1), (2.5), and (2.6).

Note that Eqs. (2.25) and (2.26) would be identical in form if $\nabla \cdot \mathbf{E} = \nabla \cdot \mathbf{H} = 0$. However, unlike $\nabla \cdot \mathbf{B}$, which is always zero because there are *no* free magnetic poles, $\nabla \cdot \mathbf{E} = \rho/\varepsilon$, which is in general nonzero because there *are* free electric charges. However, we shall now show that if any charge density exists in a conductor, it would decay extremely rapidly to zero. Taking the divergence of both sides of Eq. (2.2), we find

$$\nabla \cdot (\nabla \times \mathbf{H}) = 0 = \sigma\left(\frac{\nabla \cdot \mathbf{D}}{\varepsilon}\right) + \frac{\partial(\nabla \cdot \mathbf{D})}{\partial t}, \tag{2.27}$$

using Eqs. (2.6) and (2.24). Using Eq. (2.3), we easily find the solution of Eq. (2.27):

$$\rho = \rho_0 e^{-(\sigma/\varepsilon)t},$$

with a characteristic decay time for the charge density equal to ε/σ. For a typical metal, this time constant is on the order of 10^{-18} sec, an extremely short interval compared to the microwave period on the order of 10^{-10} sec. Therefore, $\nabla \cdot \mathbf{E}$ and $\nabla \cdot \mathbf{B}$ can be set equal to zero at microwave frequencies, making Eqs. (2.25) and (2.26) identical in form.[4]

Since Eqs. (2.25) and (2.26) are essentially identical in form, we can restrict ourselves to the solution of one of them—say Eq. (2.25) for $\mathbf{E}$—without loss of generality. Again, assume a plane wave with electric field E_x and magnetic field H_y propagating in the z-direction; i.e., let

$$E_x \propto e^{j\omega t - \gamma_0 z}, \tag{2.28}$$

where ω is the frequency and γ_0 is the propogation constant, a quantity to be determined. Substituting Eq. (2.28) into Eq. (2.25), we find

$$\gamma_0 = \alpha_0 + j\beta_0 = \pm \sqrt{-\omega^2\mu\left(\varepsilon - j\frac{\sigma}{\omega}\right)}, \tag{2.29}$$

where α_0 is the attenuation constant (nepers/m) and β_0 is the phase constant (rad/m). In the derivation of Eq. (2.29), we note that $\nabla^2 \mathbf{E} = \hat{x}\partial^2 E_x/\partial z^2$. If $\sigma = 0$, as in the case of a perfect insulator, Eq. (2.29) becomes

$$\gamma_0 = j\beta_0 = \pm j\omega\sqrt{\mu\varepsilon} = \pm j\frac{\omega}{v}, \tag{2.30}$$

---

[4] Furthermore, if $\mathbf{E}$ has a sinusoidal dependence, we find that $\nabla \cdot \mathbf{E} = 0$ since $\nabla \cdot (\nabla \times \mathbf{H}) = 0 = (\sigma + j\omega\varepsilon)\nabla \cdot \mathbf{E}$ according to Eq. (2.2).

where $v = 1/\sqrt{\mu\varepsilon}$ is the velocity of propagation in the medium. For a poor conductor, σ is small. Thus we can expand the square root in Eq. (2.29) by the binomial series and retain only terms up to linear in σ to obtain

$$\gamma_0 = \alpha_0 + j\beta_0 = \pm j\omega\sqrt{\mu\varepsilon}\left(1 - j\frac{\sigma}{2\omega\varepsilon} + \cdots\right) \simeq \pm\left(\frac{\sigma}{2}\sqrt{\frac{\mu}{\varepsilon}} + j\omega\sqrt{\mu\varepsilon}\right). \quad (2.31)$$

On the other hand, for a good conductor (σ large) we can neglect the ε term as compared to the σ/ω term so that Eq. (2.29) simplifies to

$$\gamma_0 = \alpha_0 + j\beta_0 = \pm \sqrt{j\omega\mu\sigma} = \pm(1+j)\sqrt{\frac{\omega\mu\sigma}{2}}. \quad (2.32)$$

It is interesting to note that in this case $|\alpha_0| = |\beta_0|$. Furthermore, since $E_x \propto e^{-\gamma_0 z} = e^{-\alpha_0 z_e - j\beta_0 z}$, we must use the positive sign in Eqs. (2.29) through (2.32) for waves propagated in the $+z$-direction. Conversely, the negative sign must be used for waves traveling in the $-z$-direction.

To find the relationship between **E** and **H**, we note that the field quantities must satisfy Maxwell's curl equations (2.1) and (2.2) as well as the wave equations (2.25) and (2.26). In other words, only solutions satisfying both the wave and Maxwell's equations are permissible. Although every solution that satisfies Maxwell's equations also satisfies the wave equations, the reverse is not necessarily true. This is because some information is lost when the first-order coupled Maxwell's equations are uncoupled to yield the second-order wave equations.

Returning to Eq. (2.1), we have

$$-\gamma_0 E_x = -\mu j\omega H_y, \quad (2.33)$$

so that a wave impedance, denoted by Z_0, can be defined as

$$Z_0 = \frac{E_x}{H_y} = \frac{j\omega\mu}{\gamma_0} = \sqrt{\frac{\mu}{\varepsilon - j(\sigma/\omega)}}. \quad (2.34)$$

Since Z_0, the ratio of **E** to **H**, is in general complex, there is usually a phase angle between the time dependence of the electric and magnetic vectors.

For a good conductor (σ large), we can neglect the ε term relative to the σ/ω term in Eq. (2.34). Thus we have

$$Z_0 \simeq (1+j)\frac{\omega\mu}{\sqrt{2\sigma}}. \quad (2.35)$$

It is interesting to note that as $\sigma \to \infty$, $Z_0 \to 0$. Hence $|\mathbf{E}|/|\mathbf{H}|$ is very small in a good conductor. In contrast, $|\mathbf{E}|/|\mathbf{H}|$ for a plane wave in free space is 377 ohms.

2. ELECTROMAGNETIC BOUNDARY CONDITIONS

Maxwell's equations have been postulated to apply only to points in whose neighbourhood the physical properties of the medium vary continuously. However, across any surface separating one medium from another, there are sharp changes in μ, ε, and σ on the macroscopic scale. Therefore, we may expect a corresponding occurrence of discontinuities in the electric and magnetic field quantities at these boundaries. Our aim in this section is to develop the appropriate electromagnetic boundaries conditions.

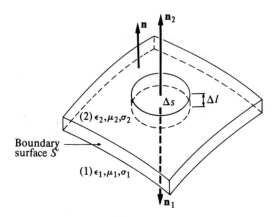

Fig. 2.1. Boundary surface for the determination of the boundary conditions of normal field components.

Replace the surface S' (Fig. 2.1) which separates medium (1) from medium (2) with a very thin transition layer of thickness Δl within which ε, μ, and σ vary rapidly but continuously from their values in medium (1) to those in medium (2). Within this layer, as within media (1) and (2), we can apply Maxwell's equations so that the field quantities and their first derivatives are continuous bounded functions of coordinate and time. Thus, the introduction of a transition layer of finite thickness enables us to treat the points inside this region as ordinary points of space about which all physical quantities vary continuously. We can then find the appropriate boundary conditions by letting Δl approach zero in our final results.

Let us draw a small right cylinder whose elements are normal to S' and whose ends lie on the surfaces of the layer so that they are separated by just the layer thickness Δl. Since $\nabla \cdot \mathbf{B} = 0$ according to Eq. (2.4), and Gauss's theorem states that

$$\oint_S \mathbf{B} \cdot \mathbf{n} \, ds = \int_V \nabla \cdot \mathbf{B} \, dv, \tag{2.36}$$

we conclude that $\oint_S \mathbf{B} \cdot \mathbf{n} \, ds = 0$ when integrated over the walls and ends of the cylinder. In Eq. (2.36), A is the area of the cylinder and V is its volume. If the area ΔS of the ends of the cylinder is made sufficiently small, we may expect $\mathbf{B}$ to have constant values over each end of the cylinder. It then follows from Eq. (2.36) and the discussion above that

$$(\mathbf{B}_1 \cdot \mathbf{n}_1 + \mathbf{B}_2 \cdot \mathbf{n}_2) \Delta S + \text{wall contribution} = 0, \tag{2.37}$$

with the contribution of the cylinder wall to the surface integral being directly proportional to Δl. As shown in Fig. 2.1, $\mathbf{n}_1$ and $\mathbf{n}_2$ are outward normals to the surface S'. In the limit as $\Delta l \to 0$, the transition layer will coincide with the surface S'. In this case, the ends of the cylinder lie just on either side of S' so that the contribution from the cylinder wall becomes vanishingly small. Thus, as $\Delta l \to 0$ and $\Delta S \to 0$, Eq. (2.37) reduces to

$$(\mathbf{B}_2 - \mathbf{B}_1) \cdot \mathbf{n} = 0, \tag{2.38}$$

where we have noted from Fig. 2.1 that $-\mathbf{n}_1 = \mathbf{n}_2 = \mathbf{n}$. According to Eq. (2.38), the *normal component of the magnetic flux density* $\mathbf{B}$ *across any surface of discontinuity is continuous*.

The boundary condition for the normal component of $\mathbf{D}$ may be derived in a similar manner. According to Eq. (2.3), we have

$$\int_V \nabla \cdot \mathbf{D}\, dv = \int_V \rho\, dv = q, \tag{2.39}$$

where q is the net charge enclosed by the volume V. Using Gauss's theorem (2.36), Eq. (2.39) becomes

$$\oint_S \mathbf{D} \cdot \mathbf{n}\, ds = q. \tag{2.40}$$

Assume that the charge is distributed throughout the transition layer with a volume charge density ρ. It then follows from Eq. (2.39) that within the small right cylinder of Fig. 2.1

$$q = \rho\, \Delta l\, \Delta S. \tag{2.41}$$

In the limit as $\Delta l \to 0$ so that the transition region coincides with the surface S', $\rho \to \infty$ as the total charge q within the cylinder remains constant because it cannot be destroyed. To circumvent this difficulty, let us introduce a new quantity, $\Omega = \rho\, \Delta l$, which is just the surface charge density in coulombs/m² (C/m²). It then follows from Eqs. (2.40) and (2.41) that

$$(\mathbf{D}_2 - \mathbf{D}_1) \cdot \mathbf{n} = \Omega. \tag{2.42}$$

We see from the above expression that the *amount of the discontinuity of the normal component of the electric displacement* $\mathbf{D}$ *across a boundary is equal to the surface charge density residing at that boundary*.

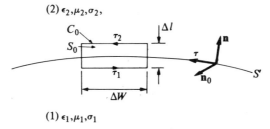

Fig. 2.2. Boundary surface for the determination of boundary conditions of tangential field components.

To study the behavior of the tangential components, we replace the cylinder of Fig. 2.1 by the rectangular path of Fig. 2.2. The rectangle has an area S_0 and a perimeter C_0 with $\mathbf{n}_0$ as its positive normal. The direction of $\mathbf{n}_0$ is determined by the direction of the circulation of C_0, using the right-hand rule.

Using Stoke's theorem, which states that $\int(\nabla \times \mathbf{E}) \cdot \mathbf{n}_0 \, ds = \int \mathbf{E} \cdot d\mathbf{l}$, Eq. (2.1) becomes

$$\int \mathbf{E} \cdot d\mathbf{l} = -\int \frac{\partial \mathbf{B}}{\partial t} \cdot \mathbf{n}_0 \, ds. \qquad (2.43)$$

From Fig. 2.2, it then follows that if the rectangle were reasonably small,

$$(\mathbf{E}_1 \cdot \boldsymbol{\tau}_1 + \mathbf{E}_2 \cdot \boldsymbol{\tau}_2)\Delta W + \text{end contributions} = -\frac{\partial \mathbf{B}}{\partial t} \cdot \mathbf{n}_0 \, \Delta W \Delta l, \qquad (2.44)$$

where $\boldsymbol{\tau}_1$ and $\boldsymbol{\tau}_2$ are the circulation vectors shown in Fig. 2.2. As the transition layer contracts to the surface S', the contribution from the end segments of the rectangle, being proportional to Δl, becomes vanishingly small. From Fig. 2.2, we can define a unit tangent vector $\boldsymbol{\tau}$ by the relation

$$\boldsymbol{\tau} = \mathbf{n}_0 \times \mathbf{n},$$

where $\mathbf{n}$ is again the positive normal to the surface S' drawn from medium (1) to medium (2). By means of the vector identity $(\mathbf{n}_0 \times \mathbf{n}) \cdot \mathbf{E} = \mathbf{n}_0 \cdot (\mathbf{n} \times \mathbf{E})$, we find that as $\Delta l \to 0$ and $\Delta W \to 0$, Eq. (2.44) becomes

$$\mathbf{n}_0 \cdot \left[\mathbf{n} \times (\mathbf{E}_2 - \mathbf{E}_1) + \lim_{\Delta l \to 0} \left(\frac{\partial \mathbf{B}}{\partial t} \Delta l \right) \right] = 0. \qquad (2.45)$$

Since the orientation of the rectangle given by the direction of $\mathbf{n}_0$ is entirely arbitrary, we find

$$\mathbf{n} \times (\mathbf{E}_2 - \mathbf{E}_1) = -\lim_{\Delta l \to 0} \frac{\partial \mathbf{B}}{\partial t} \Delta l. \qquad (2.46)$$

Making the usual assumption that the field vectors and their time derivatives are finite in magnitude, we can reduce Eq. (2.46) to

$$\mathbf{n} \times (\mathbf{E}_2 - \mathbf{E}_1) = 0, \qquad (2.47)$$

when the limit $\Delta l \to 0$ is taken. We therefore find that the *tangential component of the electric field* $\mathbf{E}$ *is continuous across the boundary*.

We can similarly deduce the boundary condition for the tangential component of $\mathbf{H}$. Instead of starting with the curl equation (2.1), which states that $\nabla \times \mathbf{E} = -(\partial \mathbf{B}/\partial t)$, we should begin with Eq. (2.2), which reads $\nabla \times \mathbf{H} = \mathbf{i} + (\partial \mathbf{D}/\partial t)$. It is then evident that the required expression can be obtained by analogy. Specifically, let us change the $\mathbf{E}$'s to $\mathbf{H}$'s and $-(\partial \mathbf{B}/\partial t)$ to $\mathbf{i} + (\partial \mathbf{D}/\partial t)$ in Eq. (2.46) to obtain

$$\mathbf{n} \times (\mathbf{H}_2 - \mathbf{H}_1) = \lim_{\Delta l \to 0} \left(\mathbf{i} + \frac{\partial \mathbf{D}}{\partial t} \right) \Delta l. \qquad (2.48)$$

Since $\partial \mathbf{D}/\partial t$ is assumed to be finite,

$$\lim_{\Delta l \to 0} \left(\frac{\partial \mathbf{D}}{\partial t} \Delta l \right) \to 0 \quad \text{as} \quad \Delta l \to 0.$$

If the current density $\mathbf{i}$ is also finite, then $\lim_{\Delta l \to 0} (\mathbf{i}\, \Delta l) \to 0$ as $\Delta l \to 0$. If, however, the total $I = \mathbf{i} \cdot \mathbf{n}_0 \,\Delta l\, \Delta W$ remains finite as $\Delta l \to 0$, then $\mathbf{i}$ may approach infinity. To circumvent this difficulty, let $\mathbf{K} = \mathbf{i}\,\Delta l$ be the surface current density as $\Delta l \to 0$. It then follows from Eq. (2.48) that

$$\mathbf{n} \times (\mathbf{H}_2 - \mathbf{H}_1) = \mathbf{K}. \tag{2.49}$$

Since $\mathbf{i} = \sigma \mathbf{E}$, it could only be infinite if σ is infinite as $\mathbf{E}$ is by necessity finite. Therefore Eq. (2.49) is applicable if the conductivity of one of the media is infinite. This assumption has an important implication. For example, to facilitate the solution of waveguide problems, waveguide walls are usually assumed to have infinite conductivity. In this case, we see from Eq. (2.49) that *for infinite conductivity, the discontinuity in the magnetic field $\mathbf{H}$ at a boundary is equal to the surface current density at the boundary.*

On the other hand, if the conductivity is assumed finite, $\mathbf{i} = \sigma \mathbf{E}$ and therefore K must be zero as $\Delta l \to 0$ since $\mathbf{E}$ is by necessity finite. It therefore follows that *the magnetic field is continuous across a boundary of finite conductivity.*

PROBLEMS

2.1 a) Starting from Maxwell's equations, derive the equation in $\mathbf{H}$ only for the case where the permittivity and conductivity are both scalars designated by ε and σ, and permeability is a tensor $\|\mu\|$.
  b) Find the relationship between the components of $\|\mu\|$ and $\mathbf{H}$ such that the equation obtained in (a) will reduce to a wave equation.
  c) If $\mu_{11} = \mu_{22} = \mu_{33}$ and $\mu_{12} = \mu_{13} = \mu_{21} = \mu_{23} = \mu_{31} = \mu_{32} = 0$, what is the spatial and time dependence of $\mathbf{H}$ for the case described in (b)?

2.2 The solution of Eq. (2.27) gives the charge density ρ as an exponentially decreasing function of time. However, footnote 4 implies that for a sinusoidally varying $\mathbf{E}$, ρ is zero. Resolve this apparent contradiction.

2.3 a) Suppose that a capacitor partially filled with a metallic slab of conductivity σ is connected across a battery via a switch. Assuming that the switch is closed at time zero, sketch the current through the circuit as a function of time.
  b) If given that the metallic slab fills the entire capacitor, sketch again the circuit current as a function of time, assuming that the switch is closed at $t = 0$.

2.4 If a uniform magnetic field increasing linearly at a time rate k from $t = 0$ is applied perpendicular to a current-carrying circuit of impedance Z, what is the change in circuit current due to the application of the field if the effective area of the circuit loop is A?

2.5 a) Consider the case where the plane surface of a perfectly conducting metal is coated with a dielectric layer of dieletric constant $\varepsilon \,(= 5\varepsilon_0)$ and thickness δ. If the electric field on the air side of the air-dielectric interface has magnitude $|E_0|$ and is oriented at 45° to the boundary, find the electric field **E** on the dielectric side of the boundary.

b) Find the tangential component of **E** and charge density at the dielectric-metal boundary.

2.6 a) Consider a parallel-plane transmission line made of perfect conductors lying in the yz-plane and separated by a distance d. Let the current flowing in the upper conductor be given by

$$I = \hat{z} I_0 \, e^{j(\omega t - \beta z)},$$

where $\hat{z}$ is a unit vector in the z-direction and β is the phase constant. Find the direction and magnitude of **H** at the inner surface of the upper and lower conductors in terms of the current.

b) Would there be a displacement current flowing in the transmission line? Explain.

CHAPTER 3

WAVEGUIDES AND CAVITIES

In this chapter we shall study the theory of waveguides and cavities and associated quantities, including the reflection coefficient, standing wave ratio, impedance, power flow, energy density, and the phase and group velocities. The treatment is essentially self-contained and emphasizes physical reasoning as well as mathemetical rigor. This chapter should be a good review for those who have studied waveguides and cavities. For those who have not, it will serve as an introduction to the more complicated structures used in microwave electronics and magnetics.

1. GENERAL SOLUTION OF WAVE EQUATION

In Chapter 2, we derived the wave equations (2.25) and (2.26) respectively for $\mathbf{E}$ and $\mathbf{H}$. If we set $\nabla \cdot \mathbf{E} = 0$ (a condition which is valid for all sinusoidal cases as explained in footnote 4 of Chapter 2) these equations reduce to

$$\nabla^2 \mathbf{E} - \gamma_0^2 \mathbf{E} = 0, \tag{3.1}$$

$$\nabla^2 \mathbf{H} - \gamma_0^2 \mathbf{H} = 0, \tag{3.2}$$

where $\gamma_0^2 = -\omega^2(\varepsilon - j\sigma/\omega)$, as first defined by Eq. (2.29), and $\mathbf{E}$ and $\mathbf{H}$ are assumed to have the time dependence $e^{j\omega t}$. Since Eqs. (3.1) and (3.2) are of the same form, we may concentrate on the solution of one of them, say Eq. (3.1), without any loss of generality.

A. Rectangular Coordinates

In rectangular coordinates, $\nabla^2 \mathbf{E} = \hat{x} \nabla^2 E_x + \hat{y} \nabla^2 E_y + \hat{z} \nabla^2 E_z$. Therefore, we can decompose Eq. (3.1) into three scalar equations which are identical in form, one for each component of $\mathbf{E}$. For example, for the z-component, we have the equation

$$\nabla^2 E_z - \gamma_0^2 E_z = 0. \tag{3.3}$$

If $\mathbf{E} \propto e^{j\omega t - \gamma z}$, Eq. (3.3) becomes

$$\frac{\partial^2 E_z}{\partial x^2} + \frac{\partial^2 E_z}{\partial y^2} + (\gamma^2 - \gamma_0^2)E_z = 0, \tag{3.4}$$

where γ is known as the propagation constant. Let the quantity $(\gamma^2 - \gamma_0^2)$ be designated by k_c^2, the square of the cutoff wave number. It then follows that $\gamma^2 = k_c^2 + \gamma_0^2$, a quantity whose physical significance we can readily appreciate. For example, for a lossless waveguide containing a lossless medium, $\gamma_0^2 = -\omega^2\mu\varepsilon$, so that

$$\gamma^2 = k_c^2 - \omega^2\mu\varepsilon. \qquad (3.5)$$

Since $E_z \propto e^{-\gamma z}$, γ is evidently real, indicating attenuation without propagation for frequencies $\omega < k_c/\sqrt{\mu\varepsilon}$. For $\omega > k_c/\sqrt{\mu\varepsilon}$, γ is imaginary, so that we have propagation without attenuation. In this respect, a waveguide is like a high-pass filter with a transition frequency ω_c from high loss to low loss equal to $k_c/\sqrt{\mu\varepsilon}$. However, we shall see later that unlike the lumped parameter high-pass filter, k_c may take on a number of discrete values, each for a particular waveguide mode. Associated with the cutoff (transition) frequency ω_c is the cutoff wavelength $\lambda_c = 2\pi/k_c$, where λ_c is the wavelength which a disturbance of frequency ω_c would have if propagated in a plane wave. It therefore follows that λ_c and ω_c are related by the equation $\lambda_c = 2\pi/\omega_c\sqrt{\mu\varepsilon}$.

If the waveguide medium is not lossless, we must generalize Eq. (3.5) to read

$$\gamma^2 = k_c^2 - \omega^2\mu\left(\varepsilon - j\frac{\sigma}{\omega}\right); \qquad (3.6)$$

where we have used Eq. (2.29). It is evident from Eq. (3.6) that in this case γ is complex, indicating propagation with attenuation. If the medium inside the waveguide is slightly absorbing, the transition from high to low loss occurs gradually over a small band of frequencies rather than at a single frequency. If the loss is sufficiently large, the transition may occur over such a broad band of frequencies that the very definition of a cutoff frequency may become meaningless.

If we let $\nabla^2_{x,y}$ equal $\partial^2/\partial x^2 + \partial^2/\partial y^2$, Eq. (3.4) becomes

$$\nabla^2_{x,y} E_z = -k_c^2 E_z. \qquad (3.7)$$

Since we are primarily interested in lossless waveguiding systems, k_c^2 would be given by Eq. (3.5). We can solve Eq. (3.7) by the well-known method of separation of variables. To do this, assume that E_z can be represented by the product of two functions, $X(x)$ and $Y(y)$:

$$E_z = X(x)Y(y). \qquad (3.8)$$

As these notations indicate, X is a function of x only and Y is a function of y only. Substituting Eq. (3.8) into Eq. (3.7), we find

$$\frac{X''}{X} + \frac{Y''}{Y} = -k_c^2. \qquad (3.9)$$

Since x and y are independent variables, the first term of Eq. (3.9) can vary independently of the second and vice versa. It then follows that the equality (3.8) can hold for all values of x and y only if the ratios X''/X and Y''/Y are constants. There are several forms for the solution of Eq. (3.8) dependent upon whether these ratios are both positive, both negative, or one positive and one negative.

If $X''/X = K_x^2$ and $Y''/Y = K_y^2$ are both positive, the solution of Eq. (3.9) is

$$X = A_1 \cosh K_x x + B_1 \sinh K_x x,$$
$$Y = C_1 \cosh K_y y + D_1 \sinh K_y y, \qquad (3.10)$$

where $K_x^2 + K_y^2 = -k_c^2$. If $X''/X = -k_y^2$ and $Y''/Y = -k_y^2$ are both negative, the solution of Eq. (3.8) is

$$X = A_2 \cos k_x x + B_2 \sin k_x x,$$
$$Y = C_2 \cos k_y y + D_2 \sin k_y y, \qquad (3.11)$$

where $-k_x^2 - k_y^2 = -k_c^2$. If $X''/X = K_x^2$ and $Y''/Y = -k_y^2$ are positive and negative respectively, the solution of Eq. (3.8) is

$$X = A_3 \cosh K_x x + B_3 \sinh K_x x,$$
$$Y = C_3 \cos k_y y + D_3 \sin k_y y, \qquad (3.12)$$

where $K_x^2 - k_y^2 = -k_c^2$. On the other hand, if $X''/X = -k_x^2$ and $Y''/Y = K_y^2$ are negative and positive respectively, the solution of Eq. (3.9) is

$$X = A_4 \cos k_x x + B_4 \sin k_x x,$$
$$Y = C_4 \cosh K_y y + D_4 \sinh K_y y, \qquad (3.13)$$

where $-k_x^2 + K_y^2 = -k_c^2$.

B. Cylindrical Coordinates

For cylindrical coordinates, Eq. (3.1) becomes

$$\frac{\partial^2 E_z}{\partial r^2} + \frac{1}{r}\frac{\partial E_r}{\partial r} + \frac{1}{r^2}\frac{\partial^2 E_z}{\partial \theta^2} = -k_c^2 E_z. \qquad (3.14)$$

Assuming that $E_z = R(r)F_\theta(\theta)$, we find

$$R'' F_\theta + \frac{R' F_\theta}{r} + \frac{F_\theta'' R}{r^2} = -k_c^2 R F_\theta. \qquad (3.15)$$

If we divide through by RF_θ and multiply through by r^2, Eq. (3.15) becomes

$$r^2 \frac{R''}{R} + \frac{rR'}{R} + k_c^2 r^2 = -\frac{F_\theta''}{F_\theta}. \qquad (3.16)$$

The left-hand side of the equation is a function of r and the right-hand side is a function of θ only. If both sides are to be equal for all values of r and θ, they must equal a constant, say n^2. Then

$$-F''_\theta = n^2 F_\theta \tag{3.17}$$

and

$$r^2 \frac{R''}{R} + \frac{rR'}{R} + k_c^2 r^2 = n^2 \tag{3.18}$$

or

$$R'' + \frac{R'}{r} + \left(k_c^2 - \frac{n^2}{r^2}\right) R = 0. \tag{3.19}$$

The solutions of Eq. (3.17) are sinusoids. Since H_z for TE (transverse electric) waves satisfies an equation identical in form to Eq. (3.14), solutions of H_z are also of this form. Solutions of Eq. (3.19) are given in terms of Bessel functions J_n and Neumann functions N_n. Thus we have

$$R = A J_n(k_c r) + B N_n(k_c r),$$
$$F_\theta = C \cos n\theta + D \sin n\theta \tag{3.20}$$

or

$$R = A_1 H_n^{(1)}(k_c r) + B_1 H_n^{(2)}(k_c r),$$
$$F_\theta = C \cos n\theta + D \sin n\theta \tag{3.21}$$

or

$$R = A_2 J_n(k_c r) + B_2 H_n^{(1)}(k_c r),$$
$$F_\theta = C \cos n\theta + D \sin n\theta, \tag{3.22}$$

where

$$H_n^{(1)}(k_c r) = J_n(k_c r) + j N_n(k_c r),$$
$$H_n^{(2)}(k_c r) = J_n(k_c r) - j N_n(k_c r); \tag{3.23}$$

$H_n^{(1)}$ and $H_n^{(2)}$ are called Hankel functions of the first and second kind. All these functions have been tabulated by Jahnke and Emde.[1]

To recapitulate, we have derived the expressions for the longitudinal component of $\mathbf{E}$, that is, E_z, by solving the wave equation (3.1) by the method of separation of variables. In rectangular coordinates, expressions for $E_z = XY$ are given by (3.10), (3.11), (3.12), or (3.13) depending on the signs for the ratios X''/X and Y''/Y. We pointed out previously in connection with Eq. (3.3) that the wave equation (3.1) can be decomposed into three scalar equations which are identical in form, one for each of the three components of $\mathbf{E}$. Therefore, the expressions given by Eqs. (3.10) through (3.13) for E_z should also be applicable for E_x and E_y. However, since there are four constants for each expression, the spatial

---

[1] E. Jahnke and F. Emde, *Tables of Functions*, Dover Publications, New York, 1943.

distribution for E_x, E_y, and E_z may be very different depending on the relative values of the constants for a given problem. Furthermore, as we shall demonstrate below, these components are not independent of each other. In fact, once E_z and H_z are specified, E_x, E_y, H_x, and H_y are likewise determined. Similar remarks can be made with regard to the solutions in cylindrical coordinates.

2. RELATIONSHIP BETWEEN LONGITUDINAL AND TRANSVERSE COMPONENTS

To find the relationship between the longitudinal and transverse components of the electromagnetic fields, we begin with Maxwell's equations (2.1) and (2.2) rewritten below:

$$\nabla \times \mathbf{E} = -\gamma_0 Z_0 \mathbf{H}, \qquad (3.24)$$

$$\nabla \times \mathbf{H} = \frac{\gamma_0}{Z_0} \mathbf{H}, \qquad (3.25)$$

where $\gamma_0 = \sqrt{-\omega^2 \mu(\varepsilon - j\sigma/\omega)}$ and $Z_0 = \sqrt{\mu/(\varepsilon - j\sigma/\omega)}$ as given respectively by Eqs. (2.29) and (2.34). Taking the cross product of $\hat{z}$ with Eq. (3.24) and using the vector identity

$$\mathbf{A} \times (\nabla \times \mathbf{B}) = \nabla(\mathbf{A} \cdot \mathbf{B}) - (\mathbf{A} \cdot \nabla)\mathbf{B} - (\mathbf{B} \cdot \nabla)\mathbf{A} - \mathbf{B} \times (\nabla \times \mathbf{A}),$$

we obtain

$$\nabla E_z + \gamma \mathbf{E} = -\gamma_0 Z_0 \hat{z} \times \mathbf{H}. \qquad (3.26)$$

Similarly, taking the cross product of $\hat{z}$ with Eq. (3.25) and again using the vector identity above, we find

$$\nabla H_z + \gamma \mathbf{H} = \frac{\gamma_0}{Z_0} \mathbf{z} \times \mathbf{E}. \qquad (3.27)$$

Taking the cross product of $\hat{z}$ with Eq. (3.27) and substituting the expression for $\hat{z} \times \mathbf{H}$ obtained from Eq. (3.26) into the resulting equation, we obtain

$$\mathbf{E}_t = -\frac{\gamma}{k_c^2} \nabla_t E_z + \frac{\gamma_0 Z_0}{k_c^2} (\hat{z} \times \nabla_t H_z), \qquad (3.28)$$

where we have decomposed $\mathbf{E}$ into $\mathbf{E}_t + \hat{z} E_z$ so that $\mathbf{E}_t$ is the component of $\mathbf{E}$ transverse to the direction of propagation. Similarly,

$$\nabla_t E_z = \hat{x}\, \partial E_z/\partial x + \hat{y}\, \partial E_z/\partial y$$

is the transverse component of the gradient of E_z, etc. Taking the cross product of $\hat{z}$ with Eq. (3.26) and substituting the expression for $\mathbf{z} \times \mathbf{E}$ obtained from Eq. (3.27) into the resulting equation, we obtain

$$\mathbf{H}_t = -\frac{\gamma}{k_c^2} \nabla_t H_z - \frac{\gamma_0}{k_c^2 Z_0} (\hat{z} \times \nabla_t E_z). \qquad (3.29)$$

Again, $\mathbf{H} = \mathbf{H}_t + \hat{z}H_z$ and $\nabla_t H_z = \hat{x}\,\partial H_z/\partial x + \hat{y}\,\partial H_z/\partial y$, etc. Thus we have succeeded in expressing the *transverse components of* **E** *and* **H** *in terms of the derivatives of their longitudinal components* as given by Eqs. (3.28) and (3.29). Since E_z and H_z as given by Eqs. (3.10) through (3.13) and (3.20) through (3.23) are solutions of the wave equation and Eqs. (3.28) and (3.29) derived from Maxwell's equations, E_t, H_t, E_z, and H_z satisfy both the wave equation and Maxwell's equations, as required.

3. TE AND TM MODES

We note from Eqs. (3.28) and (3.29) that $\mathbf{E}_t$ and $\mathbf{H}_t$ can exist if either E_z or H_z equals zero but not if both are zero. Thus we can subdivide the permissible solutions into two types.

a) A wave whose longitudinal component of the electric field E_z is equal to zero. Since the electric field of such a wave must lie entirely in the transverse plane, such a wave is customarily called a transverse electric or *TE mode*. Since H_z must be nonzero lest $\mathbf{E}_t$ and $\mathbf{H}_t$ also vanish according to Eqs. (3.28) and (3.29), TE modes are also referred to as **H**-modes, especially in physics.

b) A wave whose longitudinal component of the magnetic field H_z is equal to zero. Since its magnetic field H must correspondingly lie entirely in the transverse plane, it is designated as a transverse magnetic or *TM mode*. Physicists also refer to TM modes as *E*-modes, since E_z must be nonzero.

For TE modes ($E_z = 0$), we have from Eqs. (3.28) and (3.29)

$$\mathbf{E}_t = \frac{\gamma_0 Z_0}{k_c^2} \hat{z} \times \nabla_t H_z, \tag{3.30}$$

$$\nabla_t H_z = -\frac{k_c^2}{\gamma} \mathbf{H}_t. \tag{3.31}$$

Similarly, for TM modes ($H_z = 0$), we have from Eqs. (3.28) and (3.29)

$$\mathbf{H}_t = -\frac{\gamma_0}{k_c^2 Z_0}(\hat{z} \times \nabla_t E_z), \tag{3.32}$$

$$\nabla_t E_z = -\frac{k_c^2}{\gamma} \mathbf{E}_t. \tag{3.33}$$

We can find the relationship between $\mathbf{E}_t$ and $\mathbf{H}_t$ for TE modes by combining Eqs. (3.30) and (3.31) to get

$$\mathbf{E}_t = -Z_{\text{TE}}\,\hat{z} \times \mathbf{H}_t, \tag{3.34}$$

where

$$Z_{\text{TE}} = \frac{\gamma_0}{\gamma} Z_0 \tag{3.35}$$

is called the wave impedance for the TE modes. Taking the cross product of $\hat{z}$ and Eq. (3.34), we can transform Eq. (3.34) into

$$Z_{TE} \mathbf{H}_t = \hat{z} \times \mathbf{E}_t. \tag{3.36}$$

Similarly, for TM Modes, we can combine Eqs. (3.32) and (3.33) to get

$$Z_{TM} \mathbf{H}_t = \hat{z} \times \mathbf{E}_t, \tag{3.37}$$

where

$$Z_{TM} = \frac{\gamma}{\gamma_0} Z_0 \tag{3.38}$$

is the wave impedance of the TM modes.

Note from Eqs. (3.36) and (3.37) that since $\hat{z}$ and $\mathbf{E}_t$ are at right angles to each other, $\mathbf{H}_t$ is at right angles to $\mathbf{E}_t$ in the xy-plane and equal in magnitude to $|\mathbf{E}_t|$ divided by the characteristic impedance Z_{TE} or Z_{TM} for the TE and TM modes respectively. Of course, the relationships between the longitudinal and transverse components of the electromagnetic fields are given by Eqs. (3.30) and (3.31) for TE modes and by Eqs. (3.32) and (3.33) for TM modes.

In microwave electronics and magnetics problems, we are usually dealing with electromagnetic fields enclosed in some form of metallic waveguiding structure. It is therefore pertinent to find the specific boundary conditions on E_z and H_z at a dielectric-metallic boundary. Referring to Fig. 2.2, we shall assume that medium (1) is a perfect conductor ($\sigma = \infty$) and that medium (2) is a dielectric.[2]

From Eq. (2.47) we have

$$\mathbf{n} \times (\mathbf{E}_2 - \mathbf{E}_1) = 0. \tag{3.39}$$

It follows that

$$\mathbf{n} \times \mathbf{E}_2 = \mathbf{n} \times \mathbf{E}_1. \tag{3.40}$$

It is clear from the above equation that if we can determine the tangential component of $\mathbf{E}_1$ we can also determine the tangential component of $\mathbf{E}_2$. To find $\mathbf{E}_1$, we can use Eq. (2.25), repeated below:

$$\nabla(\nabla \cdot \mathbf{E}_1) - \nabla^2 \mathbf{E}_1 = -\mu\sigma \frac{\partial \mathbf{E}_1}{\partial t} - \mu\varepsilon \frac{\partial^2 \mathbf{E}_1}{\partial t^2}. \tag{3.41}$$

If $\mathbf{E}_1$ has a sinusoidal dependence, we find from Eq. (2.2) that $\nabla \cdot \mathbf{E}_1 = 0$, since

$$\nabla \cdot (\nabla \times \mathbf{H}_1) \equiv 0 = (\sigma + j\omega\varepsilon)\nabla \cdot \mathbf{E}_1, \tag{3.42}$$

---

[2] Although the conductivities of metals used for waveguides are not infinite, they are nevertheless sufficiently large that the field distribution within the guide, assuming infinite conductivity, closely approximates that for the actual guide. For finite conductivity, the boundary conditions would be considerably more complicated than those given here. Furthermore, because of coupling between modes as a result of losses, single-mode propagation may no longer be assumed.

where we have used Eqs. (2.6) and (2.24). Substituting Eq. (3.42) into Eq. (3.41), we have

$$\nabla^2 \mathbf{E}_1 = j\omega\mu(\sigma + j\omega\varepsilon)\mathbf{E}_1. \tag{3.43}$$

The solution of Eq. (3.43) shows that **E** decays as we go in from the surface of a metal with a characteristic decay distance, which is distance below the surface at which **E** decays to $1/e$ its value at the surface. It is equal to $1/Re\sqrt{j\omega\mu(\sigma+j\omega\varepsilon)}$, a quantity which approaches zero as σ approaches infinity. We therefore conclude that $\mathbf{E}_1$ is zero at all points inside a perfect conductor. Accordingly, Eq. (3.40) becomes

$$\mathbf{n} \times \mathbf{E}_2 = \mathbf{n} \times \mathbf{E}_1 = 0$$

at the surface of a perfect conductor. Since $\hat{z}E_z$ is always tangential to the ordinary waveguide wall, we find for TM modes

$$\boxed{E_z = 0 \text{ at surface of a perfect conductor.}} \tag{3.44}$$

Let us now turn our attention to the components of **H**. From Eq. (2.38) we have

$$\mathbf{n} \cdot \mathbf{B}_2 = \mathbf{n} \cdot \mathbf{B}_1. \tag{3.45}$$

From Eq. (2.26) we again conclude that $\mathbf{B}_1$ is zero for all points within the perfect conductor. Thus, since $\mathbf{B}_2 = \mu\mathbf{H}_2$, we have

$$\mathbf{n} \cdot \mathbf{B}_2 = \mathbf{n} \cdot \mathbf{H}_2 = 0. \tag{3.46}$$

Furthermore, according to Eq. (3.31), we find

$$\nabla_t H_{z2} = -\frac{k_c^2}{\gamma} \mathbf{H}_{t2}. \tag{3.47}$$

It follows that for the component normal to the conductor surface

$$\frac{\partial H_{z2}}{\partial n} = -\frac{k_c^2}{\gamma}(\mathbf{n} \cdot \mathbf{H}_{t2}), \tag{3.48}$$

where **n** is the surface normal. Combining Eqs. (3.46) and (3.48), we find for TE modes

$$\boxed{\frac{\partial H_z}{\partial n} = 0 \text{ at surface of a perfect conductor.}} \tag{3.49}$$

4. WAVEGUIDES

A. Rectangular Waveguides

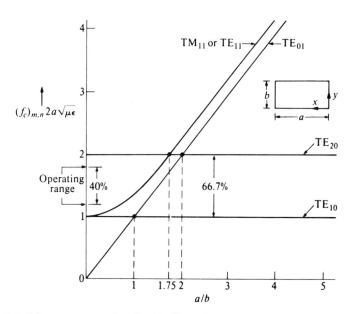

Fig. 3.1. Cutoff frequency vs. ratio of guide dimensions.

Refer now to the rectangular waveguide of Fig. 3.1. From Eq. (3.11) we have for the TM modes

$$E_z = A \sin k_x x \sin k_y y, \tag{3.50}$$

where $A = B_2 D_2$. Since $E_z = 0$ at $x = 0$ and $E_z = 0$ at $y = 0$ according to Eq. (3.44) and Fig. 3.1, A_2 and C_2 have been set equal to zero. Similarly, we have from Eq. (3.11) for TE modes

$$H_z = B \cos k_x x \cos k_y y, \tag{3.51}$$

where $B = A_2 C_2$. Since $\partial H_z/\partial x = 0$ at $x = 0$ and $\partial H_z/\partial y = 0$ at $y = 0$ according to Eq. (3.49), B_2 and D_2 have been set equal to zero.

The requirement of $E_z = 0$ at $x = a$ and at $y = b$ determines the permissible values of k_x and k_y for TM modes. Accordingly, we have from Eq. (3.50)

$$k_x a = m\pi \longrightarrow k_x = m\pi/a, \tag{3.52}$$

$$k_y b = n\pi \longrightarrow k_y = n\pi/b, \tag{3.53}$$

where m and n are integers. The requirement of $\partial H_z/\partial x = 0$ at $x = a$ and at $y = b$ leads to the *same permissible values of k_x and k_y for TE modes* as those given for TM modes.

From the definitions connected with Eq. (3.10) we have $k_c^2 = k_x^2 + k_y^2$, so that

$$(k_c)_{mn} = \sqrt{k_x^2 + k_y^2} = \sqrt{(m\pi/a)^2 + (n\pi/b)^2} \tag{3.54}$$

for both TM and TE modes. The cutoff wavelength and frequency may then be written as

$$(\lambda_c)_{mn} = \frac{2\pi}{k_c} = \frac{2}{\sqrt{(m/a)^2 + (n/b)^2}} = \frac{2ab}{\sqrt{(mb)^2 + (na)^2}}, \tag{3.55}$$

$$(f_c)_{mn} = \frac{v}{(\lambda_c)_{mn}} = \frac{k_c}{2\pi\sqrt{\mu\varepsilon}} = \frac{1}{2\sqrt{\mu\varepsilon}}\sqrt{\left(\frac{m}{a}\right)^2 + \left(\frac{n}{b}\right)^2}, \tag{3.56}$$

where v is the velocity of the wave in an unbounded medium of dielectric constant ε and permeability μ. There are, therefore, doubly infinite numbers of possible modes of each type corresponding to all the combinations of integers m and n. A TM mode with m half-wave variations in the x-direction and n half-wave variations in the y-direction is denoted as a TM_{mn} mode. Similar notation is used for the TE_{mn} modes. Note that for the TE_{mn} modes either m or n (but not both) can be zero according to Eqs. (3.51), (3.30), and (3.31). However, for the TM_{mn} modes neither m nor n can be zero because otherwise the fields would vanish completely according to Eqs. (3.50), (3.32), and (3.33). A transverse electromagnetic or TEM mode cannot propagate in a hollow waveguide, however, since by definition $E_z = H_z = 0$ for such a mode.

Thus the lowest TE mode, according to Eq. (3.55), has a cutoff wavelength of

$$(\lambda_c)_{TE_{10}} = 2a, \tag{3.57}$$

and the lowest TM mode has a cutoff wavelength of

$$(\lambda_c)_{TM_{11}} = \frac{2ab}{\sqrt{a^2 + b^2}}. \tag{3.58}$$

Note that $(\lambda_c)_{TM_{11}} < (\lambda_c)_{TE_{11}}$, so that TE_{10} represents the dominant or lowest-order mode of a rectangular guide.

For computational purposes we may put Eq. (3.56) into the form

$$(f_c)_{mn} = \frac{1}{2a\sqrt{\mu\varepsilon}}\sqrt{m^2 + [n(a/b)]^2}. \tag{3.59}$$

For TE_{10}, $(f_c)_{10} = 1/2a\sqrt{\mu\varepsilon}$ and for TE_{20}, $(f_c)_{20} = 1/a\sqrt{\mu\varepsilon}$, so that $(f_c)_{20}/(f_c)_{10} = 2$. Therefore, the maximum operating frequency range for a rectangular waveguide would be

$$\frac{2(f_c)_{10} - (f_c)_{10}}{\left(\frac{2(f_c)_{10} + (f_c)_{10}}{2}\right)} \times 100 = 66.7\%$$

if the waveguide were to propagate the dominant TE_{10} mode only. If $n \neq 0$, then $(f_c)_{mn}$ is a function of a/b.

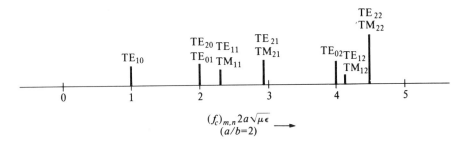

Fig. 3.2. Line spectrum of a rectangular waveguide.

We note from Fig. 3.1 that to obtain maximum operating range for the TE_{10} mode, a/b should be about 2, as is the case in practice. Note also from Fig. 3.1 that in this case

$$(f_c)_{TE_{01}} = (f_c)_{TE_{20}} < (f_c)_{TM_{11}} \text{ or } (f_c)_{TE_{11}}.$$

If we let $a/b = 2$, Eq. (3.59) becomes

$$(f_c)_{mn} = \frac{1}{2a\sqrt{\mu\varepsilon}}\sqrt{m^2 + (2n)^2} \qquad (3.60)$$

Using Eq. (3.60), we may plot $(f_c)_{mn}$ of different modes on the frequency scale as shown in Fig. 3.2.

If $f > (f_c)_{mn}$, then modes TE_{mn} or TM_{mn}, if excited, will propagate. Thus a waveguide acts like a multiple cutoff filter. Since for propagation $f > (f_c)_{TE_{10}}$, it follows that

$$\lambda < (\lambda_c)_{TE_{10}} = 2a \quad \text{or} \quad a > \lambda/2$$

if the wave is to be above cutoff. But $(f_c)_{TE_{20}} = 2(f_c)_{TE_{10}}$, so that

$$f < 2(f_c)_{TE_{10}} \quad \text{or} \quad \lambda > \frac{(\lambda_c)_{TE_{10}}}{2} = a$$

if the TE_{20} mode is to be below cutoff. Therefore, to ensure single-mode propagation, we need $\lambda/2 < a < \lambda$. Thus the cross-sectional dimensions of waveguides are usually on the order of a wavelength at microwave frequencies.

B. Cylindrical Waveguides

We now turn to the study of cylindrical waveguides. For TM modes, we have from Eq. (3.20)

$$E_z = AJ_n(k_c r)\begin{cases}\cos n\theta \\ \sin n\theta\end{cases}. \qquad (3.61)$$

The Neumann function N_n is not included because it is infinite at $r = 0$, a point contained in the waveguide. Whether $\cos n\theta$ or $\sin n\theta$ is used is a matter of choice, since one can be obtained from the other merely by a 90° rotation. Likewise, for TE modes, we have from Eq. (3.20)

$$H_z = BJ_n(k_c r) \begin{cases} \cos n\theta \\ \sin n\theta \end{cases}. \tag{3.62}$$

Again, N_n has not been included because it is infinite at $r = 0$. For TM modes, $E_z = 0$ at $r = a$ according to Eq. (3.44), so that from Eq. (3.61)

$$J_n(k_c a) = 0. \tag{3.63}$$

Since the Bessel function $J_n(x)$ has an infinite number of values of x for which it becomes zero, Eq. (3.63) may be satisfied by any one of these. Thus, if p_n is the pth root of $J_n(k_c a) = 0$, Eq. (3.63) is satisfied if

$$(k_c)_{nl} = \frac{p_{nl}}{a}. \tag{3.64}$$

Equation (3.64) defines a doubly infinite set of possible values for k_c, one for each combination of integers n and l denoted by TM_{nl}. The integer n describes the number of half-wavelength variations *circumferentially* and l describes the number of variations *radially*. The cutoff wavelength and frequency for the TM_{nl} modes are

$$(\lambda_c)_{TM_{nl}} = 2\pi/k_c = 2\pi a/p_{nl}, \tag{3.65}$$

$$(f_c)_{TM_{nl}} = k_c/2\pi\sqrt{\mu\varepsilon} = p_{nl}/2\pi a\sqrt{\mu\varepsilon}. \tag{3.66}$$

The lowest value of p_{nl} is the first root of the zero-order Bessel function denoted by p_{01} and is equal to 2.405, so that TM_{01} is the lowest TM_{nl} mode. From Eq. (3.65) we find that $\lambda_c = 2.61a$. Note that this wavelength is measured at the velocity of light ($= 1/\sqrt{\mu\varepsilon}$) in an unbounded dielectric with dielectric constant ε and permeability μ.

For TE modes, the required boundary condition is that normal derivatives of H_z be zero at $r = a$, according to Eq. (3.49). It then follows from Eq. (3.62) that

$$J'_n(k_c a) = 0. \tag{3.67}$$

If p'_{nl} is the pth root of $J'_n(k_c a) = 0$, Eq. (3.67) is satisfied when

$$(k_c)_{nl} = \frac{p'_{nl}}{a}. \tag{3.68}$$

Equation (3.68) again defines a doubly infinite number of possible TE_{nl} modes. Their cutoff wavelengths and frequencies are

$$(\lambda_c)_{TE_{nl}} = \frac{2\pi}{p'_{nl}} a, \tag{3.69}$$

$$(f_c)_{TE_{nl}} = \frac{p'_{nl}}{2\pi a\sqrt{\mu\varepsilon}} \tag{3.70}$$

The lowest value of p'_{nl} is not p'_{01} but rather p'_{11}, which is equal to 1.84. Thus the TE_{11} mode has the lowest cutoff frequency of all transverse electric modes in a given diameter of guide. From Eq. (3.69) we find that this frequency corresponds to a cutoff wavelength of $3.41a$; thus the TE_{11} mode also has the lowest cutoff frequency in a circuit guide.

The transverse components of **E** and **H** may be derived from Eqs. (3.30) and (3.31) for TE modes and from Eqs. (3.32) and (3.33) for TM modes.

From the above discussion, we see that

$$(f_c)_{TM_{01}}/(f_c)_{TE_{11}} = 2.405/1.84 = 1.308,$$

or a maximum single-mode operating range of $2(3.41 - 2.61)100/(3.41 + 2.61) = 26.6\%$ for a circular waveguide in the TE_{11} mode. This, coupled with the fact that the shape of the rectangular guide fixes the polarization of the wave, has resulted in widespread use of rectangular guides in microwave test equipment. In applications that require a circular waveguide, one must take care to ensure that the polarization does not change because of the presence of obstacles.

C. Coaxial Lines

We shall now turn to a brief discussion of coaxial lines, which are often used at the lower end of the microwave frequency spectrum. Since the center conductor carries current, transverse **H** lines may form concentric circles that enclose the current. Transverse **E** lines, being perpendicular to the **H** circles, would be radial starting and ending on charges residing on the inner and outer conductors.

In addition to the TEM mode described above, TM_{np} and TE_{np} modes can also exist in a coaxial line. We must now find the cutoff frequencies of the lowest TM and TE modes in order to adjust the dimensions of the line to avoid the excitation of these modes for a given range of operating frequencies.

From our general solution (3.20), we have for TM modes

$$A_n J_n(k_c r_i) + B_n N_n(k_c r_i) = 0,$$
$$A_n J_n(k_c r_o) + B_n N_n(k_c r_o) = 0,$$
(3.71)

or

$$N_n(k_c r_i)/J_n(k_c r_i) = N_n(k_c r_o)/J_n(k_c r_o),$$
(3.72)

since $E_z|_{r=r_i, r_o} = 0$ (see Fig. 3.3) according to Eq. (3.44).

For TE modes, the derivative of H_z normal to the two conductors must be zero at r_i and r_o, according to Eq. (3.49). Thus we have

$$C_n J'_n(k_c r_i) + D_n N'_n(k_c r_i) = 0,$$
$$C_n J'_n(k_c r_o) + D_n N'_n(k_c r_o) = 0,$$
(3.73)

or

$$N'_n(k_c r_i)/J'_n(k_c r_i) = N'_n(k_c r_o)/J'_n(k_c r_o).$$
(3.74)

Solutions of the transcendental equations (3.72) and (3.74) will give k_c as a function of r_i and r_o.[3]

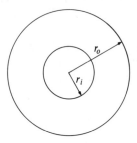

Fig. 3.3. A coaxial line.

If the radius of curvature is large, we may imagine the conductors cut radially in Fig. 3.3 and flattened into parallel planes. Recalling that for TM modes, $E_z = 0$ at $r = r_i, r_o$, we would not expect a TM_{op} mode to propagate if

$$(r_o - r_i) < p\lambda/2, \tag{3.75}$$

where p is an integer. Thus the cutoff wavelength is

$$\lambda_c = \frac{2}{p}(r_o - r_i). \tag{3.76}$$

Actual solution of Eq. (3.72) indicates that Eq. (3.76) is correct near $r_o/r_i \simeq 1$. Thus the lowest-order TM mode is TM_{01}.

For the lowest-order TE wave (TE_{11}), we would expect cutoff to occur when the average circumference of the coaxial line approximately equals one wavelength. Thus these modes will not propagate if

$$2\pi \frac{r_o + r_i}{2} < n\lambda, \tag{3.77}$$

where n is an integer. The cutoff wavelength for the TE_{1n} mode is therefore

$$\lambda_c = \frac{2\pi}{n}\left(\frac{r_o + r_i}{2}\right). \tag{3.78}$$

Calculation shows that Eq. (3.78) is valid to within 4% for values of r_o/r_i up to 5. Since

$$\frac{2\pi}{n}\left(\frac{r_o + r_i}{2}\right) > \frac{2}{p}(r_o - r_i) \tag{3.79}$$

for $p = n = 1$, TE_{11} is the lowest mode (aside from the TEM mode) found in a coaxial line.

---

[3] For the TM case, a plot of $\lambda_c/2(r_o - r_i)$ vs. r_o/r_i is given in S. Ramo et al., *Fields and Waves in Communication and Electronics*, John Wiley and Sons, New York, 1965; p. 447.

5. IMPEDANCE, REFLECTION COEFFICIENT, AND STANDING WAVE RATIO

In the last section we studied the field distribution in a waveguide in some detail. We shall now ascertain the impedance, reflection coefficient, and standing wave ratio (SWR) at a given cross-sectional plane of a waveguide that terminates in an arbitrary impedance or contains some form of discontinuity, such as irises or junctions between waveguides of different cross sections. All these obstacles have one feature in common: they set up reflected waves traveling in the negative z-direction. These waves modify the field distribution both preceding and following the obstacles, thereby setting up standing waves.

If we change the direction of propagation from $+z$ to $-z$, the sign of γ changes because the forward and reflected waves vary respectively as $e^{j\omega t - \gamma z}$ and $e^{j\omega t + \gamma z}$. Therefore, the signs of Z_{TE} and Z_{TM} also change according to Eqs. (3.35) and (3.38). This in turn produces a sign change in the relationship between the transverse components of $\mathbf{E}$ and $\mathbf{H}$ via Eqs. (3.36) and (3.37), and in the relationship between both H_t and H_z via Eq. (3.31) and E_t and E_z via Eq. (3.33). Designating the vector amplitude of the field components traveling in the $+z$- and $-z$-directions by the subscripts $+$ and $-$ respectively, we have for the total field

$$\mathbf{E} = (\mathbf{E}_{t+}e^{j\omega t - \gamma z} + \mathbf{E}_{t-}e^{j\omega t + \gamma z}) + \hat{z}(E_{z+}e^{j\omega t - \gamma z} - E_{z-}e^{j\omega t + \gamma z}), \tag{3.80}$$

$$\mathbf{H} = (\mathbf{H}_{t+}e^{j\omega t - \gamma z} - \mathbf{H}_{t-}e^{j\omega t + \gamma z}) + \hat{z}(H_{z+}e^{j\omega t - \gamma z} + H_{z-}e^{j\omega t + \gamma z}). \tag{3.81}$$

If more than one mode exists in the waveguide, the right-hand sides of Eqs. (3.80) and (3.81) should be summed over all modes. A check of the relative signs for the various terms in Eqs. (3.80) and (3.81) shows that they indeed satisfy the requirements enumerated above.

The question as to what is meant by the "characteristic" impedance of a waveguide now arises. Unlike the ordinary two-wire transmission line or the coaxial line propagating the TEM mode, the waveguide modes have transverse as well as longitudinal components of fields. In the case of a TEM mode, we can easily define an unambiguous impedance equal to the ratio of either the transverse electric field to the transverse magnetic field or that of the voltage between conductors to the current in the conductor. But it is more difficult to define a waveguide's characteristic impedance because of the presence of the longitudinal field component of Eqs. (3.80) and (3.81). The matter is further complicated by the fact that the field components are not constant over the cross section of the waveguide. For example, according to Eqs. (3.30) and (3.51), we have for the TE_{10} mode

$$\mathbf{E}_t = \hat{y}\frac{\gamma_0 Z_0}{k_c^2}\frac{\partial H_z}{\partial x} = -\hat{y}B\frac{\gamma_0 Z_0}{k_x}\sin k_x x, \tag{3.82}$$

recalling that $k_c^2 = k_x^2 + k_y^2 = k_x^2$ in this case. It follows that the voltage across the top and bottom plates of the waveguide, being equal to $-\int \mathbf{E} \cdot d\mathbf{l}$, is not a constant but rather a function of x; it is maximum at the center and zero at the side walls. Similar remarks may be made regarding the magnetic field and consequently the current distribution.

We can formulate a number of different definitions for the characteristic impedance of a waveguide, each useful for certain purposes. Schelkunoff gives three, based on the maximum voltage V, total longitudinal current I, and power P:[4]

$$Z_{V,I} = \frac{V}{I} = \frac{\pi}{2} \frac{bE_y}{aH_x}, \tag{3.83}$$

$$Z_{P,V} = \frac{V^2}{2P} = 2\frac{bE_y}{aH_x}, \tag{3.84}$$

$$Z_{P,I} = \frac{2P}{I^2} = \frac{\pi^2}{8} \frac{bE_y}{aH_x}, \tag{3.85}$$

for the TE_{10} mode in a rectangular guide. It is interesting to note that the three impedances, defined on the bases of voltage-current, power-voltage, and power-current, are all proportional to E_y/H_x or the wave impedance. This is not particularly surprising, since the power flow down the waveguide is proportional to the z-directed Poynting vector or power density $\mathbf{P} = \mathbf{E}_t \times \mathbf{H}_t = \hat{z}E_tH_t$. Thus it follows that $Z_{V,I}$, $Z_{P,V}$, and $Z_{P,I}$ differ only in multiplicative constants.

Inasmuch as we are primarily interested in the impedance at a particular plane of a waveguide relative to the guide's characteristic impedance, however it may be defined, we can concentrate on the wave impedance or the ratio between the transverse components of the field. So long as we choose a particular form of impedance, the arbitrary constants are divided out in the normalization, i.e., in the ratio of impedance/(characteristic impedance).

However, if two different waveguides are joined together, we must consider the way in which the dimensions enter into an expression for the characteristic impedance. Even in this case, we need not specify the numerical constants of Eqs. (3.83) through (3.85) that are involved, since they will cancel out in the expression for the impedance of one waveguide relative to the other so long as they are defined on the same basis. For example, given two waveguides with cross-sectional dimensions a_1, b_1 and a_2, b_2 respectively, filled by dielectrics of constants μ_1, ε_1 and μ_2, ε_2 with corresponding free-space wavelengths λ_1 and λ_2, we find from any of the three equations (3.83) through (3.85) that[5]

$$\frac{Z_1}{Z_2} = \frac{b_1 a_2}{b_2 a_1} \sqrt{\frac{\mu_1 \varepsilon_2}{\mu_2 \varepsilon_1}} \sqrt{\frac{1 - (\lambda_2/2a_2)^2}{1 - (\lambda_1/2a_1)^2}}, \tag{3.86}$$

where Z_1 and Z_2 are the characteristic impedances of waveguides 1 and 2 respectively defined on the same basis. Actually Eq. (3.86) is applicable only if $a_1 \simeq a_2$ and $b_1 \simeq b_2$. For large changes in dimensions, the real part of Z_1/Z_2 is still given by Eq. (3.86) but a shunt susceptance representing the excitation of nonpropagating modes at the discontinuity appears in the equivalent circuit.

---

[4] S. A. Schelkunoff, *Electromagnetic Waves*, D. Van Nostrand Co., New York, 1943; p. 319.

[5] G. L. Ragan, *Microwave Transmission Circuits*, M.I.T. Radiation Laboratory Series, No. 9, McGraw-Hill Book Co., New York, 1948; pp. 53–54.

We can show by a quasistatic analysis that the appearance of such a shunt susceptance is characteristic of the behavior caused by sudden changes in line or guide dimensions.[6] One must treat this situation as a rather complicated boundary value problem in electromagnetic theory in order to obtain an exact solution.[7]

To avoid the complications discussed above, we shall restrict ourselves to the transverse components $\mathbf{E}_t$ and $\mathbf{H}_t$ in calculating the reflection coefficient, SWR, etc. From Eqs. (3.80) and (3.81), we have

$$\mathbf{E}_t = \mathbf{E}_{t+}e^{j\omega t - \gamma z} + \mathbf{E}_{t-}e^{j\omega t + \gamma z}, \qquad (3.87)$$

$$\mathbf{H}_t = \mathbf{H}_{t+}e^{j\omega t - \gamma z} - \mathbf{H}_{t-}e^{j\omega t + \gamma z}. \qquad (3.88)$$

For the general case of irises and obstacles, it is difficult to proceed from Eqs. (3.87) and (3.88) because we may not be able to readily find the relationship between $\mathbf{E}_{t+}$ and $\mathbf{E}_{t-}$, etc. However, there is one type of reflection that is particularly simple to work with. It is the case in which the properties of the material filling the guide change discontinuously at $z = 0$, as shown in Fig. 3.4. In this case, $\mathbf{E}_{t+} = \mathbf{E}_{t-}$, $\mathbf{E}_{t+} = Z_1\mathbf{H}_{t+}$, and $\mathbf{E}_{t-} = Z_1\mathbf{H}_{t-}$, where Z_1 is the wave impedance of waveguide section (1). If we omit the time dependence, Eqs. (3.87) and (3.88) become

$$E_t = E_{t+}e^{-\gamma z} + E_{t-}e^{\gamma z}, \qquad (3.89)$$

$$H_t = \frac{1}{Z_1}(E_{t+}e^{-\gamma z} - E_{t-}e^{\gamma z}). \qquad (3.90)$$

At $z = 0$, we have

$$\left.\frac{E_t}{H_t}\right|_{z=0} = Z_1 \frac{1 + (E_{t-}/E_{t+})}{1 - (E_{t-}/E_{t+})} \qquad (3.91)$$

If waveguide (2) is terminated in its characteristic impedance or is infinite in length, there will be no reflected wave in section (2). It follows that $(E_t/H_t)_{z=0}$

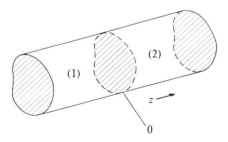

Fig. 3.4. General waveguide.

---

[6] S. Ramo et al., op. cit.; pp. 597–602.
[7] N. Marcuwitz, *Waveguide Handbook*, M.I.T. Radiation Laboratory Series, No. 10, McGraw-Hill Book Co., New York, 1948.

is equal to the characteristic impedance Z_2 of section (2). Setting the right-hand side of Eq. (3.91) equal to Z_2 and solving for the ratio E_{t-}/E_{t+}, we find

$$\rho = \frac{E_{t-}}{E_{t+}} = \frac{(Z_2/Z_1) - 1}{(Z_2/Z_1) + 1}, \qquad (3.92)$$

where the reflection coefficient ρ is the ratio of the magnitude of the reflected wave to that of the incident wave.

We can obtain the impedance at any plane of medium (1), say at $z = -l$, by setting z in Eqs. (3.89) and (3.90) equal to $-l$. In this way we find

$$\frac{Z_{-l}}{Z_1} = \frac{(Z_2/Z_1)\cosh \gamma_1 l - \sinh \gamma_1 l}{\cosh \gamma_1 l - (Z_2/Z_1)\sinh \gamma_1 l}, \qquad (3.93)$$

where we have used Eq. (3.92).

The standing wave ratio S is defined as the ratio of the maximum electric field along the line $E_{\max}$ to the minimum electrical field along the line $E_{\min}$:

$$S = E_{\max}/E_{\min}. \qquad (3.94)$$

Here $E_{\max}$ occurs at a point along the transmission line where the incidence and reflected waves interfere constructively or are adding in phase:

$$E_{\max} = |E_{t+}| + |E_{t-}|. \qquad (3.95)$$

Correspondingly, $E_{\min}$ occurs at a point along the transmission line where the incident and reflected waves interfere destructively or are adding out of phase by 180°:

$$E_{\min} = |E_{t+}| - |E_{t-}|. \qquad (3.96)$$

Combining Eqs. (3.94), (3.95), and (3.96), we have

$$S = \frac{|E_{t+}| + |E_{t-1}|}{|E_{t+}| - |E_{t-}|} \qquad (3.97)$$

Using Eq. (3.92), Eq. (3.97) becomes

$$S = \frac{1 + |\rho|}{1 - |\rho|}. \qquad (3.98)$$

The foregoing discussion applies to a single discontinuity at a plane of constant z. We can solve a problem involving more than one discontinuity by repeated application of Eq. (3.93). Unlike the more complicated cases involving irises and obstacles, these problems can be solved exactly; they also give us a good physical feel and model for the other cases.

At this point we should say a word about the special cases embodied in Eq. (3.93). First, if the transmission line or waveguide (1) is lossless, Eq. (3.93) becomes

$$\frac{Z_{-l}}{Z_1} = \frac{(Z_2/Z_1)\cos \beta_1 l + j \sin \beta_1 l}{\cos \beta_1 l + j(Z_2/Z_1)\sin \beta_1 l}, \qquad (3.99)$$

where β_1 is the phase constant. If, in addition, waveguide (2) is replaced by a load impedance Z_{-l} equal to Z_2, then Z_L, as measured at the beginning of the transmission line, is appropriately called the input impedance Z_{in}. Equation (3.99) in this case becomes

$$\frac{Z_{in}}{Z_1} = \frac{(Z_L/Z_1)\cos\beta_1 l + j\sin\beta_1 l}{\cos\beta_1 l + j(Z_L/Z_1)\sin\beta_1 l}. \tag{3.100}$$

This is a well-known formula in transmission line theory.

To summarize, we have seen that although characteristic impedances for a waveguide defined on different basis differ in multiplicative constants, they are all proportional to the wave impedance. Therefore, so long as we use normalized impedances (impedance/characteristic impedance), we need concern ourselves only with wave impedances or the correspondingly transverse field components. We have noted in this regard, from Eqs. (3.92) and (3.93), that only ratios of impedances Z_2/Z_1 and Z_{-1}/Z_1 appear. For a more rigorous treatment of this problem based on normal modes and electromagnet boundary conditions, the reader is referred elsewhere.[8]

6. RESONANT CAVITIES

A. Lumped Analogues

Before we proceed to the study of microwave resonant cavities, it is instructive to review briefly the facts connected with low-frequency lumped resonant circuits. In particular, we shall see that the concept of a complex frequency is a natural consequence of our calculation.

For a series resonant circuit composed of a resistance R, inductance L, and capacitance C, the current i is related to the driving voltage V by the following differential equation:

$$L\frac{di}{dt} + Ri + \frac{1}{C}\int i\, dt = V. \tag{3.101}$$

If, after the current is established, the impressed voltage is removed, i decays in amplitude as a function of time due to the presence of losses. Assuming $i \propto e^{j\omega t}$, we have from Eq. (3.101)

$$R + j\left(\omega L - \frac{1}{\omega C}\right) = 0. \tag{3.102}$$

If we let $1/LC = \omega_0^2$ and $\omega_0 L/R = Q$, Eq. (3.102) becomes

$$\frac{1}{Q} + j\left(\frac{\omega}{\omega_0} - \frac{\omega_0}{\omega}\right) = 0. \tag{3.103}$$

Solving for ω in Eq. (3.103), we obtain

$$\omega = \omega_0\sqrt{1 - (1/2Q)^2} + j(\omega_0/2Q). \tag{3.104}$$

---

[8] J. C. Slater, *Microwave Electronics*, D. Van Nostrand Co., New York, 1950; pp. 22–30.

Note that the frequency ω is complex. We can best ascertain the physical significance of this fact by letting $\omega = \omega_1 + j\omega_2$. Using this expression in Eq. (3.104), we find

$$\omega_0 = \sqrt{\omega_1^2 + \omega_2^2} \tag{3.105}$$

and

$$Q = \sqrt{(\omega_1/2\omega_2)^2 + \tfrac{1}{4}}. \tag{3.106}$$

Since $i \propto e^{j\omega t}$, it follows in terms of ω_1 and ω_2 that

$$i \simeq e^{-\omega_2 t}\, e^{j\omega_0\sqrt{1-(1/2Q)^2}}.$$

We can see that in contrast to the case of a purely real frequency, a frequency with a real as well as an imaginary part gives rise to an exponential decay with time of magnitude of the current.

For a parallel circuit, we can again obtain equations of the same form as Eqs. (3.103) through (3.106) if we define Q as equal to $RC\omega_0$ rather than $\omega_0 L/R$ as for the series case. Thus we see that the definitions for Q for series and parallel resonant circuits are the same when stated in terms of the real and imaginary parts of the frequency. For this reason, it is more convenient to define Q in terms of Eq. (3.106) for distributed circuits such as those encountered in microwave electronics.

B. Rectangular Cavity

Consider again the electromagnetic field in a rectangular waveguide supporting the TE_{10} mode. From Eqs. (3.51) through (3.53) and Eqs. (3.30) and (3.31) we have

$$H_z = B \cos\frac{\pi x}{a}, \tag{3.107}$$

$$H_x = B\frac{\pi}{a}\frac{\gamma}{k_c^2}\sin\frac{\pi x}{a}, \tag{3.108}$$

$$E_y = -\frac{\gamma_0 Z_0}{k_c^2}\left(\frac{\pi}{a}\right) B \sin\frac{\pi x}{a}. \tag{3.109}$$

If the direction of propagation is reversed from $+z$ to $-z$, γ and therefore H_x change sign. A rectangular cavity can be formed by closing a rectangular waveguide with two shorting plates placed a distance d apart and perpendicular to the z-axis, as shown in Fig. 3.5. In this case, waves traveling in the $+z$- and $-z$-directions combine to yield

$$H_z = (Be^{-\gamma z} + B'e^{\gamma z})\cos\frac{\pi x}{a}, \tag{3.110}$$

$$H_x = (Be^{-\gamma z} - B'e^{\gamma z})\left(\frac{\pi}{a}\right)\frac{\gamma}{k_c^2}\sin\frac{\pi x}{a}, \tag{3.111}$$

$$E_y = -(Be^{-\gamma z} + B'e^{\gamma z})\frac{\gamma_0 Z_0}{k_c^2}\left(\frac{\pi}{a}\right)\sin\frac{\pi x}{a}. \tag{3.112}$$

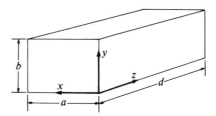

Fig. 3.5. Rectangular cavity.

Since E_y must equal zero at $z = 0$ according to Eq. (3.44), we find from Eq. (3.112) that $B' = -B$. Thus Eqs. (3.110) through (3.112) become

$$H_z = -2B \cos \frac{\pi x}{a} \sinh \gamma z, \tag{3.113}$$

$$H_x = 2B\left(\frac{a}{\pi}\right) \gamma \sin \frac{\pi x}{a} \cosh \gamma z, \tag{3.114}$$

$$E_y = B\gamma_0 Z_0 \left(\frac{a}{\pi}\right) \sin \frac{\pi x}{a} \sinh \gamma z, \tag{3.115}$$

where $k_c = \pi/a$ from Eq. (3.52).

At $z = d$, E_y must also be zero according to Eq. (3.44). Since none of the z-independent factors in Eq. (3.115) is in general zero at $z = d$, it follows that $\sinh \gamma d$ must be equal to zero. Letting $\gamma = \alpha + j\beta$ and using a trigonometric transformation, we have

$$\sinh \alpha d \cos \beta d + j \cosh \alpha d \sin \beta d = 0. \tag{3.116}$$

For Eq. (3.116) to hold, α must be zero. In this case, Eq. (3.116) reduces to

$$\sin \beta d = 0 \rightarrow \beta = l\pi/d, \tag{3.117}$$

where l is an integer. Equation (3.117) indicates that E_y must be zero at $z = 0$ and $z = d$: thus the length of the rectangular cavity must correspond to an integral number of half-wavelengths of the electric field.

Although we have implicitly assumed in the calculation above that the guidewalls are lossless, it is permissible for the dielectric filling the guide to have losses. This being the case, it is not immediately clear how α in $\gamma = \alpha + j\beta$ can be zero at the same time. The answer to this question lies in the phenomena connected with a complex frequency previously defined in the subsection on lumped analogues. Letting $\omega = \omega_1 + j\omega_2$, we find from Eq. (3.6) that

$$\gamma = \sqrt{[k_c^2 - (\omega_1^2 - \omega_2^2)\varepsilon\mu - \sigma\omega_2\mu] + j[\omega_1\mu(\sigma - 2\omega_2\varepsilon)]}. \tag{3.118}$$

If γ is imaginary, the expression in the first bracket must be negative and that in the second must be zero; that is,

$$[(\omega_1^2 - \omega_2^2)\mu\varepsilon + \sigma\omega_2\mu] > k_c^2 \qquad (3.119)$$

and

$$\omega_2 = \sigma/2\varepsilon. \qquad (3.120)$$

Equation (3.119) is a generalization of the statement that the operating frequency must be above the cutoff value for propagation to occur in a waveguide containing a lossy dielectric. Equation (3.120), on the other hand, relates the imaginary part of the frequency to the conductivity and dielectric constant of the dielectric filling the cavity.

To recapitulate, since the fields are assumed to be proportional to $e^{j\omega t - \gamma z}$, the results obtained above indicate that

$$E_y \propto \exp(-\omega_2 t)\exp(j\omega_1 t)\exp(j\{[(\omega_1^2 - \omega_2^2)\mu\varepsilon + \sigma\omega_2\mu] - k_c^2\}),$$

and likewise for H_x and H_z. Since γ is purely imaginary, there is propagation without attenuation with z. In this connection, we may recall that the standing wave in the z-direction is composed of two oppositely traveling waves. These waves travel along the $+z$- and $-z$-directions without an accompanying decrease in amplitude in spite of the fact that the dielectric filling the cavity is not lossless. Thus as the energy is absorbed with the passage of time the amplitude of the waves and that of their linear combinations must decrease with time.

In the above discussion, we have implicitly assumed that the cavity is a completely enclosed structure. In actuality, of course, this is not the case, since energy must be coupled into or away from the cavity. This coupling is usually achieved via a small hole in one wall of the cavity. The hole must be small in order to prevent the bulk of the energy from entering the cavity to reradiate and thereby lowering the total Q. Hence the total Q of the cavity is given by the expression

$$\frac{1}{Q_{\text{total}}} = \frac{1}{Q_{\text{coupling}}} + \frac{1}{Q_{\text{deielectric}}} + \frac{1}{Q_{\text{wall}}}, \qquad (3.121)$$

where Q_{total} is usually known as the loaded Q; Q_{coupling} which is related to the energy reradiated out of the coupling hole, is known as the external Q; and

$$\frac{1}{Q_{\text{dielectric}}} + \frac{1}{Q_{\text{wall}}} = \frac{1}{Q_{\text{unloaded}}}. \qquad (3.122)$$

The unloaded Q, in this sense, is the Q of the cavity "not loaded" by the presence of the coupling hole. This distinction is important to remember, since loading the cavity with a sample whose properties are to be measured would change the unloaded Q. In this case, we have

$$\frac{1}{Q_{\text{unloaded}}} = \frac{1}{Q_{\text{dielectric}}} + \frac{1}{Q_{\text{wall}}} + \frac{1}{Q_{\text{sample}}}. \qquad (3.123)$$

We turn now to calculation of the resonance frequency ω_1. Combining Eqs. (3.117) through (3.120), we find for ω_1

$$\omega_1 = \omega_0 \sqrt{1 - \left(\frac{1}{2Q}\right)^2}, \tag{3.124}$$

where

$$\omega_0 = \frac{1}{\sqrt{\mu\varepsilon}} \sqrt{\left(\frac{\pi}{a}\right)^2 + \left(\frac{l\pi}{d}\right)^2} \tag{3.125}$$

and

$$Q = \frac{\omega_0 \varepsilon}{\sigma}. \tag{3.126}$$

It is significant to note that Eq. (3.124) is of the same form as the real part of the corresponding equation (3.104) for the lumped circuit. Furthermore, it can be shown that the definitions for Q given by Eqs. (3.106) and (3.126) are equivalent; the same is true for ω_0 given by Eqs. (3.105) and (3.125). By induction, we can generalize Eq. (3.125) to read

$$\omega_0 = \frac{1}{\sqrt{\mu\varepsilon}} \sqrt{\left(\frac{m\pi}{a}\right)^2 + \left(\frac{n\pi}{b}\right)^2 + \left(\frac{l\pi}{d}\right)^2}, \tag{3.127}$$

where ω_0 is the resonance frequency of a rectangular cavity oscillating in the TE_{mnl} mode.

If the cavity walls have finite losses, the calculation of ω and Q is more complicated. However, if the losses of the walls are sufficiently small, we can determine Q_{wall} rather easily by an approximate calculation. In such a calculation, the fields in the cavity are assumed to be the same as those of lossless cavities. In particular, the component of H tangential to the perfectly conducting wall is equal to the surface current density in magnitude, according to Eq. (2.49). Knowing the surface current densities at the various cavity walls and their conductivity, we can calculate the ohmic losses.

From Eqs. (3.104) and (3.105) we see that for $Q \gg \frac{1}{2}$, $\omega_0 \simeq \omega_1$. Accordingly, $Q \simeq \omega_1/2\omega_2 \simeq \omega_0/2\omega_2$ from Eq. (3.106). This definition of Q is equivalent to

$$Q = \omega_0 U/W_L, \tag{3.128}$$

where U is the energy stored in the cavity and W_L is the average power loss. We can easily derive Eq. (3.128) from the solution i of Eq. (3.101) by noting that $U = \frac{1}{2}L|i|^2$ and $W_L = -dU/dt$. Knowing the field distribution in the cavity, we can calculate U as well as W_L. We can then determine Q from Eq. (3.128).

C. Other Types of Cavities

Besides the rectangular cavity, there are a number of cavities of other configurations; coaxial, cylindrical, spherical, reentrant, optical, etc. Although their geometry may be very different from that of the rectangular cavity, the concepts

7. VELOCITY, ENERGY, AND POWER FLOW

In the foregoing discussion we assumed that fields are harmonic in time, that is, that $\mathbf{E}$, $\mathbf{D}$, $\mathbf{B}$, and $\mathbf{H}$ are proportional to $e^{j\omega[t-(\beta/\omega)z]}$ for the lossless case. Since $(\beta/\omega)z$ must have the dimension of time, it is clear that ω/β corresponds to some sort of velocity. It turns out that this velocity, known as the phase velocity, v_p, is the velocity with which a plane of constant phase travels, since if $t - (\beta/\omega)z$ is independent of space and time, we have

$$t - \frac{\beta}{\omega} z = \text{const.} \quad (3.129)$$

Differentiating Eq. (3.129), we find

$$v_p = dz/dt = \omega/\beta. \quad (3.130)$$

From the above discussion, we see that the concept of a phase velocity applies only to fields that are periodic in space and consequently represent wave trains of infinite duration. A wave train of finite length, on the other hand, cannot be represented in a simple harmonic form; thus the term phase velocity loses its precise significance. To illustrate this point, let us consider the superposition of two sinusoidal waves, each of unit amplitude, which differ very slightly in frequency and phase constant. In this case, the resultant wave is represented by

$$\cos(\omega t - \beta z) + \cos[(\omega + \delta\omega)t - (\beta + \delta\beta)z].$$

By a trigonometric transformation, this expression becomes

$$2 \cos \tfrac{1}{2}[(\delta\omega)t - (\delta\beta)z] \cos\left[\left(\omega + \frac{\delta\omega}{2}\right)t - \left(\beta + \frac{\delta\beta}{2}\right)z\right].$$

We see that the resultant field has a phase constant essentially equal to β and oscillates at a frequency that differs negligibly from ω. It then follows that the velocity of a plane of constant phase is still given essentially by v_p of Eq. (3.130). On the other hand, it is clear that the resultant amplitude given by

$$2 \cos \tfrac{1}{2}[(\delta\omega)t - (\delta\beta)z]$$

varies slowly between the sum of the amplitudes of the component waves and zero. As a result of the interference of the two waves of slightly different frequency and phase constant, the field distribution in space and time has an envelope with a series of periodically repeated beats. In this case, the velocity with which a plane of constant amplitude travels, or group velocity v_g, is obtained from the relation

$$(\delta\omega)t - (\delta\beta)z = \text{const.} \quad (3.131)$$

Differentiating, we find

$$v_g = dz/dt = \delta\omega/\delta\beta. \quad (3.132)$$

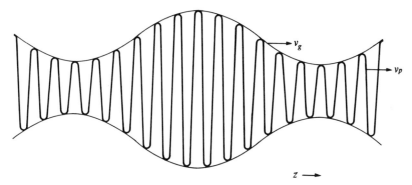

Fig. 3.6. Group and phase velocities of a wave.

The physical significance of the above results can best be illustrated by reference to Fig. 3.6. We see that the group velocity is the velocity of the envelope while the phase velocity is that of the high-frequency wave within the envelope. For $v_p > v_g$, we would expect the high-frequency wave to slip through the envelope with the passage of time.

We can construct a pulse of any shape by choosing the amplitude of the component waves as an appropriate function of ω or β and integrating by means of the Fourier integral. However, the concept of group velocity applies only to a narrow band spectrum in ω or β. As the interval $\delta\omega$ or $\delta\beta$ increases, the spread in the phase velocity of the component waves increases markedly, providing that the medium is dispersive, that is, $\omega = f(\beta)$. In this case, the packet (wave amplitude vs. ω or β) is severely deformed and the concept of group velocity becomes meaningless.[9]

To relate our discussion of v_p and v_g to waveguides, let us further examine Eq. (3.5). Solving for $\omega\sqrt{\mu\varepsilon}$, we find

$$\omega\sqrt{\mu\varepsilon} = \sqrt{k_c^2 + \beta^2}, \tag{3.133}$$

where we have set $\gamma^2 = -\beta^2$ for a lossless medium. Referring to Fig. 3.7, or

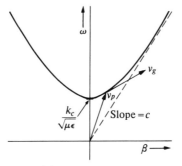

Fig. 3.7. ω–β diagram for a waveguide.

---

[9] J. A. Stratton, *Electromagnetic Theory*, McGraw-Hill Book Co., New York, 1941; pp. 331–333.

Eq. (3.133), we see that for $\beta \gg k_c$ (that is, for a guide operating way above cutoff) $\omega/\beta \simeq 1/\sqrt{\mu\varepsilon} = c$, where c is the velocity of light in the unbounded medium of dielectric constant ε and permeability μ. But for smaller values of β, the relationship between ω and β is no longer linear, as can be seen from Fig. 3.7. In other words, $v_p = \omega/\beta$ is not a constant but rather a function of ω, for the waveguide is dispersive. On the other hand, differentiating Eq. (3.5), we have

$$v_g = d\omega/d\beta = \beta/\mu\varepsilon\omega = 1/\mu\varepsilon v_p, \quad (3.134)$$

so that $v_p v_g = c^2$, a characteristic of empty waveguides. Again, we find that v_g is a function of ω. In Fig. 3.7, we see that v_p is the slope of the line from the origin to a point on the curve and v_g is the local slope at the same point.

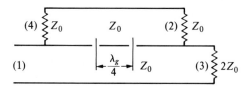

Fig. 3.8. Directional coupler.

Group velocity is often associated with the velocity of energy propagation. This identity is valid for many cases with normal dispersion $(dv_p/d\omega < 0)$, such as the waveguides discussed above, but usually invalid for systems with anomalous dispersion $(dv_p/d\omega > 0)$.[10] If v_E is the velocity of energy flow, then

$$v_E = W_T/U_{av} \quad (3.135)$$

where W_T is the average power flow in a single mode and U_{av} is the average energy stored in a unit of guide length. In this case W_T is equal to the Poynting vector **P** integrated over the cross section of the guide and **P** is defined in terms of the fields as

$$\mathbf{P} = \mathbf{E} \times \mathbf{H}. \quad (3.136)$$

For a lossless waveguide, it is clear that **P** has only a z-component, since perfectly conducting walls would consume no energy. It then follows that

$$\mathbf{P} = \hat{z} E_t H_t,$$

where E_t and H_t are transverse fields, as mentioned previously in the discussion of waveguide characteristic impedance in Section 5.

---

[10] For a careful study of the wavefront and signal velocities, see Stratton, *Electromagnetic Theory*, pp. 333–340.

PROBLEMS

3.1 Show that the TEM mode cannot propagate in an empty waveguide.

3.2 Show two methods of excitation of the TE_{30} mode in a rectangular guide. What other modes would you expect to be excited by your methods?

3.3 a) Consider two waveguides connected by a junction. The reflection coefficient looking into the junction from waveguide 1 is ρ_1 and the reflection cofficient looking into waveguide 2 is ρ_2. If $\rho_1 \neq \rho_2$ and the junction contains no nonreciprocal elements, what can you deduce about the characteristic of the junction?
b) For a unit of power entering waveguide 1, δ amount of power emerges from the junction into waveguide 2. If a unit of power enters waveguide 2, what amount of power emerges from the junction into waveguide 1?

3.4 a) Consider the directional coupler of Fig. 3.8. Explain how it works. Z_0 is the characteristic impedance of the guide and λ_g is the guide wavelength.
b) Given that one unit of power enters terminal 1, find the power dissipated in the terminating impedances at 2, 3, and 4. Assume that the coupling coefficient between the guides is 3 db and the directivities are infinite.
c) Find the VSWR at terminal 1.

3.5 a) Consider the case of two waveguides of characteristic impedance Z_{01} and Z_{03} connected by a third of characteristic impedance Z_{02} and of length l. Find l as a function of frequency ω and the relationship between Z_{01}, Z_{02}, and Z_{03} given that all power entering guide 1 emerges from guide 2 into guide 3.
b) Investigate the way in which the magnitude of the reflection coefficient at the 1–2 interface varies with frequency as it departs from the frequency for which the matching section is designed. Show that the section will be most broad-banded for the shortest matching section.

3.6 a) Consider a lossless rectangular waveguide terminated by a short. Find the impedance, reflection coefficient, and standing wave ratio at plane A–A located at a distance d in front of the short.
b) If $d = l\lambda_g/2$, where l is an integer and λ_g is the TE_{10} mode guide wavelength, the shorted waveguide can be considered a cavity with a full-size coupling hole. Calculate the resonance frequency and Q of such a cavity.

3.7 a) Expand the field distribution in the cavity described in Problem 3.6(b) in terms of its normal modes (TE_{mnl} and TM_{mnl}). Can a summation of these modes satisfy the boundary condition at plane A–A?
b) If the answer to (a) is no, what is the correct field distribution inside the cavity?

3.8 a) Consider a cubic rectangular resonator. Find the Q of the resonator for the TE_{10l} modes. Assume that the surface resisitivity of the cavity walls is R_s.
b) Sketch the Q of the resonator as a function of the mode number l and also as a function of frequency.
c) Represent the cavity by a parallel RLC resonant circuit. Find the expressions for R, L, and C for the TE_{10l} modes.

3.9 Find the approximate resonance frequency of the foreshortened lossless coaxial cavity of Fig. 3.9 as a function of a, b, l, δ, etc. For simplicity, assume that $\delta \ll l$ and that the electric field is concentrated at the gap region G while the magnetic field is distributed throughout the rest of the cavity. Contrast this frequency with the lowest resonance frequency of the TEM mode in a lossless cavity of length l.

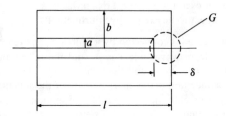

Fig. 3.9. Foreshortened coaxial cavity.

3.10 An empty rectangular waveguide with dimensions $a = 4$ cm and $b = 2$ cm is excited at a frequency 5 GHz in the fundamental mode. If the exciting source produces a maximum electric field strength of 10^4 V/m at the center of the entrance cross section, what is the power input to the guide?

3.11 a) Show mathematically that no power can propagate in a lossless waveguide below cutoff even though the electromagnetic fields inside the guide are finite.

b) If a crystal detector is inserted into the cutoff waveguide of (a), would any power be extracted? Why?

CHAPTER 4

ELECTROMAGNETIC FIELDS AND CHARGES

In this chapter we shall study the relationship between time-varying electromagnetic fields and moving charges. Fields and charges do not exist independently of each other, since fields influence the motion of charges and charges in turn give rise to fields. This relationship is contained in Maxwell's equations and the Lorentz force law. Starting with Maxwell's equations, we shall derive the inhomogeneous wave equations, the solutions of which are the so-called retarded potentials. Then, invoking the Lorentz force law, we shall consider the conservation of energy and momentum for a composite system composed of fields and charges.

1. INHOMOGENEOUS WAVE EQUATIONS

Maxwell's equations (2.1) through (2.4), repeated here, state that

$$\nabla \times \mathbf{E} = -\frac{\partial \mathbf{B}}{\partial t}, \tag{4.1}$$

$$\nabla \times \mathbf{H} = \mathbf{i} + \frac{\partial \mathbf{D}}{\partial t}, \tag{4.2}$$

$$\nabla \cdot \mathbf{D} = \rho, \tag{4.3}$$

$$\nabla \cdot \mathbf{B} = 0. \tag{4.4}$$

We can see from these equations that the electromagnetic field quantities $\mathbf{D}, \mathbf{E}, \mathbf{B}$, and $\mathbf{H}$ are functions of the current density $\mathbf{i}$ and charge density ρ and vice versa. To derive the inhomogeneous wave equations, we merely manipulate these equations with the help of the constitutive equations (2.5) and (2.6). Since $\nabla \cdot \mathbf{B} = 0$ according to Eq. (4.4) and $\nabla \cdot (\nabla \times \mathbf{A}) \equiv 0$, where $\mathbf{A}$ is any vector, we can let

$$\mathbf{B} = \nabla \times \mathbf{A}, \tag{4.5}$$

where $\mathbf{A}$ is the vector potential. Substituting Eq. (4.5) into Eq. (4.1), we have

$$\nabla \times \left(\mathbf{E} + \frac{\partial \mathbf{A}}{\partial t}\right) = 0, \tag{4.6}$$

since spatial and time derivatives are independent. Because $\nabla \times (\nabla \psi) \equiv 0$, where ψ is any scalar, Eq. (4.6) leads to

$$\mathbf{E} + \partial \mathbf{A}/\partial t = -\nabla \phi, \tag{4.7}$$

where ϕ is the scalar potential. Alternatively, we can rearrange Eq. (4.7) to read

$$\mathbf{E} = -\nabla \phi - \frac{\partial \mathbf{A}}{\partial t}, \tag{4.8}$$

where time-varying $\mathbf{E}$ is a function of both the scalar potential ϕ and vector potential $\mathbf{A}$. Taking the divergence of both sides of Eq. (4.7), we find that

$$\nabla^2 \phi + \frac{\partial}{\partial t}(\nabla \cdot \mathbf{A}) = -\frac{\rho}{\varepsilon}, \tag{4.9}$$

where we have used Eqs. (4.3) and (2.6) and assumed ε to be a scalar. It can be shown that vector $\mathbf{A}$ cannot be completely specified, however, unless both its curl and divergence are known. Since $\nabla \times \mathbf{A}$ has already been specified via Eq. (4.5), we are at liberty to specify $\nabla \cdot \mathbf{A}$. Referring to Eq. (4.9), we see that it is clearly convenient to specify $\nabla \cdot \mathbf{A}$ in terms of ϕ, since in that case Eq. (4.9), being in ϕ only, can be readily solved in terms of ρ. The most expedient functional relationship between $\nabla \cdot \mathbf{A}$ and ϕ will be evident in what follows.

From Eqs. (4.2), (2.5), and (2.6), we have

$$\nabla \times \mathbf{B} = \mu \mathbf{i} + \mu \varepsilon \frac{\partial \mathbf{E}}{\partial t}, \tag{4.10}$$

where we have assumed that both ε and μ are scalars. Using Eqs. (4.5), (4.8), and the vector identity $\nabla \times \nabla \times \mathbf{A} = \nabla(\nabla \cdot \mathbf{A}) - \nabla^2 \mathbf{A}$, Eq. (4.10) becomes

$$\nabla^2 \mathbf{A} - \mu\varepsilon \frac{\partial^2 \mathbf{A}}{\partial t^2} = -\mu\mathbf{i} + \nabla\left(\nabla \cdot \mathbf{A} + \mu\varepsilon \frac{\partial \phi}{\partial t}\right). \tag{4.11}$$

In solving Eq. (4.11), it is clearly expedient to let[1]

$$\nabla \cdot \mathbf{A} = -\mu\varepsilon \frac{\partial \phi}{\partial t}, \tag{4.12}$$

Since then the last term in Eq. (4.11) vanishes and we have an equation in $\mathbf{A}$ only. Adopting the relation (4.12), we find that Eqs. (4.9) and (4.11) become

$$\nabla^2 \phi - \mu\varepsilon \frac{\partial^2 \phi}{\partial t^2} = -\frac{\rho}{\varepsilon}, \tag{4.13}$$

$$\nabla^2 \mathbf{A} - \mu\varepsilon \frac{\partial^2 \mathbf{A}}{\partial t^2} = -\mu\mathbf{i}. \tag{4.14}$$

Note that Eqs. (4.13) and (4.14) are identical in form and Eqs. (4.9) and (4.11)

---

[1] Equation (4.12) is known as the Lorentz condition.

contain only ϕ and **A** respectively with the introduction of relation (4.12). Equations (4.13) and (4.14) are known as the *inhomogeneous wave equations* with solutions

$$\phi = [\phi]_t = \int_V \frac{[\rho]_{t-r/v'} d\tau}{4\pi\varepsilon r}, \tag{4.15}$$

$$\mathbf{A} = [\mathbf{A}]_t = \int_V \frac{\mu[\mathbf{i}]_{t-r/v'} d\tau}{4\pi r}, \tag{4.16}$$

where $v' = 1/\sqrt{\mu\varepsilon}$ is the phase velocity of the wave in the medium. The quantity $[\phi]_t$, usually written ϕ for simplicity, represents the potential of some point P at a distance r from the time-varying charge density ρ. Note that to compute ϕ at time t, we must use the charge density ρ at an earlier time $t - (r/v')$ because it takes a time r/v' for the effect of any changes in the charge density to reach point P, as depicted in Fig. 4.1. We can, of course, apply the same reasoning to **A**. Thus ϕ and **A** are appropriately known as *retarded potentials*.

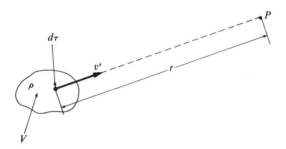

Fig. 4.1. Quantities in the calculation of retarded potentials.

2. MOTION OF CHARGES IN E.M. FIELDS

Maxwell's equations (4.1) through (4.4) indicate that the field quantities **E, H, D,** and **B** are determined by the charge density ρ and current density **i**. The scalar potential ϕ and vector potential **A**, which are related to **E, H, D,** and **B**, are also dependent on ρ and **i** via the inhomogeneous wave equations (4.13) and (4.14). One must not surmise from this, however, that ρ and **i** can be specified independently of **E, H, D,** and **B**. Although the fields are functions of charges and currents, the forces acting on charges are in turn determined by the fields according to the Lorentz force law,

$$\mathbf{F} = q(\mathbf{E} + \mathbf{v} \times \mathbf{B}), \tag{4.17}$$

where **F** is the force acting on a charge q moving with velocity **v** through an electric field **E** and magnetic field **B**. Combining Eq. (4.17) with Newton's second law,

$$\mathbf{F} = \frac{d}{dt}(m\mathbf{v}), \tag{4.18}$$

we find that

$$q(\mathbf{E} + \mathbf{v} \times \mathbf{B}) = \frac{d}{dt}(m\mathbf{v}). \tag{4.19}$$

For given values of **E** and **B**, the velocity **v** of charge q can be determined from the above equation. We must emphasize again, however, that in most cases **E** and **B** cannot be specified independently of ρ and **i**.

In microwave electronics devices, we are likely to deal with a large number of charges in an electron beam rather than with a single charge. It is therefore more convenient to speak of a charge density ρ rather than a charge q. If the velocity of ρ is $\mathbf{v}_\rho$, then

$$\mathbf{i}_\rho = \rho \mathbf{v}_\rho. \tag{4.20}$$

We can easily obtain Eq. (4.20) by noting that the total current I is equal to the rate of change of charge:

$$I = dq/dt. \tag{4.21}$$

It thus follows that the current density **i** (current I per unit area A) is given by

$$i = \frac{d(q/A)}{dt}. \tag{4.22}$$

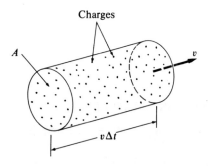

Fig. 4.2. Motion of charges *in vacuo*.

In Fig. 4.2, all charges within the indicated volume emerge from the right surface in time Δt; in the time interval Δt, the charges at the left surface of the volume at $t = 0$ arrive at the right surface. Hence the charge density ρ is given by

$$\rho = \frac{\Delta q}{A v \, \Delta t}, \tag{4.23}$$

where Δq is the charge within the volume $A(v\,\Delta t)$. Equation (4.20) then follows from Eq. (4.22) and (4.23) and Fig. 4.2. In this derivation, we have assumed that all charges move with the same velocity $\mathbf{v}$ so that $\mathbf{v} = \mathbf{v}_\rho$. Similarly, $\mathbf{i}$ is denoted by $\mathbf{i}_\rho$, the convection current.

Equation (4.20) gives the relationship between the current density $\mathbf{i}_\rho$ and the velocity of the charge density $\mathbf{v}_\rho$ for charges *in vacuo* and therefore applies to electron beams such as those in microwave tubes (Klystrons, traveling wave tubes, etc.). In a conductor the motion of charges (free electrons) is influenced by the potential environment set up by the atoms and ions in the interior of the conductor. In the classical picture, the particles lose energy to the atoms of the medium through collisions. A detailed analysis of the individual trajectories of a large number of electrons using the Lorentz force law (4.17) would be hopelessly complicated. In the wave-mechanical picture, the wave packets representing electrons are reflected by nonperiodic potential energy gradients and lose energy to the recoiling atoms.[2] However, it is often possible to describe the flow of charge particles in terms of an average drift velocity $\mathbf{v}_d$ which is in turn proportional to the electric field $\mathbf{E}$, that is, $\mathbf{v}_d = \pm \mu \mathbf{E}$, at a given temperature. The plus and minus signs are for positive and negative charges respectively, while μ (not to be confused with the permeability) is known as the mobility of the material. Replacing ρ and $\mathbf{v}_\rho$ of Eq. (4.20) respectively by $\pm |\rho|$ and $\pm \mu\mathbf{E}$, we obtain Ohm's law:

$$\mathbf{i}_c = \sigma \mathbf{E}, \tag{4.24}$$

where $\sigma = |\rho|\mu$ is the conductivity of the material in ohms per meter. Thus the conduction current, $\mathbf{i}_c$, is proportional to the electric field $\mathbf{E}$ for most materials. For electron beams *in vacuo*, therefore, $\mathbf{i}$ of Eq. (4.2) is equivalent to $\rho \mathbf{v}_\rho$ of Eq. (4.20). Likewise, for the case of conduction in materials, $\mathbf{i}$ is given by $\sigma \mathbf{E}$ of Eq. (4.24).

Since current is defined as the rate of change of charge, it seems reasonable to expect that $\mathbf{i}$ and ρ of Eqs. (4.2) and (4.3) respectively are related and therefore cannot be independently specified. We can easily find this relationship by taking the divergence of both sides of Eq. (4.2) to yield

$$\nabla \cdot \mathbf{i} + \frac{\partial \rho}{\partial \tau} = 0, \tag{4.25}$$

in which we have also used Eq. (4.3) and the vector identity $\nabla \cdot (\nabla \times \mathbf{A}) \equiv 0$, where $\mathbf{A}$ is any vector. Equation (4.25), known as the equation of continuity, states that the divergence (spatial) of the current density is equal to the negative time rate of change of charge density. Simultaneous solution of Maxwell's equations (4.1) through (4.4) and Eqs. (4.20), (4.24), and (4.25) yields the required solution to the entire problem.

---

[2] If the potential is periodic, there will be no energy transfer between the drifting electrons and the atoms of the crystal. However, because of impurities, imperfections, dislocations, and lattice vibrations of the crystal, the potential is always somewhat nonperiodic.

3. ENERGY RELATIONS AND POYNTING'S VECTOR

In the foregoing discussion, we showed how to determine fields, charges, and currents by solving a self-consistent set of equations. Of course, it is energy, not fields, that actuates our instruments. Thus our ultimate objective in solving practical microwave electronics problems must be the determination of energy and power flow via the field quantities. We shall now show how to derive the desired energy relation from Maxwell's equations.

Taking the dot product of $\mathbf{H}$ and both sides of Eq. (4.1) and the dot product of $\mathbf{E}$ and both sides of Eq. (4.2), we have

$$\mathbf{H} \cdot \nabla \times \mathbf{E} = -\mathbf{H} \cdot \frac{\partial \mathbf{B}}{\partial t}, \tag{4.26}$$

$$\mathbf{E} \cdot \nabla \times \mathbf{H} = \mathbf{E} \cdot \mathbf{i} + \mathbf{E} \cdot \frac{\partial \mathbf{D}}{\partial t}. \tag{4.27}$$

Subtracting Eq. (4.27) from Eq. (4.26) and using the vector identity $\mathbf{A} \cdot (\nabla \times \mathbf{B}) - \mathbf{B} \cdot (\nabla \times \mathbf{A}) = \nabla \cdot (\mathbf{B} \times \mathbf{A})$, where $\mathbf{A}$ and $\mathbf{B}$ are any two vectors, we find

$$-\mathbf{H} \cdot \frac{\partial \mathbf{B}}{\partial t} - \mathbf{E} \cdot \frac{\partial \mathbf{D}}{\partial t} - \mathbf{E} \cdot \mathbf{i} = \nabla \cdot (\mathbf{E} \times \mathbf{H}). \tag{4.28}$$

If we integrate over the volume V, Eq. (4.28) becomes

$$\int_V \left(\mathbf{H} \cdot \frac{\partial \mathbf{B}}{\partial t} + \mathbf{E} \cdot \frac{\partial \mathbf{D}}{\partial t} + \mathbf{E} \cdot \mathbf{i} \right) dv = -\oint_S (\mathbf{E} \times \mathbf{H}) \cdot d\mathbf{s}, \tag{4.29}$$

where we have also used the Gauss theorem (2.36). Equation (4.29) is the most general energy relation in that it is applicable to materials with constant as well as time-varying properties. However, if μ and ε are independent of time, as is usually the case, Eq. (4.29), with the aid of the constitutive equations (2.5) and (2.6), becomes

$$\int_V \left[\frac{\partial}{\partial t} \left(\frac{\mu H^2}{2} \right) + \frac{\partial}{\partial t} \left(\frac{\varepsilon E^2}{2} \right) + \mathbf{E} \cdot \mathbf{i} \right] dv = -\oint_S (\mathbf{E} \times \mathbf{H}) \cdot d\mathbf{s}, \tag{4.30}$$

where we note that $\mathbf{H} \cdot (\partial \mathbf{B}/\partial t) = (\partial/\partial t)(\mu H^2/2) = (\partial/\partial t)(\mathbf{B} \cdot \mathbf{H}/2)$, etc. We must now interpret the meaning of each term in Eq. (4.30).

As we shall show later, the first term in Eq. (4.30) represents the rate of change of stored energy in the magnetic field. Similarly, the second term represents the rate of change of stored energy in the electric field. The third term is the familiar ohmic term and represents energy loss in the form of heat per unit time. The net increase in stored energy and heat loss must be supplied externally. It therefore follows that the right-hand side of Eq. (4.30) represents the energy flow *into* the volume V across the bounding surface S per unit time. Alternatively, the energy flow W out of the volume per unit time (power) is

$$W = \oint_S \mathbf{P} \cdot d\mathbf{s}, \tag{4.31}$$

where

$$\mathbf{P} = \mathbf{E} \times \mathbf{H} \tag{4.32}$$

is the *Poynting vector*. It is clear from the derivation of Eq. (4.30) that Eq. (4.31) will always correctly represent the power flow across the bounding surface. It does not necessarily follow, however, that $\mathbf{P}$ is a vector giving the magnitude and direction of actual power flow density at a given point in space, although such an interpretation is convenient. The reason for this ambiguity is that various spatial distributions of $\mathbf{P}$ can give rise to the same W, given by Eq. (4.31).

4. ENERGY, STRESS, AND MOMENTUM IN E.M. FIELDS

We have asserted, in connection with Eq. (4.30), that $\mu H^2/2$ and $\varepsilon E^2/2$ represent the energy density of electric and magnetic fields respectively. In this section we shall try to justify this interpretation.

Consider first the case of an electrostatic field. The potential difference between two points is

$$\phi = \int \mathbf{E} \cdot d\mathbf{l}. \tag{4.33}$$

Substituting the Lorentz force law (4.17) into Eq. (4.33), we have

$$\phi = \frac{1}{q} \int \mathbf{F} \cdot d\mathbf{l}. \tag{4.34}$$

We see from this equation that the potential difference ϕ is equal to the work done by the electrostatic force per unit charge in moving the charge from one point to another. It follows that the incremental electrostatic energy dE_e expended in moving an infinitesimal charge dq through a potential difference ϕ is

$$dE_e = \phi \, dq. \tag{4.35}$$

But we have from Eq. (4.3)

$$\int \mathbf{D} \cdot d\mathbf{s} = \int \sigma \, ds \tag{4.36}$$

or

$$D_n = \sigma, \tag{4.37}$$

where σ is the surface charge density and D_n is the component of $\mathbf{D}$ normal to the surface. In the derivation of Eq. (4.36), we have used Gauss's theorem, Eq. (2.36), to convert the volume integral into a surface integral. From Eqs. (2.6), (4.33), and

(4.37), it is evident that ϕ is proportional to q, that is, $q = C\phi$, where C is the capacitance. Inserting this relation into Eq. (4.35) and integrating, we find

$$E_e = \int \frac{q}{C} dq = \frac{q^2}{2C} = \frac{q\phi}{2}. \tag{4.38}$$

It now follows that the incremental energy dE_e is

$$dE_e = \tfrac{1}{2}(\mathbf{D} \cdot d\mathbf{s})(\mathbf{E} \cdot d\mathbf{l}) = \tfrac{1}{2}(\mathbf{D} \cdot \mathbf{E})\, dv = \frac{\varepsilon}{2} E^2\, dv, \tag{4.39}$$

where we have also used Eqs. (4.33) and (4.36). We can interpret Eq. (4.39) to mean that the energy density in an electrostatic field is given by $\varepsilon E^2/2$, as mentioned in the last section. Although this interpretation is plausible, we must remember that only

$$E_e = \int \tfrac{1}{2} \mathbf{D} \cdot \mathbf{E}\, dv \tag{4.40}$$

is unique in that it correctly gives the total energy of the system. In an analogous manner, we can show that

$$dE_m = \frac{\mu}{2} H^2\, dv. \tag{4.41}$$

We derived Eqs. (4.39) and (4.41) for electrostatic and magnetostatic fields respectively. The interpretation of $\varepsilon E^2/2$ and $\mu H^2/2$ as energy densities can be extended to quasi-static cases in which displacements and variations occur so slowly that they are equivalent to a sequence of stationary states. In thermodynamic terms, such changes are said to be reversible. For the dynamic case, such interpretations are still permissible, but their justification is more subtle and involves the fact that fields and potentials are propagated with a finite velocity.[3]

Since an electromagnetic field possesses energy propagating at a finite velocity, it must also carry some momentum with it. Starting with Maxwell's equations, we can show that in vacuum

$$\int_S {}^2\mathbf{S} \cdot \mathbf{n}\, ds = \int_V \rho \mathbf{E}\, dv + \int_V \mathbf{i} \times \mathbf{B}\, dv + \frac{\partial}{\partial t} \int_V \varepsilon_0 \mathbf{E} \times \mathbf{B}\, dv, \tag{4.42}$$

where $\mathbf{n}$ is surface normal and ${}^2\mathbf{S}$ is a second-rank tensor identified with stresses in the electromagnetic field.[4] The left-hand side of Eq. (4.42) represents the net force transmitted across the closed surface S bounding a region of volume V. The first and second terms on the right-hand side of Eq. (4.42) are electric and magnetic forces respectively [see the Lorentz force law (4.17)]. To interpret the last term of Eq. (4.42), consider the case in which there are no charges or currents within the closed surface S. In this case, Eq. (4.42) becomes

$$\int_S {}^2\mathbf{S} \cdot \mathbf{n}\, ds = \frac{\partial}{\partial t} \int_V \varepsilon_0 \mathbf{E} \times \mathbf{B}\, dv. \tag{4.43}$$

---

[3] J. A. Stratton, *Electromagnetic Theory*, McGraw-Hill Book Co., New York, 1941; pp. 133–134.
[4] *Ibid.*, pp. 97–100.

According to this equation, there is a force on a volume element of empty space. This paradox can be resolved if we interpret

$$\mathbf{G}_{electro} = \int_V \mathbf{g}_{electro}\, dv = \int_V \varepsilon_0\, \mathbf{E} \times \mathbf{B}\, dv \qquad (4.44)$$

as the momentum carried by an electromagnetic field. A dimensional analysis of Eq. (4.44) shows that **G** has indeed the dimension of a momentum, i.e., kg-m/sec. On the other hand, the force exerted on charged matter within the surface S is, from Eq. (4.42),

$$\int_V (\rho \mathbf{E} + \mathbf{i} \times \mathbf{B})\, dv = \frac{d}{dt} G_{mech}, \qquad (4.45)$$

where G_{mech}, which is analogous to $G_{electro}$ for the e.m. field, is the total linear momentum of matter, i.e. of the moving charges. Substituting Eqs. (4.44) and (4.45) into Eq. (4.43), we have a statement of the law of conservation of momentum for a system composed of charges and field within a bounded region:

$$\int_S {}^2\mathbf{S} \cdot \mathbf{n}\, ds = \frac{d}{dt}(G_{mech} + G_{electro}). \qquad (4.46)$$

If the surface S is extended to enclose the entire field, the left-hand side of Eq. (4.46) must vanish, so that

$$G_{mech} + G_{electro} = \text{const.} \qquad (4.47)$$

Thus there is inertia associated with an electromagnetic field, similar to the case of ordinary matter.

Proceeding from Eq. (4.42), we can show that the total force transmitted by an electromagnetic field across a closed surface S *in vacuo* is

$$\mathbf{F} = \int_S \left[\varepsilon_0 (\mathbf{E} \cdot \mathbf{n})\mathbf{E} + \frac{1}{\mu_0}(\mathbf{B} \cdot \mathbf{n})\mathbf{B} - \tfrac{1}{2}(\varepsilon_0 E^2 + \frac{1}{\mu_0} B^2)\mathbf{n} \right] ds. \qquad (4.48)$$

If the medium is ordinary material rather than free space, Eq. (4.48) remains correct for the static case, providing that ε_0 and μ_0 are replaced respectively by ε and μ of the medium. For an electromagnetic field in a material medium, we must interpret the right-hand side of Eq. (4.48), after replacing ε_0 by ε and μ_0 by μ, in the sense of Eq. (4.46), i.e., not as the force exerted by the field on the matter within S but as the inward flow of momentum per unit time through S. For an electromagnetic field, therefore, we have

$$\int_S \left[\varepsilon (\mathbf{E} \cdot \mathbf{n})\mathbf{E} + \mu(\mathbf{H} \cdot \mathbf{n})\mathbf{H} - \tfrac{1}{2}(\varepsilon E^2 + \mu H^2)\mathbf{n} \right] ds = \frac{d}{dt}(G_{mech} + G_{electro}). \qquad (4.49)$$

PROBLEMS

4.1 Show by substitution that expression (4.15) for the retarded potential ϕ is the solution for the inhomogeneous wave equation (4.13).

4.2 Suppose you suspect that an electromagnetic field exists in some region of space. Devise an experiment to measure the field.

4.3 a) Consider a coaxial line structure in which the line propagates the fundamental mode. A ring of electrons coaxial with the conductor is injected at $t = 0$ into the line at $z = 0$ with axial velocity $\hat{z}u_0$ where $\hat{z}$ is a unit vector in the $+ z$-direction. Sketch the path of the beam as a function of z.

b) If $u_0 = 3 \times 10^8$ cm/sec, characteristic impedance of the coax $= 50$ ohms, and power carried by the coax $= 100$ mW, find the ratio of the magnetic force divided by the electric force acting on the electrons.

4.4 a) Given that **E** and **H** are the complex vectors representing the electric and magnetic fields respectively in the time-periodic case, show that the average Poynting vector $\bar{\mathbf{P}}$ is

$$\bar{\mathbf{P}} = \tfrac{1}{2} \operatorname{Re}(\mathbf{E} \times \mathbf{H}^*).$$

b) Consider the case of a round wire of radius a carrying a dc current I_0. Show by means of the Poynting vector that the total power flow into the conductor is $I_0^2 R$ per unit length, where R is the resistance of the conductor per unit length.

4.5 Calculate the electromagnetic stress exerted on the walls of a rectangular cavity of linear dimensions a, b, and d resonating in the TE_{10l} mode.

CHAPTER 5

SPACE-CHARGE EFFECTS

In this chapter we shall consider the interaction between stationary and moving charges due to Coulomb forces, emphasizing the coulomb repulsion between electrons such as those present in an electron beam. We shall consider relativistic effects so that we can apply our results to the study of microwave particle accelerators, since the electrons in a linear accelerator travel at almost the velocity of light.

1. INTERACTION BETWEEN STATIONARY CHARGES

A. Coulomb's Force Law for Point Charges

Consider two point charges as shown in Fig. 5.1(a). The force **F** acting on these charges is given by Coulomb's inverse square law:

$$\mathbf{F} = \hat{a}_r \frac{q_1 q_2}{4\pi\varepsilon r^2}, \qquad (5.1)$$

where $\hat{a}_r$ is a unit vector pointing from one charge directly away from the other as shown in Fig. 5.1(a). Thus, we see from Eq. (5.1) that if q_1 and q_2 are of the same sign, **F** is in the direction of $\hat{a}_r$; that is, *charges of the same sign repel each other with equal and opposite force*. Conversely, if q_1 and q_2 are of opposite signs, **F** is in the direction of $-\hat{a}_r$; that is, *charges of opposite signs attract each other* with equal and opposite force. Equation (5.1) is in the rationalized MKS system of

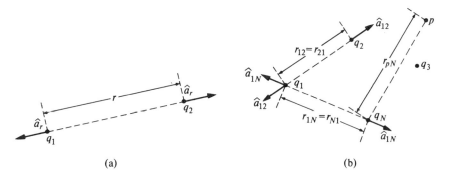

Fig. 5.1. (a) Coulomb interaction between two point charges. (b) Coulomb interaction between N point charges.

practical units. For free space, the dielectric constant ε, designated by ε_0, is equal to 8.854×10^{-12} F/m. For a medium with a relative dielectric constant equal to ε_r, ε in Eq. (5.1) is equal to $\varepsilon_r \varepsilon_0$.

For a complex of N charges, the force acting on the mth charge is by induction,

$$\mathbf{F}_m = \sum_{n=1}^{N}{}' \hat{a}_{mn} \frac{q_m q_n}{4\pi\varepsilon r_{mn}^2}, \qquad (5.2)$$

where $\sum_{n=1}^{N}{}'$ means that the summation should be carried out for $n = 1$ to $n = N$ with the case $m = n$ omitted. For example, the force acting on q_1 is obtained from Eq. (5.2) as

$$\mathbf{F}_1 = \hat{a}_{12} \frac{q_1 q_2}{4\pi\varepsilon r_{12}^2} + \hat{a}_{13} \frac{q_1 q_3}{4\pi\varepsilon r_{13}^2} + \cdots \hat{a}_{1N} \frac{q_1 q_N}{4\pi\varepsilon r_{1N}^2}. \qquad (5.3)$$

The meaning of this equation becomes clear if we refer to Fig. 5.1(b); the resultant force acting on q_1 is the sum of all forces of repulsion or attraction due to all other charges in the complex.

From Eq. (5.2), we see that the force approaches infinity as the distance between charges approaches zero. Therefore, if the Coulomb forces of repulsion were the only forces acting on the protons in a nucleus, the nucleus would tend to fly apart because protons are very close together in the nucleus. This of course, is not the case. As protons come closer and closer together to form the nucleus, other attractive forces due mesons come into play and hold the nucleus together. This situation is depicted by the force vs. distance plot in Fig. 5.2. Note that here F is positive for a force of attraction and negative for a force of repulsion.

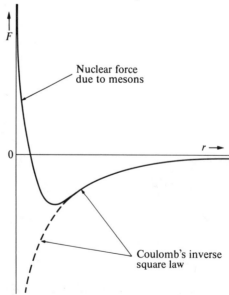

Fig. 5.2. Force between particles vs. particle separation distance.

B. Electric Field Resulting from a Charge Complex

If a test charge Δq is placed at some point p in the charge complex of Fig. 5.1(b), a force **F** will act on it. The associated electric field is defined as the force per unit charge acting on the test charge as the magnitude of the test charge Δq approaches zero:

$$\mathbf{E} = \lim_{\Delta q = 0} \frac{\mathbf{F}}{\Delta q}. \tag{5.4}$$

If Δq were not vanishingly small, it would also produce an electric field which would modify the resultant field due to the charges in the complex. In that case, the electric field obtained would not be a true measure of the effects due to the charge complex. It follows from Eq. (5.2) that the electric field at point p is simply

$$\mathbf{E}_p = \sum_{n=1}^{N'} \hat{a}_{pn} \frac{q_n}{4\pi\varepsilon r_{pn}^2}, \tag{5.5}$$

where r_{pn} is the distance between point p and charge q_n.

C. Electrostatic Potential Resulting from a Charge Complex

From Maxwell's equation (2.1) we see that for an electrostatic field

$$\nabla \times \mathbf{E} = 0. \tag{5.6}$$

It follows that **E** can be represented as the gradient of a scalar potential:

$$\mathbf{E} = -\nabla\phi. \tag{5.7}$$

Therefore, if we know **E** at some point p due to a given charge, we can determine ϕ. Since ϕ is a scalar, the total potential due to a complex of charges is simply the algebraic sum of the potentials due to the individual charges. To obtain the total electric field using Eq. (5.5), we must perform a vector addition. Thus ϕ_p, the total potential at point p, is simply

$$\phi_p = \sum_{n=1}^{N} \phi_n, \tag{5.8}$$

where ϕ_n is the scalar electric potential due to charge q_n. Once we have obtained the total potential, we can find the total electric field at point p from Eq. (5.7).

2. TRANSFORMATIONS OF RELATIVITY

Moving charges constitute currents which in turn give rise to magnetic fields. Thus, whereas stationary charges give rise only to electrostatic fields, moving charges may well appear in the form of both electric and magnetic fields in the laboratory frame of reference. That is, electric and magnetic fields observed in a given region are not uniquely defined but depend on the state of motion of the observer. If the relative velocity between charges and observer is comparable to the velocity of light, as is usually the case in microwave particle accelerators, the special theory of relativity must be invoked. Thus we shall digress briefly to study the pertinent aspects of relativity.

A. Lorentz Transformation

Consider two coordinate systems moving relative to each other at a constant velocity as depicted in Fig. 5.3. As shown here, the primed and unprimed axes are respectively parallel, the origins are coincident at $t = 0$, and the x- and x'-axes slide along each other at a relative velocity **u**. It is clear from the figure that a point (x', y', z') in the moving frame can be expressed in terms of the coordinates (x, y, z) of the stationary frame and the velocity u as follows:

$$x' = x - ut,$$
$$y' = y, \qquad (5.9)$$
$$z' = z.$$

In these equations, known as the *Galilean transformations*, we have implicitly assumed the identity of times t and t' measured in the two frames. However, according to the theory of relativity, this assumption turns out to be invalid for u comparable to the velocity of light c. In what follows, we shall therefore derive a set of transformation equations—the *Lorentz transformations*—using the theory of relativity. In the limit of small u/c, of course, we would expect the Lorentz transformation to reduce to the Galilean transformation given by Eq. (5.9).

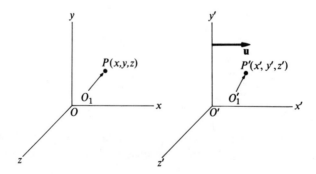

Fig. 5.3. Coordinate systems in constant relative motion.

Let us first define an event for an observer O as something that happens at a point $P(x, y, z)$ at time t with observer O stationed at the origin of the unprimed coordinate system. The *same* event will occur for observer O' stationed at the origin of the primed system at a point $P'(x', y', z')$ at time t'. Now, if a pulse of light (from a point source) is sent out at the instant that O and O' coincide, that is, at $t = 0$, an auxiliary observer O_1 stationed at a *fixed* point (x, y, z) records the event of this light pulse passing him as having occurred at time t. It then follows that observer O can characterize this event by the equation

$$x^2 + y^2 + z^2 = (ct)^2. \tag{5.10}$$

Correspondingly, O' has also stationed an auxiliary observer O'_1 at a fixed point (x', y', z'). If O_1 and O'_1 just happen to coincide at the instant the light pulse passes, O'_1 would record the event as having occurred at a time t'. Therefore, O' can characterize this event by writing

$$(x')^2 + (y')^2 + (z')^2 = (ct')^2. \tag{5.11}$$

In Eqs. (5.10) and (5.11), we have implicitly assumed that the velocity of light c is the same when measured in both the primed and unprimed reference frames in spite of the fact that there is relative velocity between the frames. This assumption constitutes one of the two fundamental postulates of relativity. As expressed by Einstein,[1] the postulate states:

> *Light is always propagated in empty space with a definite velocity c which is independent of the state of motion of the emitting body.*

The constancy of c has been substantiated by a number of experiments, the best known being that of Michelson and Morley.[2]

The second postulate of relativity holds that all physical laws are of the same form in all inertial frames. Einstein states:[3]

> *The same laws of electrodynamics and optics will be valid for all frames of reference for which the equations of mechanics hold good.*

According to this postulate, the transformation equations which link the observations in coordinate xyz to those in $x'y'z'$ must be such that O can derive Eq. (5.11) from Eq. (5.10) and vice versa for O' since they are describing the same event. As we will show in Appendix A, x' and t' must be linearly related individually to x and t.

After evaluating the coefficients involved in these relationships, we find

$$\begin{aligned} x' &= \kappa(x - ut), \\ y' &= y, \\ z' &= z, \\ t' &= \kappa[t - (ux/c^2)], \end{aligned} \tag{5.12}$$

---

[1] A. Einstein, "On the Electron Dynamics of Moving Bodies," *Ann. Phys.* **17,** 891 (1905). An English translation can be found in *The Principles of Relativity*, Dover Publications, New York, 1958.
[2] A. A. Michelson and E. W. Morley, "On the Relative Motion of the Earth and the Luminiferous Ether," *Am. J. Sci.*, Series III **34,** 333 (November 1887).
[3] Einstein, *Electron Dynamics of Moving Bodies.*

where $\kappa = (1 - u^2/c^2)^{-1/2}$. Inverting the equations of (5.12), we find

$$x = \kappa(x' + ut'),$$
$$y = y'$$
$$z = z',$$
$$t = \kappa[t' + (ux'/c^2)].$$
(5.13)

Equations (5.12) and (5.13) are known as the Lorentz transformations.[4] As we anticipated, (5.12) reduces to (5.9) for small values of u/c.

B. Velocity Transformation

By implicitly differentiating the expressions (5.12) and taking proper ratios, we can obtain the velocity transformation equations:

$$v'_x = \frac{v_x - u}{1 - (uv_x/c^2)},$$
$$v'_y = \frac{v_y}{\kappa[1 - (uv_x/c^2)]},$$
(5.14)
$$v'_z = \frac{v_z}{\kappa[1 - (uv_x/c^2)]},$$

where $v'_x = dx'/dt'$, etc. Similarly, by implicitly differentiating (5.13) and taking proper ratios, we obtain the inverse transformations;

$$v_x = \frac{v'_x + u}{1 + (uv'_x/c^2)},$$
$$v_y = \frac{v'_y}{\kappa[1 + (uv'_x/c^2)]},$$
(5.15)
$$v_z = \frac{v'_z}{\kappa[1 + (uv'_x/c^2)]},$$

where $v_x = dx/dt$, etc.

C. Force Transformation

According to the second postulate of relativity, the Lorentz equations properly

---

[4] H. A. Lorentz, "Electromagnetic Phenomena in a System Moving with Any Velocity Less than That of Light," *Proc. Amst. Acad.* **6**, 809 (1904). Reprinted in English in *The Principles of Relativity*, Dover Publications, New York, 1958; pp. 11–34.

transform all physical laws. Let us derive the force transformation laws that transform the force law in the unprimed system,

$$\mathbf{F} = \frac{d}{dt}(m\mathbf{v}), \tag{5.16}$$

to that of the primed system,

$$\mathbf{F}' = \frac{d}{dt'}(m'\mathbf{v}'), \tag{5.17}$$

and vice versa. With the help of Eqs. (5.14) and (5.15), Eq. (5.17) becomes

$$F'_x = \frac{dt}{dt'} \frac{d}{dt}[\kappa(v_x - u)m],$$

$$F'_y = \frac{dt}{dt'} \frac{d}{dt}[mv_y], \tag{5.18}$$

$$F'_z = \frac{dt}{dt'} \frac{d}{dt}[mv_z],$$

where we have used the relation

$$m' = \kappa\left(1 - \frac{uv_x}{c^2}\right)m, \tag{5.19}$$

to be derived by the reader. From the last expression of Eq. (5.12) we can show that the ratio dt/dt' is $1/\kappa(uv_x/c^2)$. Substituting this expression into Eq. (5.18) and comparing the results with Eq. (5.16), we obtain the force transformation equations:

$$F'_x = F_x - \frac{uv_y/c^2}{1 - (uv_x/c^2)} F_y - \frac{uv_z/c^2}{1 - (uv_x/c^2)} F_z,$$

$$F'_y = \frac{F_y}{\kappa[1 - (uv_x/c^2)]}, \tag{5.20}$$

$$F'_z = \frac{F_z}{\kappa[1 - (uv_x/c^2)]}.$$

By applying these transformations to the Lorentz force law (4.17), we can now find the transformation laws for electric and magnetic fields.

D. Electric and Magnetic Field Transformation

If we solve expressions (5.20) for F_x, F_y, and F_z, we find

$$F_x = F'_x + \kappa \frac{uv_y}{c^2} F'_y + \kappa \frac{uv_z}{c^2} F'_z,$$

$$F_y = \kappa \left(1 - \frac{uv_x}{c^2}\right) F'_y, \quad (5.21)$$

$$F_z = \kappa \left(1 - \frac{uv_x}{c^2}\right) F'_z.$$

Now, suppose that the electrostatic field and magnetostatic flux density due to time-independent sources in $x'y'z'$, as observed by an observer O' stationary in $x'y'z'$, are respectively $\mathbf{E}'(x'y'z')$ and $\mathbf{B}(x'y'z')$. If a charge q moves through $x'y'z'$ at a velocity $\mathbf{v}'$, observer O' can say that the force acting on q is

$$\mathbf{F}' = q(\mathbf{E}' + \mathbf{v}' \times \mathbf{B}'), \quad (5.22)$$

in accordance with the Lorentz force law (4.17).

In component form, Eq. (5.22) reads

$$F'_x = q(E'_x + v'_y B'_z - v'_z B'_y),$$

$$F'_y = q(E'_y + v'_z B'_x - v'_x B'_z), \quad (5.23)$$

$$F'_z = q(E'_z + v'_x B'_y - v'_y B'_x).$$

Using Eq. (5.14) to express the components of $\mathbf{v}'$ in terms of the components of $\mathbf{v}$ (the velocity of the charge relative to xyz) and substituting the resulting equations in Eq. (5.21), we have

$$F_x = q\left[E'_x + \frac{\kappa u}{c^2}(v_y E'_y + v_z E'_z) + \kappa(v_y B'_z - v_z B'_y)\right],$$

$$F_y = q\left[\kappa\left(1 - \frac{uv_x}{c^2}\right)E'_y + v_z B'_x - \kappa(v_x - u)B'_z\right], \quad (5.24)$$

$$F_z = q\left[\kappa\left(1 - \frac{uv_x}{c^2}\right)E'_z - v_y B'_x + \kappa(v_x - u)B'_y\right].$$

Now, if the Lorentz force law is invariant under the Lorentz transformation, observer O stationary in xyz can write

$$\mathbf{F} = q(\mathbf{E} + \mathbf{v} \times \mathbf{B}), \quad (5.25)$$

where $\mathbf{E}(x, y, z)$ and $\mathbf{B}(x, y, z)$ are fields as observed by observer O stationary in xyz. We can see from Eq. (5.25) that if $\mathbf{v} = 0$ we have

$$F_x = qE_x,$$

$$F_y = qE_y, \quad (5.26)$$

$$F_z = qE_z.$$

Comparing Eq. (5.26) with Eq. (5.24) with $\mathbf{v} = 0$, we find the transformation equations for the electric field:

$$\boxed{\begin{aligned} E_x &= E'_x, \\ E_y &= \kappa(E'_y + uB'_z), \\ E_z &= \kappa(E'_z - uB'_y). \end{aligned}} \qquad (5.27)$$

Similarly, comparing Eqs. (5.24) and (5.25) with the help of Eq. (5.26), we find the transformation equations for $\mathbf{B}$:

$$\boxed{\begin{aligned} B_x &= B'_x, \\ B_y &= \kappa\left(B'_y - \frac{u}{c^2} E'_z\right), \\ B_z &= \kappa\left(B'_z + \frac{u}{c^2} E'_y\right). \end{aligned}} \qquad (5.28)$$

We can easily find the inverse transforms from Eqs. (5.27) and (5.28):

$$\boxed{\begin{aligned} E'_x &= E_x, \\ E'_y &= \kappa(E_y - u\,B_z), \\ E'_z &= \kappa(E_z + u\,B_y), \end{aligned}} \qquad (5.29)$$

and

$$\boxed{\begin{aligned} B'_x &= B_x, \\ B'_y &= \kappa\left(B_y + \frac{u}{c^2} E_z\right), \\ B'_z &= \kappa\left(B_z - \frac{u}{c^2} E_y\right). \end{aligned}} \qquad (5.30)$$

It should be noted that the sources which produce these fields are of a restricted class, being time independent in $x'y'z'$. However, this restriction can be removed by a simple extension.[5]

---

[5] R. S. Elliott, *Electromagnetics*, McGraw-Hill Book Co., New York, 1966; p. 270.

3. INTERACTION BETWEEN MOVING CHARGES

It is illuminating to apply the field transformation laws of the last section to an examination of repulsion forces between charges in an electron beam. For simplicity, let us consider just two electrons on the x-axis separated by a distance Δx and moving with a velocity **u** in the positive x-direction as shown in Fig. 5.4(a).

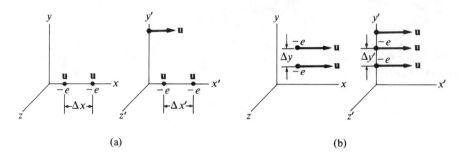

Fig. 5.4. (a) Electrons on x-axis moving along x-axis. (b) Electrons on y-axis moving along x-axis.

The electrons within the $x'y'z$ frame are at rest since $x'y'z'$ moves at a velocity **u** relative to frame xyz. Thus we know that the electric field at one electron due to the presence of the other is

$$E'_x = \frac{e}{4\pi\varepsilon_0(\Delta x')^2}. \tag{5.31}$$

Equation (5.31) follows directly from Eqs. (2.6) and (2.40). The repulsion force also follows from the Lorentz force law (4.17) and Eq. (5.31);

$$F'_x = eE'_x = \frac{e^2}{4\pi\varepsilon_0(\Delta x')^2}. \tag{5.32}$$

From the first Eq. (5.12), we can show (see Appendix A) that $\Delta x' = \kappa \Delta x$. Inserting this relationship into Eqs. (5.31) and (5.32), we find

$$E_x = \frac{1}{\kappa^2} \frac{e}{4\pi\varepsilon_0(\Delta x)^2}, \tag{5.33}$$

$$F_x = \frac{1}{\kappa^2} \frac{e^2}{4\pi\varepsilon_0(\Delta x)^2}, \tag{5.34}$$

where we have also used the first Eq. (5.29), $E'_x = E_x$.

If the electrons are located on a line perpendicular to the direction of motion, i.e., on the y'-axis shown in Fig. 5.4(b), the second of equations (5.27) shows that

$$E_y = \kappa E'_y = \kappa \frac{e}{4\pi\varepsilon_0(\Delta y)^2}, \tag{5.35}$$

since $B'_z = 0$ because the electrons are stationary in $x'y'z'$. In deriving Eq. (5.35), we have again used Eqs. (2.6) and (2.40) and the second of equations (5.12). Furthermore, it follows from the second of equations (5.21) that $F_y = F'_y/\kappa$ since $v_x = u$ for the case under consideration. This relation, along with Eqs. (4.17) and (5.35), yields

$$F_y = \frac{1}{\kappa} \frac{e^2}{4\pi\varepsilon_0(\Delta y)^2}. \qquad (5.36)$$

Note that here F_y is not equal simply to the electric force eE_y, since there is also a magnetic force due to the motion of the electrons. Note also that E_x, F_x, E_y, and F_y given by Eqs. (5.33) through (5.36) are quantities as seen by a stationary observer.

It is also interesting to observe that F_x and F_y both approach zero as v approaches the velocity of light c. This implies that the longitudinal and transverse forces of repulsion between electrons in an electron beam traveling at a relativistic velocity approach zero as $v \to c$, as far as the stationary observer is concerned. Consequently, except for the first section in which the electrons are accelerated, the electron beam in a high-energy microwave electron accelerator needs no electrostatic or magnetostatic focusing. We will give a more detailed account of this phenomenon later in our study of microwave particle accelerators.

PROBLEMS

5.1 a) Two negative charges, each of strength $-q$, are located at $x = +1$ and $x = -1$. Find the locations $+y_0$ and $-y_0$ of the positive charges, each of strength $+q$, such that the force acting on each of them due to the presence of the other three charges is zero.
b) Find the electric flux density at a distance $\mathbf{r}$ from the origin.
c) Show that if $|\mathbf{r}| > 1$, the electric flux density integrated over a spherical surface is zero.

5.2 An observer in the unprimed system measures the time interval between the passing of his origin by the two ends of a meter stick in the primed system. He then multiplies this time interval by u to obtain the length in the unprimed system. Derive the Lorentz contraction from this procedure.

5.3 Obtain the time dilatation by measuring the distance in the unprimed system between two events occurring at the origin of the primed system and dividing this distance by u to obtain the time interval in the unprimed system.

5.4 Is the ultimate rise time of an oscilloscope limited by the finite velocity of light?

5.5 a) Sketch the electric field intensity about a stationary charge.
b) Sketch the electric field intensity about a charge moving at a velocity comparable to that of light.

5.6 Derive Eq. (5.19).

CHAPTER 6

COMPARISON BETWEEN CONVENTIONAL AND MICROWAVE TUBES

In this chapter we shall first examine the high-frequency limitations of conventional vacuum tubes. We will show that the degradation in their performance at high frequencies is due to three factors; lead inductance effects, transit time effects, and gain-bandwidth product limitations. The study of these phenomena will then lead us to means of overcoming them and the invention of microwave tubes.

1. DEFICIENCY OF CONVENTIONAL VACUUM TUBES AT HIGH FREQUENCIES

At high frequencies, the performance of the conventional vacuum tube is impaired by the effects of (1) lead inductance, (2) electron transit time, and (3) gain-bandwidth product limitation. Briefly stated, the leads from electrodes to the external terminals of the tube may have sufficient inductance at high frequencies to cause resonance effects with the electrode capacitances and coupling between electrodes. Because the grid is negatively biased with respect to the cathode, the velocity of the electrons in transit between the cathode and grid is small. Consequently, a high-frequency modulation voltage can easily cause the electrons to oscillate back and forth in the cathode-grid space or to return to the cathode if the transit time of the electrons in the absence of modulation is large compared to the period of the modulation signal. In other words, as an electron proceeds from the cathode to the grid, its velocity is sufficiently small so that alternately accelerating and decelerating forces will act upon it as the modulation voltage changes in polarity with time. Finally, if the plate circuit is resonant, then the gain-bandwidth product is a constant dependent only on the tube and circuit parameters. Thus a higher gain can be obtained only at the expense of a smaller bandwidth and vice versa. We shall now examine each of these three effects in more detail.

2. LEAD INDUCTANCE EFFECTS

As we shall demonstrate, the input capacitance of a triode is much larger than that of a pentode. To minimize the shunting effect of the input capacitance on the applied signal, pentodes rather than triodes are used at higher frequencies. Therefore, we shall first examine the lead inductance effects of pentodes. By a slight modification, we can apply our results to triodes.

Consider the circuit shown in Fig. 6.1(a). In this figure, R_f is the bias resistor that provides the negative bias on the grid while C_f is sufficiently large to serve as an effective short circuit for the ac component of the current. Similarly, R_{sg} drops the plate supply voltage to the value appropriate for operating the screen grid while the screen grid bypass capacitance C_{sg} is sufficiently large to provide the effective short circuit for the ac component of the screen grid current. The equivalent circuit for Fig. 6.1(a) is given in Fig. 6.1(b). Here L_g and L_k represent the lead inductances in the grid and cathode circuits respectively. Note that we have assumed, for simplicity, that the cathode and suppressor are internally connected by zero impedance and that C_{sg} also neutralizes any screen grid lead inductance.

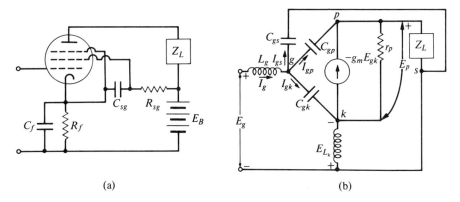

Fig. 6.1. (a) Pentode circuit. (b) Equivalent circuit for (a).

From Fig. 6.1(b) and assuming that the amplified currents are much larger than the charging currents, we have

$$E_p = -g_m E_{gk} \frac{r_p(Z_L + j\omega L_k)}{r_p + Z_L + j\omega L_k}, \qquad (6.1)$$

and

$$E_{gp} = E_{gk} - E_p. \qquad (6.2)$$

Substituting Eq. (6.2) into Eq. (6.1), we find

$$E_{gp} = \left(1 + g_m r_p \frac{Z_L + j\omega L_k}{r_p + Z_L + j\omega L_k}\right) E_{gk}. \qquad (6.3)$$

The grid current I_g is the sum of three terms:

$$I_g = I_{gp} + I_{gk} + I_{gs} = j\omega C_{gp} E_{gp} + j\omega C_{gk} E_{gk} + j\omega C_{gs}(E_{gk} - E_{L_k}), \qquad (6.4)$$

$$E_{gk} = E_g - I_g j\omega L_g + E_{L_k}, \qquad (6.5)$$

and

$$E_{L_k} = E_p \frac{j\omega L_k}{Z_L + j\omega L_k}. \qquad (6.6)$$

Lead inductance effects

Substituting E_p from Eq. (6.1) into Eq. (6.6), we find

$$E_{L_k} = -g_m r_p \frac{j\omega L_k}{r_p + Z_L + j\omega L_k} E_{gk}. \tag{6.7}$$

Combining Eqs. (6.3), (6.4), (6.5), and (6.7), we obtain E_{gk} in terms of E_g:

$$E_{gk} = \frac{E_g}{1 - \omega^2 L_g [C_{gk} + C'_{gs} + C_{gp}(1 + A'\cos\theta_{A'} + jA'\sin\theta_{A'})] + \frac{j\omega L_k g_m r_p}{r_p + Z_L + j\omega L_k}} \tag{6.8}$$

where

$$C'_{gs} = C_{gs}\left(1 + g_m r_p \frac{j\omega L_k}{r_p + Z_L + j\omega L_k}\right), \tag{6.9}$$

$$A' \angle \theta_A = g_m r_p \frac{Z_L + j\omega L_k}{r_p + Z_L + j\omega L_k}. \tag{6.10}$$

It follows that the input admittance Y_g between the grid and cathode terminals is

$$Y_g = \frac{I_g}{E_g} = \frac{j\omega[C_{gk} + C'_{gs} + C_{gp}(1 + A'\cos\theta_{A'} + jA'\sin\theta_{A'})]}{1 - \omega^2 L_g[C_{gk} + C'_{gs} + C_{gp}(1 + A'\cos\theta_{A'} + jA'\sin\theta_{A'})] + \frac{j\omega L_k g_m r_p}{r_p + Z_L + j\omega L_k}} \tag{6.11}$$

For pentodes, the term containing C_{gp} is typically small compared to $C_{gk} + C'_{gs}$; because of the shielding effect of the screen, C_{gp} is a stray capacitance. In this case, Eq. (6.11) simplifies to

$$Y_g = \frac{j\omega(C_{gk} + C'_{gs})}{1 - \omega^2 L_g(C_{gk} + C'_{gs}) + \frac{j\omega L_k g_m r_p}{r_p + Z_L + j\omega L_k}}. \tag{6.12}$$

In order to appreciate the effects of L_g and L_k on tube performance, let us set first one and then the other equal to zero. If we let $L_g = 0$, Eq. (6.12) becomes

$$Y_g = \frac{j\omega(C_{gk} + C'_{gs})}{1 + \frac{j\omega L_k g_m r_p}{r_p + Z_L + j\omega L_k}}. \tag{6.13}$$

Now, noting that

$$j\omega L_k g_m r_p/(r_p + Z_L + j\omega L_k) = E_{L_k}/E_{gk}$$

and assuming that

$$E_{L_k} \ll E_{gk} \quad \text{and} \quad r_p \gg (Z_L + j\omega L_k),$$

we can simplify Eq. (6.13) to

$$Y_g \simeq \omega^2 g_m L_k (C_{gk} + C_{gs}) + j\omega(C_{gk} + C_{gs}), \tag{6.14}$$

where Eq. (6.9) has also been used. We see from Eq. (6.14) that in addition to the expected input capacitance $(C_{gk} + C_{gs})$ there is an input resistance R_g given by

$$R_g = 1/\omega^2 g_m L_k(C_{gk} + C_{gs}), \qquad (6.15)$$

varying as $1/\omega^2$. Thus at high frequencies energy is drawn from the signal source because of the coupling together of the grid and plate circuits caused by the cathode lead inductance L_k.

If L_k is set equal to zero, Eq. (6.12) becomes

$$Y_g \simeq \frac{j\omega(C_{gk} + C_{gs})}{1 - \omega^2 L_g(C_{gk} + C_{gs})}, \qquad (6.16)$$

where we have again used Eq. (6.9). It is interesting to note from Eq. (6.16) that at a frequency

$$\omega = \sqrt{1/L_g(C_{gk} + C_{gs})}, \qquad (6.17)$$

$Y_g \to \infty$ and $Z_g \to 0$ or resonance occurs. At this frequency, the input signal is short-circuited and becomes ineffective in inducing changes in the plate. Thus we conclude that the grid lead inductance L_g could give rise to a resonance phenomenon with the electrode capacitances at a signal frequency given by Eq. (6.17).

Finally, for a triode instead of a pentode, C_{gs} in Eq. (6.11) would be equal to zero. But C_{gp}, in the absence of the screening effect of the screen grid, would be larger, on the order of C_{gk}, rather than being much smaller. Since, as mentioned previously, triodes are usually employed at lower frequencies, we can neglect L_g and L_k for the sake of simplicity. Setting L_g and L_k equal to zero in Eq. (6.11), we immediately obtain

$$Y_g \simeq -\omega C_{gp} A' \sin \theta_{A'} + j\omega[C_{gk} + C_{gp}(1 + A' \cos \theta_{A'})] \qquad (6.18)$$

Thus, in this case the input resistance is

$$R_g = -\frac{1/\omega C_{gp}}{A' \sin \theta_{A'}}, \qquad (6.19)$$

where R_g can be either positive or negative. In the former case, energy is fed from the grid into the plate circuit, whereas in the latter the reverse occurs. More important, the input capacitance is

$$C_g = C_{gk} + C_{gp}(1 + A' \cos \theta_{A'}), \qquad (6.20)$$

which, because of the term containing the amplification factor A', may be much larger than C_g of the pentode given by Eq. (6.14). This confirms our statement that pentodes rather than triodes are used at higher frequencies due to their smaller input capacitance.

There are several ways to minimize lead and electrode capacitance effects.[1] First, the tube may be scaled to smaller size. The accompanying reduction in

---

[1] F. E. Terman, *Radio Engineering*, McGraw-Hill Book Co., New York, 1947; pp. 189–190.

lead length and electrode area decreases both lead inductance and electrode capacitance. However, as the frequency increases, the necessary reduction in tube size to minimize the effects of lead inductance and electrode capacitance may be such as to render the task technologically impractical.

Second, large or double wires may be used to decrease both lead resistance and inductance, and separating the leads can largely eliminate the capacitance between various lead wires. These remedies can yield tubes that operate satisfactorily up to about 200 GHz. Higher frequencies require a more imaginative approach, involving the arrangement of tube electrodes so that they represent extensions of the external circuit. For example, the electrode structure of a lighthouse tube fits in naturally with coaxial line circuits. Such tubes can operate at a low microwave frequency of 1000 GHz. Their maximum usable frequency is limited not so much by lead inductance and electrode capacitance effects as by electron transit time considerations, which we shall now examine.

3. TRANSIT TIME EFFECTS

In order to evaluate the effects of electron transit time, we must first examine electronic motion in an electric field. Consider, for example, the case of an electron of charge $-e$ emitted from electrode A and moving toward electrode B as shown in Fig. 6.2.[2]. The electrodes are assumed to be planar and interconnected by a battery of voltage V. While the electron is in transit, electric field

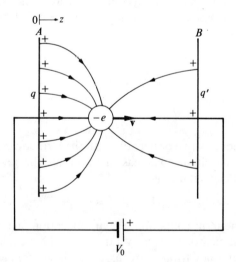

Fig. 6.2. Electron in transit between electrodes.

---

[2] Since the position of an electron must be given in terms of wave function in quantum mechanics rather than by a definite trajectory, we should really be considering the case of a bundle composed of a large number of electrons so that classical mechanics is applicable.

lines emanating from positive charges residing on electrodes A and B converge on the electron as depicted in Fig. 6.2. Since field lines tend to terminate on the closer conductor, we would expect the number of E-lines terminating on electrode A to decrease with time as the electron moves from left to right. Correspondingly, the number of E-lines terminating on electrode B increases with time. From Maxwell's equations (2.3) and (2.6) and by means of the divergence theorem $\oint \nabla \cdot \mathbf{D} \, dv = \oint \mathbf{D} \cdot d\mathbf{s}$, we have

$$E = \sigma_s/\varepsilon \qquad (6.21)$$

at the conductor surface. Here σ_s is the surface charge density. In deriving Eq. (6.21) we also noted that the electric field is perpendicular to the surface of the perfect conductor according to Eq. (3.44).

Since E at the electrodes is changing with time while the electron is in transit, it follows from Eq. (6.21) that the charge density is also changing with time. Inasmuch as the induced positive charges on electrode A decrease with time, we must conclude that an induced current must flow from electrode A through the external circuit to electrode B while the electron is in transit. Since the total charge within the space bounded by electrodes A and B must be zero, the total charges on A and B ($q + q'$) must be equal to e at any time during electron transit. When the electron arrives at B, $q \to 0$ while $q' \to e$; thus a total charge q has been transferred from A to B via the external circuit. At the instant of electron arrival, e and $-e$ neutralize each other at electrode B and induced current flow ceases. This may seem surprising, since at first thought we might (erroneously) expect no current to flow until the electron strikes electrode B and flows back to A via the external circuit.

We can compute induced charge $q'(t)$ on B by equating the work done in transferring the induced positive charge from A to B via the external circuit to the energy gained by the electron in moving from A to B. The electron is acted upon by a uniform field V_0/d (not shown in Fig. 6.2) so that the work W done on it by the field in moving it to z is

$$W = (eE)z = eV_0 z/d, \qquad (6.22)$$

where we have used the Lorentz force law (4.17). On the other hand, the battery does an amount of work equal to $q'V_0$ in transferring the induced charge from 0 to z. Equating this quantity to W of Eq. (6.22), we find

$$q'V_0 = eV_0 z/d \qquad (6.23)$$

or

$$q' = ez/d. \qquad (6.24)$$

The induced current i is therefore given by

$$i = \frac{dq'}{dt} = \frac{e}{d}\frac{dz}{dt} = \frac{ev}{d}. \qquad (6.25)$$

As we shall now see, Eq. (6.25) is a fundamental relationship in the study of high-frequency devices.

To express i explicitly as a function of time, we must find $v(t)$. To accomplish this we combine Newton's second law and the Lorentz force law (4.17) to get

$$eE = m\frac{d^2z}{dt^2}. \tag{6.26}$$

If $-e$ starts from A ($z = 0$) with zero velocity, we find from Eq. (6.26) that

$$v(t) = \frac{eE}{m}t. \tag{6.27}$$

Combining Eqs. (6.25) and (6.27), we finally obtain

$$i(t) = \frac{e^2 E}{md}t. \tag{6.28}$$

In this case, $i(t)$ increases linearly with time until the electron reaches B, at which time i drops abruptly to zero, consistent with the above physical reasoning.

In an actual vacuum tube (refer again to Fig. 6.2) there is a time-varying signal voltage superimposed upon the dc voltage V_0 that modifies the path of the electrons. Indeed, since electrode B represents the negatively biased control grid, V_0 may be very small or even reversed in sign. Therefore, the transit time of the electron (say in the absence of the signal) may be comparable to the period if the signal frequency is sufficiently high, say 1 GHz. In other words, in the conventional vacuum tube, electrons are first density-modulated and then accelerated. The density modulation by the signal voltage is performed in the cathode-grid region where the velocity of the electron is relatively small because of the retarding effect of the negative grid potential. As a consequence, the alternately accelerating and decelerating signal potential can significantly affect the velocity of the electron. If the frequency of the signal is sufficiently high, an electron may well be subjected to alternately accelerating and decelerating forces on its way to the grid if the transit time of the electron (in the absence of the signal, say) is comparable to the signal period. Once the electron passes the grid (of a triode, say) it is quickly accelerated to the anode by the high anode voltage.

It is interesting to determine the equivalent impedance presented to the grid circuit due to transit time effects, as we have done for lead inductance effects. To begin with, the grid, like other electrodes· can have changing charges induced on it as it is traversed by an electron stream. The relative phase angle between the signal voltage and the induced grid current indicates whether or not there is energy transferred from the signal source to the electron stream. This can occur even when the grid is sufficiently negative to attract no electrons.

Consider a triode with a negatively biased grid upon which is superimposed a small alternating voltage. As the instantaneous grid potential becomes less negative, the number of electrons flowing to the plate increases. However, the electron density is proportionately greater on the anode side, since the finite transit time delays the arrival of the electrons at the plate. The excess of approaching over receding electrons causes the induced charge on the grid to change with time and the grid to draw current in spite of the fact that it is negatively biased and captures no electrons from the stream. Conversely, somewhat later in the cycle the instantaneous grid voltage becomes more negative and the excess of receding electrons causes an induced current to flow out of the grid. We would then expect the magnitude of the fundamental component of this current I_g flowing in and out of the grid to be proportional to the number of electrons involved, i.e., to the tube transconductance g_m, signal voltage E_g, transit time τ, and frequency f:

$$I_g = C_1 g_m E_g \tau f, \tag{6.29}$$

where C_1 is a constant. It follows that the admittance Y_g presented to the grid is

$$Y_g = I_g/E_g = C_1 g_m \tau f. \tag{6.30}$$

From the above discussion, we might surmise that the transit time effect would merely give rise to a reactive grid current and grid admittance. This is not quite the case, however, since the induced grid current is not in quadrature with the signal voltage. Consider, for example, the instant when the applied voltage is at maximum value. Because of finite electron transit time, there is still a disproportionately larger number of approaching electrons on the cathode side of the grid as compared with the receding ones on the anode side. In other words, there is still finite and positive, rather than zero, induced grid current at the crest of the signal voltage cycle. This gives rise to a real component of the grid admittance representing power loss to the grid. The associated energy loss, of course, gives rise to the acceleration of electrons in the stream.

The real part of Y_g is of particular interest. If θ is the angle by which the fundamental component of the induced grid current fails to lead the grid voltage by $\pi/2$ rad, we have

$$G_g = Y_g \sin \theta, \tag{6.31}$$

where G_g is the grid conductance due to electron transit time effects. For small θ, $\sin \theta \simeq \theta$ which is in turn proportional to the product of the frequency and electron transit time. Equations (6.30) and (6.31) therefore combine to yield

$$G_g = C_2 g_m \tau^2 f^2. \tag{6.32}$$

Although we have neglected space-charge effects in derivating Eq. (6.32), their inclusion would merely change the numerical constant C_2.[3]

From the above discussion, it seems reasonable to surmise that transit time effects can be minimized by *first accelerating and then velocity-modulating* the electron stream. This is indeed the remedy offered in microwave tubes such as Klystrons. In these devices the beam is first modulated by an r.f. signal as it

---

[3] D. O. North, "Analysis of the Effects of Space Charge on Grid Impedance," *Proc. Inst. Radio Engrs.* **24**, 108–136 (1936).

traverses the input cavity gap of very small spacing. During the beam's subsequent drift along the drift tube, its velocity modulation is converted into density modulation. This in turn gives rise to electron bunches which induce an r.f. current at the output cavity gap. We will study these processes in detail in Section 1 of Chapter 7.

4. GAIN-BANDWIDTH PRODUCT LIMITATION

In the case of ordinary vacuum tubes, the output is usually a tuned circuit in order to maximize output voltage. Figure 6.3 shows the equivalent circuit of a

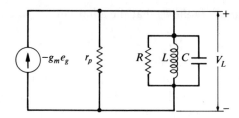

Fig. 6.3. Equivalent circuit of a pentode with tuned circuit load.

pentode. Here r_p and R are the plate and load resistances respectively and L is the tuning inductance for the stray capacitance C.[4] We can easily find the load voltage V_L from Fig. 6.3 as

$$V_L = - \frac{g_m e_g}{G + j[\omega C - (1/\omega L)]}, \qquad (6.33)$$

where $G = 1/r_p + 1/R$. It follows that the amplifier gain A_r at the resonance frequency $\omega = 1/\sqrt{LC}$ is simply

$$A_r = |V_{Lr}/e_g| = g_m/G. \qquad (6.34)$$

To find the $\frac{1}{2}$ power bandwidth of the circuit, we set $|V_L|$ of Eq. (6.33) equal to $|V_{Lr}|/\sqrt{2}$, where $|V_{Lr}|$, the resonant load voltage, is given by Eq. (6.34). The result shows that the $\frac{1}{2}$ power points occur at frequencies where

$$\omega C - \frac{1}{\omega L} = G. \qquad (6.35)$$

The roots of this quadratic equation are

$$\omega_2 = (G/2C) + \sqrt{(G/2C)^2 + 1/LC} \qquad (6.36)$$

and

$$\omega_1 = - (G/2C) + \sqrt{(G/2C)^2 + 1/LC}. \qquad (6.37)$$

---

[4] Figure 6.3 could schematically represent also the output gap of the Klystron amplifier of Fig. 7.1.

Subtracting ω_1 from ω_2, we obtain the $\frac{1}{2}$ power bandwidth $\Delta\omega$:

$$\Delta\omega = \omega_2 - \omega_1 = G/C. \tag{6.38}$$

Thus the gain-bandwidth product of the circuit of Fig. 6.3 is

$$A_r\Delta\omega = g_m/C, \tag{6.39}$$

where we have used Eqs. (6.34) and (6.38).

It is interesting to note that since the gain-bandwidth product is a function of the transconductance and stray capacitance only, it is a constant independent of frequency. This means that for a given tube an increase in bandwidth can be obtained only at the expense of a lower gain.

The remedy for this restriction is to use a nonresonant circuit. In this case, the bandwidth would not be restricted by the Q of the circuit. The gain, of course, is correspondingly lower and cannot be increased for lumped circuits used in conventional vacuum tubes. To obtain a high overall gain in microwave tubes, however, one may use a distributed and extended circuit, e.g., a helix or a periodically loaded waveguide used in traveling wave tubes. Because of the extended electron beam-circuit interaction for the whole length of the propagating structure ($\gg$ wavelength), one can obtain a quite respectable overall gain over a broad bandwidth in spite of the fact that the gain per unit wavelength is small.

PROBLEMS

6.1 a) In the equivalent circuit of Fig. 6.1(b), a number of electrode capacitances and lead inductances are omitted. Insert these capacitances and inductances in the appropriate places.
  b) Under what conditions is it justifiable to neglect the capacitances and inductances mentioned in part (a)?

6.2 Consider an electron moving from cathode to anode as shown in Fig. 6.2. Show that the sum of convection and displacement currents due to the electron is equal to the induced current due to the electron. For simplicity, assume that the electron starts from rest at the cathode at $t = 0$.

6.3 Consider the case of a triode composed of a cathode, grid, and anode with a resistance R and a dc voltage V_0 connected in series between the grid and anode. Let the beam current i crossing the grid be given by $i = I_0 + I_1 \sin \omega t$, where I_0 is the dc component. Assuming that the electrons have negligible velocity while passing through the grid, find the difference in the average power dissipated in the anode without modulation ($I_1 = 0$) and with modulation ($I_1 \neq 0$).

6.4 a) If an electron follows a traveling wave in such a way as to see the same field at all times, is its velocity equal to the group or phase velocity of the wave?
  b) Given that an electron beam is injected axially into one end of a rectangular waveguide of length l with velocity u_0, compute the direction and magnitude of the electron velocity at the waveguide exit. Assume that the waveguide is supporting the fundamental mode.

CHAPTER 7

CAVITY-TYPE MICROWAVE TUBES

In this chapter we shall study the behavior of microwave tubes using resonant cavity structures. These devices include the Klystron amplifier and oscillator and the magnetron oscillator. Due to the use of high-Q resonant cavities, these amplifiers are characterized by high gain and small bandwidth or, in the case of oscillators, by small beam power required to generate oscillations. As a prelude to the study of these devices, we will examine the phenomenon of velocity and density modulation, which is fundamental to the understanding of cavity-type microwave tubes.

1. VELOCITY AND DENSITY MODULATION

In Section 3 of Chapter 6, we studied briefly the phenomenon of velocity and density modulation in a Klystron amplifier. We showed that transit time effects which degrade vacuum tube performance at high frequencies can be overcome by first accelerating and then velocity-modulating the electron stream. This is indeed the remedy offered in microwave tubes such as Klystrons. Consider, for example, the two-cavity Klystron amplifier shown in Fig. 7.1. The electrons are first accelerated by the dc voltage V_0 and then modulated by the signal voltage $V_1 \sin \omega t$ across the gap of the buncher cavity. After the electrons emerge from the buncher cavity, they drift along in the field-free tube until they arrive at the

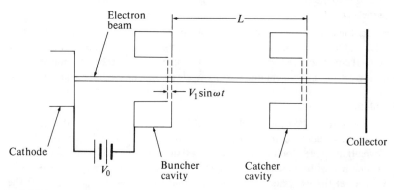

Fig. 7.1. A Klystron amplifier.

catcher cavity. Due to the presence of the modulation voltage, an electron emerging from the buncher cavity will have a velocity dependent upon the value of the instantaneous voltage during its transit across the gap. This spread in velocity will be converted into density modulation. Electron bunches are formed in the drift tube as the faster electrons catch up with slower ones that left the buncher gap earlier. The bunches induce a voltage across the gap of the catcher cavity as they pass through it on their way to the collector. Note that so long as the transit time across the cavity gap is small compared to the period of the modulating signal, ordinary transit time effects are essentially eliminated. Because the electron velocity at the buncher is high as a result of dc preacceleration, it is technologically possible to have a cavity gap of sufficiently small width such that the transit time through it is small compared to a period. In contrast, the electron velocity in the ordinary vacuum tube is quite small in the cathode-grid region within which the electron stream is density-modulated.

Let us proceed to a quantitative analysis of the Klystron amplifier of Fig. 7.1. By the law of conservation of energy, the velocity of the electrons u_0 at the entrance of the buncher gap is

$$u_0 = \sqrt{2eV_0/m}, \tag{7.1}$$

where e and m are respectively charge magnitude and mass of the electron. For simplicity, let us assume for the moment that the transit time across the buncher gap is zero. Again applying the law of conservation of energy, we have for the electrons at the gap exit

$$\tfrac{1}{2} m u_1^2 = e(V_0 + V_1 \sin \omega t_1), \tag{7.2}$$

where u_1 is the velocity of the electrons after they emerge from the buncher gap and t_1 is the time at which an electron crosses the gap. Solving for u_1 in Eq. (7.2), we find

$$u_1 = \sqrt{(2e/m) V_0 [1 + (V_1/V_0) \sin \omega t_1]}. \tag{7.3}$$

If V_1/V_0 is small, as is usually the case, Eq. (7.3) reduces by binomial expansion to

$$u_1 \simeq u_0 \left(1 + \frac{V_1}{2V_0} \sin \omega t_1 \right), \tag{7.4}$$

where we have used Eq. (7.1). We can see from Eq. (7.4) that the exit velocity of the electron is to a first approximation proportional to the r.f. voltage $V_1 \sin \omega t_1$, aside from an additive time-independent term.

Before we analyze the Klystron problem further mathematically, let us discuss how density modulation could be accomplished in the field-free drift tube region. Since there is a spread in electron velocity after the electrons emerge from the buncher, we would expect the relative position of the electrons to be a function of the axial coordinate and time. To be more specific, let us consider the situation

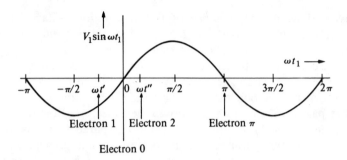

Fig. 7.2. Velocity modulation.

depicted in Fig. 7.2. The exit velocity of electron 0, which crosses the gap at $t_1 = 0$, is clearly u_0 since the modulation voltage is zero at $t_1 = 0$. Electron 1, which crosses the gap at time t', will have a velocity u'_1, which is less than u_0 because of the negative modulation voltage. Correspondingly, u''_1 of electron 2, which crosses the gap at time t'', will be larger than u_0. Since Electron 1 has a lower velocity but leaves earlier than electron 0, it will tend to fall back toward electron 0. Similarly, since electron 2 has a higher velocity but left later than electron 0, it will tend to catch up with electron 0. A little reflection will show that the electrons that cross the gap during the time interval $-(\pi/2\omega) < t_1 < 0$ will tend to fall back toward electron 0, while the electrons that cross the gap during the time interval $0 < t_1 < \pi/2\omega$ will tend to catch up with electron 0. Thus an electron bunch will be formed about the electron that crosses the gap when the modulation voltage is zero and going from negative to positive. Similarly, electrons that leave between the intervals $\pi/2\omega < t_1 < \pi/\omega$ and $\pi/\omega < t_1 < 3\pi/2\omega$ will diverge from electron π, which crosses the gap at $t_1 = \pi/\omega$. This diffusion of electrons forms what is known as an "antibunch." We therefore conclude that electron bunches will form about dc electrons crossing the gap at

$$\omega t_1 = n(2\pi), \quad (7.5)$$

where n is an integer including zero.

From the above discussion, we would expect the electron or current density in the drift region to be a function of the axial distance from the buncher cavity. In particular, we are interested in determining the current i_2 at the catcher gap because it induces a voltage across the gap of the cavity. We can find i_2 by invoking the law of conservation of charge.

2. THE KLYSTRON AMPLIFIER

During a time interval Δt_1, a charge q_1 passes the buncher gap. If I_0 is the dc current, we have the relation

$$q_1 = I_0 \, \Delta t_1. \quad (7.6)$$

This same amount of charge q_1 also passes the output gap at a later time in an interval Δt_2, which may or may not be equal to Δt_1. Thus

$$q_1 = i_2 \Delta t_2, \tag{7.7}$$

where i_2 is the current at the catcher gap. Combining Eqs. (7.6) and (7.7), we find

$$i_2 = I_0 \frac{\Delta t_1}{\Delta t_2}. \tag{7.8}$$

In passing to the limit of infinitesimally small Δt's, $\Delta t_1/\Delta t_2$ becomes $|dt_1/dt_2|$. Therefore, to determine i_2 we must first determine dt_1/dt_2. Now t_2, the arrival time at the catcher, is simply related to t_1, the departure time at the buncher, as follows:

$$t_2 = t_1 + \frac{L}{u_1} = t_1 + \frac{L}{u_0\sqrt{1 + (V_1/V_0)\sin \omega t_1}}, \tag{7.9}$$

where L, as shown in Fig. (7.1), is the buncher-catcher separation and Eq. (7.3) has also been used. If we again assume that $V_1/V_0 \ll 1$, Eq. (7.9) simplifies to

$$t_2 \simeq t_1 + \frac{L}{u_0}\left(1 - \frac{V_1}{2V_0}\sin \omega t_1\right). \tag{7.10}$$

Differentiating Eq. (7.10) and combining the resulting equation with Eq. (7.8), we find the expression for i_2:

$$i_2 = \frac{I_0}{|1 - (\omega L/u_0)(V_1/2V_0)\cos \omega t_1|}. \tag{7.11}$$

Figure 7.3 is a plot of i_2 of Eq. (7.11) as a function of ωt_1. For definiteness, $(\omega L/u_0)(V_1/2V_0)$ has been set equal to 0.5, 1.0, 1.5. Observe that there are pulses of current, one per cycle, centered about the electron that crosses the input gap at $\omega t_1 = n2\pi$, in accordance with Eq. (7.5) obtained above from physical reasoning. The current is clearly rich in harmonics. Since the output cavity is usually tuned to the input frequency, a Fourier analysis of $i_2(t)$ to find the amplitude of the fundamental component is in order.

Expanding $i_2(t)$ in a Fourier series, we have

$$i_2 = A_0 + \sum_n (A_n \cos n\omega t_2 + B_n \sin n\omega t_2), \tag{7.12}$$

using t_2 instead of t_1 in order to relate i_2 to the arrival time t_2 at the output gap. In the usual fashion, the coefficients A_n and B_n are given by the integrals

$$A_0 = \frac{1}{2\pi}\int_{-\pi}^{\pi} i_2 d(\omega t_2),$$

$$A_n = \frac{1}{\pi}\int_{-\pi}^{\pi} i_2 \cos n\omega t_2 \, d(\omega t_2), \tag{7.13}$$

$$B_n = \frac{1}{\pi}\int_{-\pi}^{\pi} i_2 \sin n\omega t_2 \, d(\omega t_2). \tag{7.14}$$

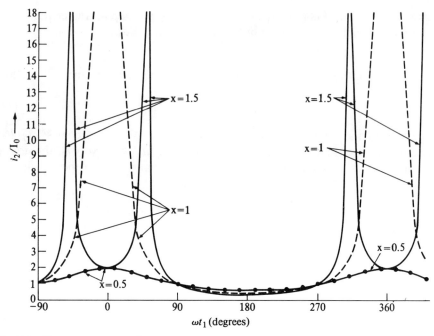

Fig. 7.3. Output current vs. time.

Using Eqs. (7.8) and (7.10), we can convert these integrals into functions of t_1 instead of t_2:

$$A_0 = \frac{1}{2\pi} \int_{-\pi}^{\pi} I_0 \, d(\omega t_1),$$

$$A_n = \frac{1}{\pi} \int_{-\pi}^{\pi} I_0 \cos n\omega \left(t_1 + \frac{L}{u_0} - \frac{LV_1}{2u_0 V_0} \sin \omega t_1 \right) d(\omega t_1), \quad (7.15)$$

and

$$B_n = \frac{1}{\pi} \int_{-\pi}^{\pi} I_0 \sin n\omega \left(t_1 + \frac{L}{u_0} - \frac{LV_1}{2u_0 V_0} \sin \omega t_1 \right) d(\omega t_1). \quad (7.16)$$

Noting that

$$\cos(x + y) = \cos x \cos y - \sin x \sin y,$$

and

$$\sin(x + y) = \sin x \cos y + \cos x \sin y,$$

we can readily evaluate integrals (7.15) and (7.16) to yield

$$A_0 = I_0,$$

$$A_n = 2I_0 J_n \left(\frac{nV_1 \omega L}{2u_0 V_0} \right) \cos \frac{n\omega L}{u_0}, \quad (7.17)$$

and

$$B_n = 2I_0 J_n \left(\frac{nV_1 \omega L}{2u_0 V_0} \right) \sin \frac{n\omega L}{u_0}, \quad (7.18)$$

where J_n is the nth-order Bessel function. Substituting these expressions back into Eq. (7.12), we finally obtain the equation for the output current i_2:

$$i_2 = I_0 + \sum_n 2I_0 J_n\left(\frac{nV_1\omega L}{2u_0 V_0}\right) \cos n\omega\left(t_2 - \frac{L}{u_0}\right). \qquad (7.19)$$

Note that we have derived Eq. (7.19) by neglecting coulomb repulsion between charges. The transverse motion of electrons can be greatly reduced by a strong axial magnetic field, but we cannot eliminate the longitudinal repulsion forces, except at extreme relativistic speeds. Thus the velocity of the electrons in the drift tube will not be a constant, as was implicitly assumed. Because the inclusion of coulomb repulsion forces will only complicate our calculations without changing the essential features of our results, we shall leave this consideration until later. If V_1/V_0 is not $\ll 1$, we must use graphical methods and Eq. (7.9) instead of Eq. (7.10) to obtain i_2.

If the output cavity is tuned to the input frequency, as is usually the case, all components of i_2 except the one corresponding to $n = 1$ will be of negligible amplitude. Thus Eq. (7.19) simplifies to

$$(i_2)_\omega = 2I_0 J_1\left(\frac{V_1}{2V_0}\frac{\omega L}{u_0}\right)\cos \omega\left(t_2 - \frac{L}{u_0}\right). \qquad (7.20)$$

An examination of Eq. (7.20) shows that i_2 is clearly not a monotonically decreasing function of the electron drift angle $\omega L/u_0$ in radians, in contrast to the transit time effects expected of conventional vacuum tubes. Indeed, there is an optimum value of L, other things being constant, which gives rise to the maximum possible value of i_2. An examination of the plot of $J_1(x)$ shows that $[J_1(x)]_{max}$ occurs at $x = 1.84$. Thus we find

$$L_{optimum} = 1\cdot 84\frac{2V_0}{V_1}\frac{u_0}{\omega}. \qquad (7.21)$$

Note, however, that in deriving Eq. (7.19) for i_2 we have implicitly neglected the transit time through the input and output gaps. If the gap transit times are considered, other effects such as beam coupling and beam loading must be taken into account. We will discuss these effects in detail in the next section.

The power output of the amplifier is $(I_2)_\omega (V_2)_\omega/2$, where $(I_2)_\omega$ and $(V_2)_\omega$ are respectively the beam current and gap voltage at frequency ω across the output gap. The magnitude of $(V_2)_\omega$ depends on the r.f. component of the beam current at the output gap and cavity parameters. However, it cannot exceed $V_0 + V_1 \simeq V_0$, the dc accelerating potential. To extract as much energy as possible from the electron beam it is clearly advisable for all electrons to cross the gap when the output cavity voltage is maximum retarding. If the cavity voltage is V_0, all electrons will be stopped at the gap, indicating that all of their energy has been extracted. It follows that no voltage higher than V_0 can exist in the gap, since the beam is the sole source of energy for the excitation of the output cavity. In practice, of course, it is impossible for all electrons to arrive at the output gap at the same time for a given cycle. This would require zero transit time across cavity gaps, sawtooth voltage modulation, no space-charge repulsion, etc. These details are left as a problem for the reader.

Since the dc beam power is equal to $V_0 I_0$, the conversion efficiency η of the tube is simply

$$\eta = \frac{(V_2)_\omega (I_2)_\omega}{2 V_0 I_0}. \tag{7.22}$$

If $(I_2)_\omega$ is maximized according to Eq. (7.21) and $(V_2)_\omega$ is set equal to V_0 in accordance with the foregoing discussion, Eqs. (7.20) and (7.22) yield a maximum conversion efficiency of 58%. In this calculation we have made use of the fact that $[J_1(x)]_{max} = J_1(1.84) = 0.58$. In practice, the conversion efficiency of Klystron amplifiers is considerably smaller, in the range of 15 to 30%.

If we place a floating cavity between the input and output cavities, we can increase the conversion efficiency by arranging the transit angles between cavities so that more electrons enter the catcher during the retarding half-period than would otherwise. The intermediate cavity has the effect of further bunching the beam by the gap voltage, which is induced by the r.f. component of the beam current.

Klystrons with more than three cavities may be staggered-tuned to obtain amplification over a bandwidth of a few percent. The resonance frequencies of the intermediate cavities are offset from the mid-band frequencies and their Q's are adjusted by loading to predetermined values. The analysis of these multicavity problems is similar to that presented for the two-cavity Klystron, although necessarily more complicated. Therefore, we shall not study these cases further here, but instead turn to a detailed examination of electronic behavior in a cavity gap in the next section.

3. ELECTRON-FIELD INTERACTION

In order for the voltage across a cavity gap to modulate the electron beam traversing it, there must be some interaction between the electric field across the gap and the electrons in the beam.[1] In general, this field-electron interaction leads to an exchange of energy between the cavity field and the beam. In this section we shall examine the detailed features of this interaction, leading eventually to the resulting representation of the cavity gap as an equivalent circuit.

First, if we take the divergence of both sides of Eq. (2.2), we find

$$\nabla \cdot \left(\mathbf{i} + \frac{\partial \mathbf{D}}{\partial t} \right) = 0, \tag{7.23}$$

---

[1] In many instances, such as the case in point, the magnetic force acting on an electron is much smaller than the electric force. Therefore we can neglect the second term of the Lorentz force given by Eq. (4.17).

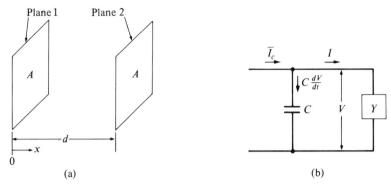

Fig. 7.4. (a) Cavity gap. (b) Equivalent circuit for (a).

since $\nabla \cdot (\nabla \times \mathbf{A}) \equiv 0$ where $\mathbf{A}$ is any vector. If the gap is formed by two perfectly conducting parallel planes of area A as shown in Fig. 7.4(a),[2] and if fringing effects are neglected, Eq. (7.23) can be integrated to read

$$I_c + A\varepsilon_0 \frac{\partial E}{\partial t} = I, \qquad (7.24)$$

where $I_c = iA$ is the total convection current due to the motion of charges, $A\varepsilon_0 \, \partial E/\partial t$ is the total displacement current, and I is the total current, convection plus displacement. Due to the geometry of the system, I_c, E, and I are all x-directed. Multiplying both sides of Eq. (7.24) by dx/d and integrating from 0 to d, we obtain the average value of the driving (convection) current $\bar{I}_c$ as

$$\bar{I}_c = \frac{1}{d}\int_0^d I_c \, dx = I + C\frac{dV}{dt} = VY + C\frac{dV}{dt}, \qquad (7.25)$$

where $C = \varepsilon_0 A/d$ is the capacitance of the gap and $V = -\int_0^d E \, dx$ is the voltage across the gap. Thus we can draw an equivalent circuit, shown in Fig. 7.4(b), with $VY = I$, where Y is the external admittance presented to the total current I.[3] Equation (7.25) is appropriately known as the circuit equation. We can determine the voltage from Eq. (7.25) for a given Y if $\bar{I}_c$ is known. However, to find $\bar{I}_c$ we

---

[2] It is clear that if an electron beam traverses the cavity gap, coaxial holes must exist on the two plates. These holes may be fitted with grids or may be gridless. In any event, the fields in the region near the axis of the holes will be essentially x-directed, as assumed here.

[3] Since V can be analyzed into its Fourier components $V_n e^{jn\omega t}$ [see Eq. (7.27)], a Y can always be defined for each component of V. Thus, in general,

$$\bar{I}_c = \sum_n \left\{ Y_n V_n e^{jn\omega t} + C\frac{d}{dt}\left[(V_n e^{jn\omega t})\right] \right\}.$$

must first determine the motion of the electrons in the gap. To accomplish this, we use the electronics equation (4.19):

$$q\mathbf{E} = \frac{d}{dt}(m\mathbf{u}), \qquad (7.26)$$

where q, m, and $\mathbf{u}$ are the charge, mass, and velocity of the electrons. Since $\mathbf{E}$ is related to the gap voltage V, Eqs. (7.25) and (7.26) are not independent of each other but must be solved simultaneously. In other words, the convection current gives rise to the voltage across the gap, and the gap voltage in turn influences the motion of the electrons.

If the voltage across the gap equals $\sum_n V_n e^{jn\omega t}$, then Eq. (7.26) becomes

$$\ddot{x} = \sum_n \frac{e}{m} \frac{V_n}{d} e^{jn\omega t}, \qquad (7.27)$$

where e is the magnitude of the electronic charge. If the injection velocity at $x = 0$ is given, we can integrate Eq. (7.27) to yield the electronic position x as a function of time t. If Fig. 7.4(a) represents a buncher gap, the injected velocity at $x = 0$ will have only a dc component. If it represents a catcher gap, the injected velocity will contain r.f. as well as dc components. For the sake of generality, we shall assume that the entrance velocity at $x = 0$ is given by

$$u_1 = u_0 + \sum_n u'_n e^{jn\omega t_1}, \qquad (7.28)$$

where t_1 is the time at which the electron crosses the $x = 0$ plane (plane 1) and u_0 is the dc beam velocity. We can now integrate Eq. (7.27) with the help of Eq. (7.28) to yield

$$\dot{x} = \sum_n \left[\frac{e}{m} \frac{V_n}{djn\omega} \left(e^{jn\omega t} - e^{jn\omega t_1} \right) + u_0 + u'_n e^{jn\omega t_1} \right]. \qquad (7.29)$$

Integrating this expression again, we have

$$x = -\sum_n \left\{ \frac{eV_n}{mdn^2\omega^2} \left[e^{jn\omega t} - e^{jn\omega t_1} - je^{jn\omega t_1}(n\omega t - n\omega t_1) \right] \right.$$
$$\left. + \frac{u_0 + u'_n e^{jn\omega t_1}}{n\omega}(n\omega t - n\omega t_1) \right\}, \qquad (7.30)$$

since $x = 0$ at $t = t_1$.

To find the convection current I_c in the gap, we must invoke the law of conservation of charge. Since charge is conserved, whatever charges cross the plane at $x = 0$ must also eventually cross the plane x. However, since modulation by the gap voltage changes the velocity of the electrons as they traverse the gap, they may take different time intervals to cross a given incremental distance about planes located at 0 and x. More explicitly, we have

$$I_c \Delta t = I_{c1} \Delta t_1, \qquad (7.31)$$

where I_c is the convection current at plane x and I_{c1} is the injected current at $x = 0$ containing both a dc and an r.f. component, that is,

$$I_{c1} = I_0 + \sum_n I'_{cn} e^{jn\omega t_1}, \tag{7.32}$$

where I_0 is the dc beam current. In the limit when the time intervals approach zero, Eq. (7.31) gives

$$I_c = I_{c1} \left| \frac{dt_1}{dt} \right|. \tag{7.33}$$

Thus, with the injection current I_{c1} given, we can find I_c if we know the ratio dt_1/dt. The latter can be found by explicit differentiation of Eq. (7.30). We would expect dt_1/dt and therefore I_c to be a function of x. By averaging this result over the entire gap, from plane 1 to plane 2, we can obtain the average value of I_c or $\bar{I}_c$.

Differentiating Eq. (7.30) with respect to t_1 and substituting the resulting expression for dt/dt_1 into Eq. (7.33), we obtain

$$I_c = \sum_n (I_0 + I'_{cn} e^{jn\omega t_1}) \tag{7.34}$$

$$\times \left| \frac{-j\dfrac{eV_n}{mdn\omega}(e^{jn\omega t} - e^{jn\omega t_1}) + u_0 + u'_n e^{jn\omega t_1}}{\dfrac{eV_n}{mdn\omega}(n\omega t - n\omega t_1)e^{jn\omega t_1} + u_0 + u'_n[1 - j(n\omega t - n\omega t_1)]e^{jn\omega t_1}} \right|,$$

also using Eq. (7.32). As it stands, Eq. (7.34) is an exact expression for I_c. If r.f. effects are small compared to dc effects, we can simplify this result considerably. More explicitly, let us rewrite Eq. (7.34) to read

$$I_c = \sum_n (I_0 + I'_{cn} e^{jn\omega t_1}) \tag{7.35}$$

$$\times \left| \frac{1 + \dfrac{u'_n}{u_0} e^{jn\omega t_1} - j\dfrac{eV_n/mdn\omega}{u_0}(e^{jn\omega t} - e^{jn\omega t_1})}{1 + \dfrac{u'_n}{u_0}[1 - j(n\omega t - n\omega t_1)]e^{jn\omega t_1} + \dfrac{eV_n/mdn\omega}{u_0}(n\omega t - n\omega t_1)e^{jn\omega t_1}} \right|.$$

This exact expression for I_c may be greatly simplified if $u'_n/u_0 \ll 1$ and $(eV_n/mdn\omega)/u_0 \ll 1$. Dividing the numerator by the denominator and neglecting terms containing $(u'_n/u_0)^2$, $(eV_n/mdn\omega)^2/u_0^2$, $(eV_n/mdn\omega)u'_n/u_0^2$, and higher, we find from Eq. (7.35) that

$$I_c \simeq \sum_n \left\{ (I_0 + I'_{cn} e^{jn\omega t_1}) \right. \tag{7.36}$$

$$\left. + I_0 \left[\left(j\frac{u'_n}{u_0} - \frac{eV_n/mdn\omega}{u_0} \right)(n\omega t - n\omega t_1)e^{jn\omega t_1} - j\frac{eV_n/mdn\omega}{u_0}(e^{jn\omega t} - e^{jn\omega t_1}) \right] \right\}$$

We see from Eq. (7.36) that the convection current I_c is a function of t and t_1.

86 Electron-field interaction

However, from a consideration of the geometry of Fig. 7.4(a), t and t_1 are related by the expression

$$t - t_1 = \frac{x}{\dot{x}} = \frac{x}{\left[\sum_n \frac{e}{m} \frac{V_n}{dj n\omega} (e^{jn\omega t} - e^{jn\omega t_1}) + u_0 + u'_n e^{jn\omega t_1}\right]}, \quad (7.37)$$

where Eq. (7.29) has been used. Consistent with the assumptions made to derive Eq. (7.36), by retaining only those terms in Eq. (7.37) up to the first power in the r.f. quantities we find

$$t_1 \simeq t - \sum_n \frac{x}{u_0}\left[1 - \frac{u'_n}{u_0} e^{jn\omega t_1} - \frac{eV_n/mdjn\omega}{u_0}(e^{jn\omega t} - e^{jn\omega t_1})\right]. \quad (7.38)$$

Substituting Eq. (7.38) into Eq. (7.36) for t_1 and again neglecting higher-order terms, we finally find

$$I_c(x,t) = I_0 + \sum_n \left\{ I'_{cn} e^{-jn\omega x/u_0} e^{jn\omega t} + jI_0 \left(\frac{u'_n}{u_0}\right)\left(\frac{n\omega x}{u_0}\right) e^{-jn\omega x/u_0} e^{jn\omega t} \right. \quad (7.39)$$

$$\left. - I_0 \left(\frac{eV_n/mdn\omega}{u_0}\right)\left[\frac{n\omega x}{u_0} e^{-jn\omega x/u_0} + j\left(1 - e^{-jn\omega x/u_0}\right)\right] e^{jn\omega t}\right\}.$$

As expected for the case of small signal approximation, the convection current I_c, aside from a dc term I_0, is a linear combination of r.f. components, each proportional to $e^{jn\omega t}$. The first term in this equation is clearly the dc beam current, while the second is the phase-shifted r.f. injection current. The third and fourth terms are the injection velocity and gap voltage modulation respectively. Thus Eq. (7.39) tells us that there is a distribution of convection currents within the gap. Physically, this means that because of entrance r.f. current and velocity, and velocity modulation resulting from the gap voltage, there is a spectrum of densities and velocities for the electrons in the gap at any given time.

We note again from Eq. (7.39) that $[I_c(x,t) - I_0]$ is a linear combination of terms proportional to $e^{jn\omega t}$. Therefore, we may focus our attention on only one of the Fourier components with no loss of generality. Furthermore, since the buncher and catcher cavities are usually tuned to the same frequency, say ω, the other r.f. components of $I_c(x,t)$ will be smaller by a factor of approximately $Q(\sim 10^3$ to 10^4 at microwave frequencies). Therefore, to a very good approximation, we can concern ourselves only with the component proportional to $e^{j\omega t}$ in the discussion to follow.

Inasmuch as $I_c(x,t) - I_0$ is approximately proportional to $e^{j\omega t}$, the relative velocities of electrons at various values of x are independent of time. It follows that to find the average value of the convection current we need merely integrate $I_c(x)$ from 0 to d and divide the result by the gap spacing d. Before proceeding to find $\bar{I}_c(x)$, let us denote the second, third, and fourth terms on the right-hand side of Eq. (7.39) (omitting the factor $e^{j\omega t}$) by $\bar{I}_c(I_c')$, $\bar{I}_c(u')$, and $\bar{I}_c(V)$ respectively:

$$I_c(x) - I_0 = \bar{I}_c(I_c') + \bar{I}_c(u') + \bar{I}_c(V). \tag{7.40}$$

Now, from Eqs. (7.39) and (7.40) we find

$$\bar{I}_c(I_c') = \frac{I_{c1}'}{d}\int_0^d e^{-j\omega x/u_0}\,dx = I_{c1}'\frac{\sin(\omega d/2u_0)}{\omega d/2u_0}e^{-j\omega d/2u_0}. \tag{7.41}$$

Thus we see that, because various phases of the current component attributed to the injected r.f. current I_c' are present, the effect of I_c' is reduced by a factor $\sin(\omega d/2u_0)/(\omega d/2u_0)$ known as the *beam coupling coefficient*. In addition, the phase of this component of the average convection current is shifted by an angle corresponding to the middle of the gap relative to the phase of the injection r.f. current, as expected.

By the same averaging procedure used for Eq. (7.41), we can find the average values of $\bar{I}_c(u')$ and $\bar{I}_c(V)$ as

$$\bar{I}_c(u') = -I_0\left(\frac{u_1'}{u_0}\right)\left[\left(\cos\frac{\omega d}{2u_0} - \frac{\sin(\omega d/2u_0)}{\omega d/2u_0}\right) - j\sin\frac{\omega d}{2u_0}\right]e^{-j\omega d/2u_0} \tag{7.42}$$

and

$$\bar{I}_c(V) = -jI_0\left(\frac{2eV_1/md\omega}{u_0}\right)\left[\cos\frac{\omega d}{2u_0} - \frac{\sin(\omega d/2u_0)}{\omega d/2u_0}\right]e^{-j\omega d/2u_0}. \tag{7.43}$$

As is the case with $\bar{I}_c(I_c')$, we see that there are also amplitude reductions and phase shifts connected with $\bar{I}_c(u')$ and $\bar{I}_c(V)$. The effect that gives rise to the terms exhibited by Eq. (7.43) is referred to as *beam loading*. Since $\bar{I}_c(V)$ is proportional to V_1, we can define an admittance $Y_V = -\bar{I}_c(V)/V_1$ by means of Eq. (7.43), which yields

$$G(V) = \frac{I_0}{2V_0}\frac{\sin(\omega d/2u_0)}{\omega d/2u_0}\left[\cos\left(\frac{\omega d}{2u_0}\right) - \frac{\sin(\omega d/2u_0)}{\omega d/2u_0}\right], \tag{7.44}$$

$$B(V) = \frac{I_0}{2V_0}\frac{\cos(\omega d/2u_0)}{\omega d/2u_0}\left[\cos\left(\frac{\omega d}{2u_0}\right) - \frac{\sin(\omega d/2u_0)}{\omega d/2u_0}\right], \tag{7.45}$$

where $G(V) + jB(V) = Y(V)$.

For self-consistency, the $\bar{I}_c$ of Eq. (7.40) derived from the electronics equation (7.27) must equal the $\bar{I}_c$ of the circuit equation (7.25). Equating Eq. (7.25) to Eq. (7.40) (after averages are taken), we obtain

$$\bar{I}_c(I_c') + \bar{I}_c(u') = VY(V) + C\frac{dV}{dt} + VY. \tag{7.46}$$

This equation is represented as an equivalent circuit in Fig. 7.5. Note that to the right of the dashed line the equivalent circuit is the same as that of Fig. 7.4(b). Thus the calculation following the electronics equation (7.27) merely delineates the sources of the average convection current $\bar{I}_c$. Note also that the external (i.e., external to the cavity gap) admittance Y is due to the inductance of the cavity or magnetic field residing in the rest of the cavity and the load connected to the cavity output and cavity ohmic loss.

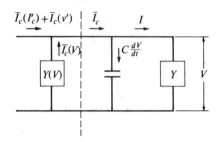

Fig. 7.5. Current sources and equivalent circuit for a cavity gap.

4. THE KLYSTRON OSCILLATOR

A. A Two-Cavity Oscillator

If a portion of the catcher-cavity output is fed back into the input of the buncher cavity in the proper phase, a Klystron oscillator will result. Figure 7.6 depicts this possibility. We can easily show that for the voltage polarities indicated, the gain with feedback $A_f = V_{\text{out}}/V_{\text{in}}$ is related to the forward gain $A_0 = V_{\text{out}}/V'_{\text{in}}$ by

$$A_f = \frac{A_0}{1 - \beta A_0}. \tag{7.47}$$

Thus, if the characteristics of the amplifier and feedback circuit are such that $1 - \beta A_0$ equals zero, oscillation results since A_f is infinite and no input is necessary to obtain a finite output.

Let us first determine the amplitude and phase of the forward-gain A_0. For simplicity, assume that the transit time across the buncher and catcher gaps is

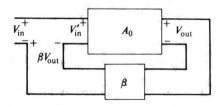

Fig. 7.6. Schematic diagram for a feedback circuit.

negligible. Equations (7.41), (7.42), (7.44), and (7.45) then reduce respectively to

$$\bar{I}_c(I'_c) = I'_{c_1},$$
$$\bar{I}_c(v') = 0,$$
$$G(V) = 0,$$
$$B(V) = 0.$$
(7.48)

It follows from Fig. 7.5 that if the catcher cavity is at resonance—that is, if $\omega C = -B$—$V_{out} = \bar{I}_c/G$. Identifying $\bar{I}_c$ with the Klystron output current $(i_2)\omega$ of Eq. (7.20), we readily find

$$V_{out} = \frac{2I_0 J_1\left(\frac{|V'_{in}|}{2V_0}\frac{\omega L}{u_0}\right) e^{j\omega(t_2 - L/u_0)}}{G},$$
(7.49)

using the complex notation for $\cos \omega[t_2 - (L/u_0)]$. However, to compute the complex open-loop gain $A_0 = V_{out}/V'_{in}$, we must scrutinize the phase of this voltage, V_{out}, relative to that of V'_{in}. From Fig. 7.3, we see that the center of the electron bunch or the maxima of V_{out} occur at $\omega t_1 = n2\pi$, where n is an integer including zero. But as far as the output gap is concerned, the bunch centered about the electron that crosses the buncher gap at $t_1 = 0$ will take a time interval of L/u_0 to arrive at the output gap, where L is the buncher-catcher separation and u_0 is the dc beam velocity. Correspondingly, there is an additional phase shift of $\omega L/u_0$ rad for the output voltage. (Although this phase shift is important to the study of oscillations, it was not germane to the discussion of Klystron amplifiers in Section 2.) We can now show that for a small amplitude of oscillation, that is, $(V_1 \omega L/2V_0 u_0) \ll 1$,

$$A_0 = \frac{V_{out}}{V'_{in}} = \frac{I_0 |V'_{in}| \omega L e^{j\omega L/u_0} e^{j2\pi n}}{-j2V_0 u_0 G |V'_{in}|},$$
(7.50)

where we have made use of the fact that $V'_{in} = |V'_{in}| \sin \omega t_1 = Re(-j|V'_{in}| e^{i\omega t_1})$ [see Eq. (7.2)]. Note also that $J_1(x) \simeq x/2$ for $x \ll 1$. Accordingly, we find the amplitude and phase of A_0 from Eq. (7.50) as

$$A_0 = \frac{I_0 \omega L}{2V_0 u_0 G} \bigg/ \frac{\omega L}{u_0} + \frac{\pi}{2} + 2n\pi,$$
(7.51)

where n is an integer. From Eq. (7.47), we see that oscillation would occur if $1 - \beta A_0 = 0$. Accordingly, Eq. (7.50) leads to the amplitude and phase requirement for β:

$$|\beta| = \frac{2V_0 u_0 G}{I_0 \omega L},$$
(7.52)

$$\underline{/\beta} = -2n\pi - \frac{\omega L}{u_0} - \frac{\pi}{2}.$$
(7.53)

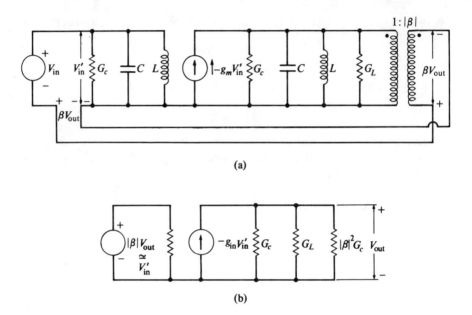

Fig. 7.7. (a) Equivalent circuit for a two-cavity Klystron oscillator at any frequency. (b) Equivalent circuit for (a) at resonance and in the oscillation state.

Some physical interpretation of Eqs. (7.52) and (7.53) is in order. To begin with, G represents the total conductance presented to the ω component of the catcher current. As Fig. 7.7 shows, there are three contributions to G. Besides the catcher cavity and load conductances, there is the conductance of the feedback circuit as seen at the output with V_{in} shorted. A consideration of this equivalent gives

$$G = (1 + |\beta|^2)G_c + G_L. \tag{7.54}$$

Substituting Eq. (7.54) into Eq. (7.52), we obtain an expression for I_0 in terms of $|\beta|$:

$$I_0 = \frac{2V_0 u_0}{|\beta|\omega L}[(1 + |\beta|^2)G_c + G_L]. \tag{7.55}$$

For maximum power transfer to the load, G_L should be set equal to $(1 + \beta^2)G_c$, which in this case is the equivalent source impedance. Making this assumption in Eq. (7.55) and then differentiating I_0 with respect to $|\beta|$ and setting the resulting expression equal to zero, we find the optimum value of $|\beta|$ to be 1. It then follows from Eq. (7.55) that the starting current $(I_0)_s$ is given by

$$(I_0)_s = 8V_0 u_0 G_c/\omega L \tag{7.56}$$

To minimize the starting current, we see from this equation that, other things being constant, we should use high-Q microwave cavities. The conductance G across the output gap accounts for the catcher and buncher-cavity losses and the load, but we have implicitly neglected other losses, such as those due to beam coupling and beam loading and those in the cavity couplings and feedback line. In practice, the value of I_0 required for stable oscillation is several times the value given by Eq. (7.56).

In order to maintain stable oscillations for all conditions of loading, we must strongly couple the buncher and catcher cavities. If the coupling coefficient of these cavities is larger than the critical value, there will be two peaks in the amplitude response vs. frequency curve; oscillations will exist only at or near the peaks of these amplitude curves. The corresponding phase shifts at these two peaks are of course different, thus requiring different values of transit time to satisfy the phase condition (7.53). In other words, $/\beta$ of that equation is dependent on not only the length of the coupling line but also the relative tuning of the cavities. Typically, the peak separation is very pronounced as a result of tight loading. Indeed, cavity tuning and adjustment of the feedback line can have a pronounced effect upon the relative amplitude and separation of the peaks. However, the peak separation is usually small compared to the resonance frequency ω of the cavity when they are uncoupled, so it is permissible to use ω in the above expressions.

B. Reflex Oscillator

Although the two-cavity Klystron oscillator just described is the natural outgrowth of the Klystron amplifier, it is not the only type of oscillator. Indeed, the most commonly used microwave laboratory oscillator is the reflex oscillator. This oscillator has only one cavity; feedback is provided by a turnabout electron beam. The electron beam traverses the gap twice by means of a reflector electrode. Thus the cavity bunches the beam on its first passage and catches it on its return passage. Figure 7.8 is the schematic diagram of the reflex oscillator.

Let us first assume that, because of the presence of noise, oscillations exist in the cavity so that a voltage $V_1 \sin \omega t_1$ appears across the cavity gap. As we shall demonstrate later, the frequency ω for stable oscillation, although very close to the resonance frequency of the cavity in the absence of the electron beam, can be changed by a few percent by varying the beam voltage V_0 and reflector voltage V_R. This procedure is known as *electronic tuning*. Coarse frequency tuning, on the other hand, is accomplished by changing the gap spacing of the cavity.

In the presence of the gap voltage $V_1 \sin \omega t$, the electron beam will emerge from the cavity toward the reflector with a velocity distribution given by Eq. (7.3), repeated here:

$$\left.\frac{dx}{dt}\right|_{\substack{x=0 \\ t=t_1}} = u_1 = \sqrt{(2e/m)(V_0 + V_1 \sin \omega t_1)}. \tag{7.57}$$

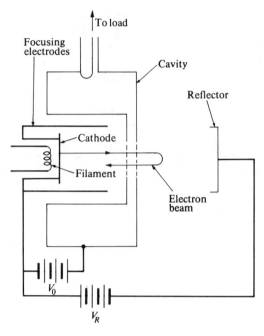

Fig. 7.8. Schematic diagram of a reflex Klystron oscillator.

Given this information as the initial condition in the solution of the equation of motion for electrons [see Eq.(4.19)],

$$-eE = m\frac{d^2x}{dt^2}, \tag{7.58}$$

we find

$$x = -\frac{e(V_R + V_0)}{2md}(t - t_1)^2 + \sqrt{\frac{2e}{m}}(V_0 + V_1 \sin \omega t_1)(t - t_1), \tag{7.59}$$

using the relationship $E = (V_R + V_0)/d$. To find the round-trip transit time of the electron $t_R - t_1$, we merely let $x = 0$ and $t = t_R$ in Eq. (7.59), yielding

$$t_R - t_1 = T_0\sqrt{1 + (V_1/V_0)\sin \omega t_1}, \tag{7.60}$$

where $T_0 = 4d\sqrt{mV_0/2e}/(V_0 + V_R)$ is the transit time for an electron that crosses the gap when the oscillation voltage is zero. We have seen that in a two-cavity oscillator an electron bunch is formed about an electron that crosses the gap when the modulation voltage is zero and going from negative to positive. In the reflex oscillator an electron bunch is formed about an electron that crosses the gap when the voltage is zero and going from positive to negative. The physical reason for this difference is that the transit time decreases with increasing beam voltage for the amplifier while it increases for the reflex oscillator. In the latter case, other factors being constant, an increase in the beam voltage increases the electron velocity, thereby projecting it further into the anode-reflector space with a consequent increase in the round-trip transit time.

If the returning electron beam is to contribute a maximum amount of energy to the oscillation, it must cross the gap when the gap field is maximum retarding. In this way, a maximum amount of kinetic energy is extracted from the returning electron beam to substain the cavity oscillation. With a little reflection and the help of Fig. 7.2, we can see that for maximum energy transfer to occur $\omega(t_R - t_1)$, corresponding to the center of the bunch, must equal $(n - \frac{1}{4}) 2\pi$, where $n = 1, 2, 3, \ldots$. It follows from Eq. (7.60) that for given values of V_0 and V_R there is some oscillation frequency ω that satisfies

$$\omega T_0' = 4d\omega \frac{\sqrt{(m/2e)V_0}}{V_0 + V_R} = (n - \tfrac{1}{4}) 2\pi, \tag{7.61}$$

where T_0' is the transit time of the electron that crosses the gap at zero oscillation voltage and at values of V_0 and V_R which satisfy the above relation. Thus we have found the relationship between the oscillation frequency ω and the various voltages.

Returning to Eq. (7.60), we find for $V_1 \ll V_0$

$$t_R - t_1 \simeq T_0\left(1 + \frac{V_1}{2V_0} \sin \omega t_1\right). \tag{7.62}$$

Except for the difference in sign in front of V_1, this expression for t_R is exactly the same as that for t_2, the arrival time at the catcher for the electrons in a Klystron amplifier, as given by Eq. (7.10). Invoking the law of charge conservation as before [see Eqs. (7.6) through (7.8)], we find that $i_R = I_0 \mid dt_1/dt_R \mid$, where i_R is the current of the returning beam at the cavity gap. Combining this expression for i_R and Eq. (7.62), we find

$$i_R = \frac{I_0}{\mid 1 + \omega T_0(V_1/2V_0) \cos \omega t_1 \mid}. \tag{7.63}$$

Again, this expression for i_R is the same as that for i_2 given by Eq. (7.11) except for the difference in sign in front of the term containing V_1. It follows that i_R vs. ωt_1 for various values of $\omega T_0 V_1/2V_0$ will have the same shapes as i_2 vs. ωt_1 shown in Fig. 7.3 except that the pulses will be centered at $n\pi$, where n is a positive odd integer.

Decomposing i_R into its Fourier components and evaluating the coefficients involved, we find

$$i_R = I_0 + \sum_n 2I_0 J_n\!\left(n\omega T_0 \frac{V_1}{2V_0}\right) \sin\left[n\omega(t_R - T_0)\right], \tag{7.64}$$

which is similar in form to Eq. (7.19) for a Klystron. For an oscillator with a high Q-cavity, all components of i_R except that corresponding to $n = 1$ will be of negligible amplitude.

94 The Klystron oscillator

We can now determine the conditions required for sustained oscillation in terms of the cavity-gap equivalent circuit of Fig. 7.5. For simplicity and consistency with the foregoing analysis, we shall assume that the transit time across the gap is negligible compared to the period of the oscillation. It follows from Eqs. (7.42), (7.44), and (7.45) that in this case $Y(V) \to \infty$, $\bar{I}_c(u') \to 0$, and the driving current of the circuit is given by i_R of Eq. (7.64) with $n = 1$. In terms of T_0 and T_0', this equation can be rewritten to read

$$i_R = 2I_0 J_1\left(\omega T_0 \frac{V_1}{2V_0}\right) \sin(\omega t_R - \omega T_0' - \phi), \tag{7.65}$$

where

$$\phi = \omega T_0 - \omega T_0' \tag{7.66}$$

is an angle which represents the deviation of the transit angle from that corresponding to maximum retardation of an electron on its return trip. When $T_0 = T_0'$, $\phi = 0$, and i_R is in phase with the oscillation voltage V_1. Therefore, we can conveniently define an admittance Y_R on the basis of Eq. (7.65) to account for the effect of the electron beam as follows:

$$Y_R = \frac{2I_0 J_1(\omega T_0(V_1/2V_0))}{V_1}(-\cos\phi + j\sin\phi). \tag{7.67}$$

Note that we obtained Y_R by dividing $-i_R$ in complex notation by V_1. Since i_R represents the driving current flowing into the parallel circuit composed of C and Y, as shown in Fig. 7.5, Y_R represents a source of supply of r.f. current. Indeed, when $\phi = 0, 2\pi, 4\pi, \ldots$, Eq. (7.67) shows that Y_R is a purely negative resistance or a source of power. On the other hand, if $\phi = \pi, 3\pi, 5\pi, \ldots$, Y_R is real and positive. For intermediate values of ϕ, Y_R is complex. This situation is depicted in Fig. 7.9(a), where $\text{Re}(Y_R)$ and $\text{Im}(Y_R)$ are plotted against ϕ.

To complete our analysis of the reflex oscillator, we need to find Y, which represents the admittance connected across the cavity gap in Fig. 7.5. To begin

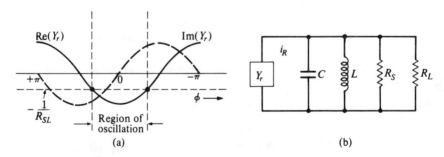

Fig. 7.9 (a) Admittance due to the electron beam as a function of transit angle deviation (b) Equivalent circuit for the reflex oscillator cavity.

with, the magnetic field in the rest of the resonant cavity represents an inductance in parallel with the capacitance of the gap. Likewise, we may represent the cavity loss and useful load by parallel resistances R_S and R_L respectively. All of this is summarized schematically by the equivalent circuit of Fig. 7.9(b).

From Fig. 7.9(a) and (b), we can readily see that oscillation will occur if $\text{Re}(Y_R)$ is negative and has an absolute value larger than or equal to $1/R_{SL}$, where R_{SL} represents the parallel combination R_S and R_L. However, if

$$\text{Re}(Y_R) < 0 \quad \text{and} \quad |1/\text{Re}(Y_R)| > (1/R_{SL}),$$

V_1 will grow until

$$|1/\text{Re}(Y_R)| = (1/R_{SL}),$$

at which point stable oscillation is established. This behavior is embodied in the nonlinear nature of Y_R as given by Eq. (7.67). To be more specific, for stable oscillation, we have from Fig. 7.9(a) and (b):

$$\frac{2I_0 J_1(\omega T_0(V_{1S}/2V_0))}{V_{1S}} \cos \phi = \frac{1}{R_{SL}}. \tag{7.68}$$

For given values of V_0 and V_R, T_0 and ϕ are determined and Eq. (7.68) yields the stable oscillation amplitude V_{1S}. Furthermore, for circuit resonance to occur, the imaginary part of the total impedance of the circuit represented by Fig. 7.9(b) must equal zero, or

$$-\frac{1}{\omega L} + \omega C + \text{Im}(Y_R) = 0, \tag{7.69}$$

where Y_R corresponds to oscillation voltage V_{1S} given by Eq. (7.68). Thus at $\phi = 0$ the oscillation frequency ω is simply the resonance frequency of the cavity in the absence of the beam, that is, $\omega_0 = 1/\sqrt{LC}$. Since $\text{Im}(Y_R)$ can be either positive or negative in the oscillation region, the actual oscillation frequency ω can be either larger or smaller than ω_0. With a little reflection and the use of Eqs. (7.66), (7.68), and (7.69), we can find the output and frequency deviation plot of Fig. 7.10 as a function of V_R with V_0 constant. Since T_0 and ϕ are functions of

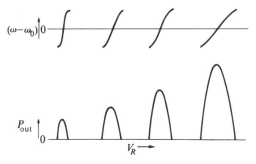

Fig. 7.10. (a) Deviation of reflex oscillator oscillating frequency from cavity resonance frequency. (b) Power output of oscillator vs. reflector voltage.

the beam voltage V_0 as well as the reflector voltage V_R, ω can also be changed by changing V_0. This means of changing the oscillation frequency ω by changing electrode voltages is called *electronic tuning*, a phenomenon that does not occur in low-frequency oscillators.

5. APPLICATION OF KLYSTRONS

Because of their relatively high inherent shot noise, Klystron amplifiers are not suitable for extremely low-level linear operation. However, because of their high gain at low signal levels [see Eq. (7.20)], we can use a chain of Klystron voltage amplifiers in cascade to amplify a low-level signal to the desired value. In this case, only the last stage is driven to saturation and acts as a power amplifier. In crystal-controlled microwave transmitters, for example, milliwatts of power at the desired microwave frequency are first generated by the crystal and then amplified by a chain of Klystron amplifiers culminating in a high power output from the last stage for transmission via the antenna.

Frequency translation can be accomplished for Klystrons by modulating the beam voltage as a function of time. For example, if the beam voltage V_0 is not constant but instead of the form $V_0(1 + kt)$, where k is a constant, we can easily see from Eq. (7.20) that the output frequency is $\omega[1 + k(L/u_0)]$ for $kt \ll 1$. In practice, of course, the linear increase or decrease of beam voltage, aside from a dc bias, cannot continue indefinitely. Therefore, a periodic linear sawtooth waveform is used.

The use of periodic instead of continuous modulation brings up the question of phase coherence. The signal output during every modulation period must be in phase with that of every other modulation period in order to give a large output power at a single output frequency. Phase coherence is assured if the magnitude of the sawtooth voltage is such that the difference in transit angle between the maximum and minimum voltage points is an integral multiple of 2π, at least for low modulation frequencies. When phase coherence exists, the output is perfectly periodic except for a number of either "missing" or "overlapping" disturbances in each modulation period. These interruptions in periodicity are caused by the fly-back of the sawtooth. Their effect on the degree of suppression of the carrier and the undesired sidebands is usually small, however, compared to that of finite fly-back time, etc. Frequency translation by means of linear-sawtooth modulation of transit time devices such as Klystrons and traveling wave tubes has been studied in detail.[4]

Amplitude, phase, and pulse modulation of Klystrons can be accomplished by changing the beam voltage or current. Since, according to Eq. (7.20), the output current is proportional to the beam current I_0, linear amplitude modulation can be accomplished by amplitude modulating I_0. Similarly, if I_0 is pulse modulated, the output will be composed of microwave pulses.

---

[4] R. C. Cummings, "The Servodyne Frequency Translator," *Proc. IRE* **45**, 175 (1956).

Phase modulation can be accomplished by amplitude modulating the beam voltage. Let the beam voltage V_b be given by

$$V_b = V_0 + V_m \sin \omega_m t, \qquad (7.70)$$

where V_m is the modulation amplitude and ω_m is the modulation frequency. The transit time T for the electron that crosses the buncher gap at zero gap voltage, i.e., the electron at the center of the bunch, is then

$$T \simeq \frac{L}{u_0} - \frac{L}{u_0} \frac{V_m}{2V_0} \sin \omega_m t. \qquad (7.71)$$

The derivation of this equation is similar to that of Eq. (7.10). If ω_1 is the input frequency, the phase angle, θ_2, of the bunched current at the catcher gap referred to the buncher gap, is simply $\omega_1(t + T)$, or

$$\theta_2 = \omega_1 t + \frac{\omega_1 L}{u_0} - \frac{\omega_1 L V_m}{2 u_0 V_0} \sin \omega_m t. \qquad (7.72)$$

It follows that the output frequency, being equal to $d\theta_2/dt$, is given by

$$\omega_2 = \omega_1 - \frac{\omega_1 \omega_m L V_m}{2 u_0 V_0} \cos \omega_m t. \qquad (7.73)$$

In phase modulation terminology, the modulation index m_p—the angle in radians through which the phase is displaced at the peak of the modulation cycle away from the phase that would exist if there were no modulation—is

$$m_p = \omega_1 L V_m / 2 u_0 V_0, \qquad (7.74)$$

where we have used Eq. (7.73)

Reflex oscillators are widely used in the laboratory for microwave measurements. Typically, their outputs range from milliwatts to hundreds of milliwatts. Since their output power and frequency can be varied to a limited extent by changing the beam and reflector voltages (electronic tuning), they are extremely flexible devices. Coarse frequency tuning involves changing the cavity gap spacing or the gap capacitance C by means of a diaphragm and micrometer arrangement.

PROBLEMS

7.1 a) Refer to the Klystron amplifier shown in Fig. 7.1. Frequently, it is necessary to cool the collector, which is heated by electron bombardment. Given that the resonance frequency of the cavity ω is 10^{10} cps, $V_0 = 5000$ V, $I_0 = 0.1$ A, the area of the cavity gap $A = 1$ cm^2, gap spacing $d = 1$ mm, and Q of the cavity including the effect of the load = 3000, compute the power dissipated in the collector. For simplicity, assume negligible power extraction at harmonic frequencies and neglect the effect of finite transit time across the gaps. Assume that $(I_2)_\omega$ corresponds to its maximum possible value.

b) Describe a way to extract more power at frequency ω from the beam by applying a voltage of appropriate magnitude and polarity to the collector.

Problems

7.2 a) Assume that an electron beam of dc current I_0 is first accelerated by a voltage V_0 and then by a sawtooth $V_1(t)$ of peak-to-peak amplitude V_1 and period T. Find the current i_2 at a distance z from the modulation gap. For simplicity, assume that the transit time across the modulation gap is negligible and $V_1 \ll V_0$.
  b) In practice, how would you establish a nearly sawtooth-like voltage across a cavity gap?

7.3 Consider the Klystron input gap shown in Fig. 7.1. The transit time through the gap is not necessarily small compared to the r.f. modulation period $2\pi/\omega$, nor is V_1 necessarily small compared to V_0. Let d be the gap width and calculate the exit electron velocity spectrum u_1.

7.4 If a third floating cavity is placed between the buncher and catcher of a Klystron amplifier, it is possible to increase its conversion efficiency. Analyze the behaviour of such a three-cavity amplifier.

7.5 Find the optimum value of $|\beta|$ and the minimum value of I_0 required for stable oscillation to occur in a two-cavity Klystron oscillator. Assume that the two cavities are not identical and the coupling line is lossless. Also assume that maximum power is transferred to the output resistive load.

7.6 a) How are electrons in the reflected beam collected in a reflex Klystron?
  b) Is some form of beam focusing necessary in the anode-reflector space for, say, an x-band Klystron? Elaborate.

7.7 a) What is the difference between frequency and phase modulations?
  b) Devise a method of frequency modulation for a Klystron amplifier and derive the modulation expression corresponding to Eq. (7.73) for phase modulation.

CHAPTER 8

MICROWAVE DEVICES WITH PERIODIC STRUCTURES

In the last chapter we studied the behavior of microwave tubes of the cavity type. These tubes, although capable of overcoming the lead inductance and transit time effects of ordinary vacuum tubes, are limited in bandwidth because of the gain-bandwidth product restriction characteristic of resonant structures. In this chapter we shall demonstrate how to obtain a large gain over a rather broad bandwidth by letting the electron beam interact with the fields of a nonresonant periodic structure over an extended length. Devices with nonresonant configurations include the traveling wave tube (TWT), backward wave oscillator (BWO), magnetron, and some particle accelerators.

In an empty waveguide, the phase velocity, or the velocity with which a point of constant phase travels, is inevitably greater than the velocity of light c. Since electrons cannot travel faster than the velocity of light due to relativistic considerations, there can be no cumulative interaction between an electron beam and the electromagnetic fields of an empty waveguide. Therefore, microwave devices such as the traveling wave amplifier and linear electron accelerator must have a structure that can sustain waves whose phase velocities are smaller than the velocity of light. One obvious means of accomplishing this is by loading the waveguide with a dielectric. If the dielectric completely fills the waveguide and the waveguide is far above cutoff, the phase velocity is $c/\sqrt{\varepsilon/\varepsilon_0}$, where ε is the dielectric constant. Thus if a phase velocity equal to $0.1c$ is desired, ε should be $100\ \varepsilon_0$. Dielectrics with a dielectric constant of this magnitude would probably be sufficiently lossy to render the structure impractical.

A better approach to the problem is to use propagating structures whose physical attributes are periodically varying with distance in the direction of propagation z. The fields in such structures can be Fourier resolved into spatial harmonics in z. Some of these harmonics have phase velocities which in general are less than c; therefore, they can propagate in synchronism with an electron beam.

In this chapter we shall first study the passive properties of these periodic structures. We shall then consider their properties in the presence of an electron beam in conjunction with our discussion of active microwave devices such as the TWT and BWO. (We might mention here in passing that, because of the periodic attributes of such structures, stop bands and pass bands may appear in the frequency domain.)

1. PERIODIC STRUCTURES

A. Introduction

Let us first examine the problem of mode propagation in an empty waveguide. We can obtain the expression for the phase velocity v_p from Eqs. (3.5) and (3.130) as

$$v_p = \frac{c}{\sqrt{1 - (\lambda/\lambda_c)^2}}, \qquad (8.1)$$

where λ and λ_c are the free space and cutoff wavelength respectively for a given waveguide mode. Thus v_p is inevitably larger than the velocity of light at a finite frequency. On the other hand, according to the theory of relativity, the velocity of energy or mass propagation must be less than the velocity of light. If an electron beam is continuously accelerated or decelerated it must see essentially the same axial r.f. electric field as it traverses the propagating structure. In other words, the electron must ride along a point of constant phase. Now, since the phase velocity v_p is the velocity with which a point of constant phase travels, we see that v_p must be nearly equal to the electronic velocity v_0. Since $v_p > c$ for the case in point and $v_0 < c$ because of relativistic effects, it is clear that synchronism between the wave and an electron beam can never occur in an empty waveguide.

One obvious way to decrease the phase velocity is to completely fill the waveguide with a dielectric of dielectric constant ε. Equation (8.1) then becomes

$$v'_p = \frac{c/\sqrt{\varepsilon/\varepsilon_0}}{\sqrt{1 - (\lambda/\lambda_c)^2}}, \qquad (8.2)$$

where v'_p is the phase velocity of the dielectric-filled guide. Thus if v'_p is $0.1c$ for $(\lambda/\lambda_c)^2 \ll 1$, ε must be about 100 ε_0. As mentioned previously, a dielectric with a constant of this magnitude is usually too lossy to be of practical interest at microwave frequencies. Of course, the losses can be reduced by filling the guide only partially with the dielectric, but the reduction in phase velocity is then correspondingly smaller. Thus, other factors being equal, an electron beam with higher v_0 is required.

A better approach to the v_p-reduction problem is to use periodic structures. In this case, the phase velocity of some of the spatial harmonics in z obtained by Fourier analysis of the waveguide field may be smaller than the velocity of light c. In what follows, we shall study the fundamental nature of such structures.

B. Floquet's Periodicity Theorem

The study of the behavior of periodic structures is based principally on Floquet's periodicity theorem, which states:

> *The electromagnetic field distribution at an arbitrary plane of a periodic structure for a given mode of oscillation at a given frequency can differ, by at most a complex constant, from that at planes an integral multiple of a period away from the reference plane.*

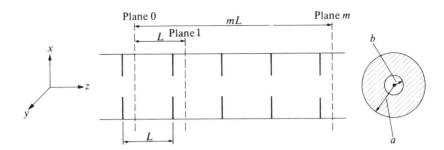

Fig. 8.1. Periodically loaded cylindrical waveguide.

This statement can be easily understood with reference to Fig. 8.1. For definiteness, let structure be a periodically disk-loaded cylindrical waveguide. Consider first the electromagnetic field distribution at cross-sectional plane 0. This field must be a solution of Maxwell's equations that obeys the boundary conditions of the structure. If the structure is infinitely long, we may think of the left and right ends of the structure joining at infinity so as to render the structure continuous. It is then evident that the cells are indistinguishable from each other. In particular, the cross-sectional field distribution at planes 0, 1, or m must have the same x- and y-dependence and the fields in cells 0, 1, and m must also satisfy the same boundary conditions. However, for propagating waves, the phases at planes 0, 1, and m may not be the same. Furthermore, if the walls and irises of the structure are not perfectly conducting, the amplitudes of the fields at these planes are also in general different because of unavoidable attenuation. Thus we conclude that the fields at planes 0, 1, and m, etc., can differ from one another by at most a complex constant, in accordance with Floquet's theorem.

Let us now derive the second-order differential equation in **E** for the case where the parameters μ, ε, and σ are possibly functions of the spatial coordinates. Taking the curl of Eq. (2.1), we have

$$\nabla \times \nabla \times \mathbf{E} = -j\omega(\nabla\mu \times \mathbf{H} + \mu\nabla \times \mathbf{H}), \qquad (8.3)$$

using Eq. (2.5) and the vector identity $\nabla \times (\phi\mathbf{A}) = \nabla\phi \times \mathbf{A} + \phi\nabla \times \mathbf{A}$. Substituting the expressions for **H** and $\nabla \times \mathbf{H}$ in terms of **E** from Eq. (2.1) and (2.2) respectively, we find the second-order differential equation in **E**:

$$\nabla \times \nabla \times \mathbf{E} - \omega^2\mu\varepsilon\left(1 + \frac{\sigma}{j\omega\varepsilon}\right)\mathbf{E} = \frac{\nabla\mu}{\mu} \times (\nabla \times \mathbf{E}), \qquad (8.4)$$

where we have made use of Eqs. (2.6) and (2.24). We may note here that if μ were not a function of the spatial coordinates, that is, if $\nabla\mu = 0$, Eq. (8.4) would be identical in form to Eq. (2.25), providing the vector identity $\nabla \times \nabla \times \mathbf{A} = \nabla(\nabla \cdot \mathbf{A}) - \nabla^2\mathbf{A}$ was used. However, because of the possible spatial dependence of ε, it is not expedient to employ this vector identity here.

In principle, at least, we can solve Eq. (8.4) and a similar equation for $\mathbf{H}$,

$$\nabla \times \nabla \times \mathbf{H} - \omega^2 \mu \varepsilon \left(1 + \frac{\sigma}{j\omega\varepsilon}\right)\mathbf{H} = \left(\frac{\nabla\sigma + j\omega\nabla\varepsilon}{\sigma + j\omega\varepsilon}\right) \times (\nabla \times \mathbf{H}), \quad (8.5)$$

and determine the electromagnetic fields for a given periodic structure by invoking the appropriate boundary conditions. The physical configurations of most periodic structures in microwave electronics, however, are sufficiently complicated that we must use approximations to solve Eqs. (8.4) and (8.5). Before we examine the details of such structures, it would be beneficial to study some of their common features.

C. Spatial Harmonics

As a consequence of Floquet's theorem, we can resolve $\mathbf{E}$ into partial waves commonly known as spatial harmonics. Refer again to Fig. 8.1 and assume that we have found the solution to the general wave equation (8.4) which satisfies the boundary conditions. If the solution at $z = z_0$ is given by $\mathbf{E}(r, \theta, z_0)$, it follows from Floquet's theorem that at $z = z_0 + L$ the solution is related to $\mathbf{E}(r, \theta, z_0)$ by

$$\mathbf{E}(r, \theta, z_0 + L) = C\mathbf{E}(r, \theta, z_0), \quad (8.6)$$

where C is an arbitrary constant. Equation (8.6) follows from Floquet's theorem since it requires that the fields at z_0 and $z_0 + L$ differ by at most a complex constant at a given frequency. Similarly, we have

$$\mathbf{E}(r, \theta, z_0 + 2L) = C\mathbf{E}(r, \theta, z_0 + L) = C^2\mathbf{E}(r, \theta, z_0), \quad (8.7)$$

where the constant C in the middle expression is the same as that used in Eq. (8.6) since all cells are physically indistinguishable. Consequently, the ratio of the fields at $z_0 + L$ and z_0 must equal that at $z_0 + 2L$ and $z_0 + L$, that is,

$$\frac{\mathbf{E}(r, \theta, z_0 + L)}{\mathbf{E}(r, \theta, z_0)} = \frac{\mathbf{E}(r, \theta, z_0 + 2L)}{\mathbf{E}(r, \theta, z_0 + L)}.$$

By induction, we can generalize Eq. (8.7) to read

$$\mathbf{E}(r, \theta, z_0 + mL) = C^m \mathbf{E}(r, \theta, z_0), \quad (8.8)$$

where m is an integer. For waves propagating in the positive z-direction of a passive structure, it is clear that $\mathbf{E}(r, \theta, z_0 + mL)$ must be less than or equal to $\mathbf{E}(r, \theta, z_0)$. It then follows that $|C|$ must be less than 1. Of course, the converse is true for waves propagating in the negative z-direction. This, coupled with the fact that $\mathbf{E}(r, \theta, z_0 + mL)$ differs from $\mathbf{E}(r, \theta, z_0)$ solely due to attenuation and phase shift, means that the most general expression for C is[1]

$$C = \exp -\left(\gamma_0 + j\frac{2\pi n}{L}\right)L, \quad (8.9)$$

---

[1] Note that the $\gamma_0, \alpha_0, \beta_0$ used here are not the same as those defined by Eq. (2.29) for a plane wave in an infinite medium.

where n is an integer, positive or negative, and the propagation constant $\gamma_0 = \alpha_0 + j\beta_0$ may be a function of frequency but not of spatial coordinates.

We shall now expand $\mathbf{E}(r, \theta, z)$ as a Fourier series of spatial harmonics of fundamental period L. If losses are present, there will in general be coupling between these harmonics which complicates our analysis without the benefit of much additional insight. Accordingly, we shall assume that the structure is lossless, so that $\alpha_0 = 0$ and $\gamma_0 = j\beta_0$. In order to satisfy the relations (8.8) and (8.9), a typical mode of propagation must be of the form

$$\mathbf{E}_n(r, \theta, z) = \mathbf{A}_n \exp\left[-j\left(\beta_0 + \frac{2\pi n}{L}\right)z\right]. \tag{8.10}$$

Note that

$$\frac{\mathbf{E}_n(r, \theta, z_0 + mL)}{\mathbf{E}_n(r, \theta, z_0)} = e^{-j\beta_0 mL}$$

which is in accord with Floquet's theorem. To represent the actual field $E(r, \theta, z)$ in the propagating structure at a given frequency, we must sum up all possible waves of the form (8.10). Thus

$$\mathbf{E}(r, \theta, z) = \sum_n \mathbf{E}_n(r, \theta, z) = \sum_{n=-\infty}^{\infty} \mathbf{A}_n \exp\left[-j\left(\beta_0 + \frac{2\pi n}{L}\right)z\right], \tag{8.11}$$

or the phase constant β_n of the nth mode is

$$\beta_n = \beta_0 + \frac{2\pi n}{L}. \tag{8.12}$$

Equation (8.12) shows that the field in a periodic structure can be expanded as an infinite series of waves, all at the same frequency but with different phase velocities v_{pn}. Specifically, from Eqs. (3.130) and (8.12), we find

$$v_{pn} = \frac{\omega}{\beta_n} = \frac{\omega}{\beta_0 + (2\pi n/L)}, \quad n = \ldots -2, -1, 0, 1, 2 \ldots. \tag{8.13}$$

It is important to note here that v_{pn} decreases for increasingly higher values of n; for this discussion let both β_0 and n be positive. Therefore, it appears possible for a wave of suitable n to have a phase velocity less than the velocity of light c, in contrast to the empty waveguide case where v_p is always larger than c, as given by Eq. (8.1). It also follows that synchronization and interaction between the wave and electron beam are possible, giving rise to the possibility of realization of active microwave devices.

On the other hand, the group velocity, being equal to $d\omega/d\beta_n$ according to Eq. (3.132), is just equal to

$$v_g = d\omega/d\beta_0, \tag{8.14}$$

which is independent of n.

D. ω–β Diagram

The most important task in a problem involving ω–β diagrams is to determine β_n as a function of ω. Once $\beta_n(\omega)$ is found, v_{pn} and v_g, for example, can be computed using Eqs. (8.13) and (8.14). In this way, we can find which wave is suitable for a particular microwave electronics device. According to Eq. (8.12), $\beta_n(\omega)$ and $\beta_0(\omega)$ would be identical were it not for the presence of the frequency-independent term $2\pi n/L$. Therefore, a determination of $\beta_0(\omega)$ is tantamount to a determination of $\beta_n(\omega)$. Since a periodically loaded guide is certainly not the simplest possible propagating structure, determining β_0 as a function of ω will not be particularly easy. We shall discuss such calculations later in this chapter. At the moment, we are more interested in the general nature of the ω–β diagram for a periodic structure than in the details themselves.

First, *if there is no anisotropic material present, ω must be an even function of β_0*, for, in the absence of such material, the propagation characteristics must be reciprocal, i.e., independent of the direction of propagation. Second, *ω must be a periodic function of β_0 with period $2\pi/L$*. We can easily prove this statement as follows. Suppose β_0 increases by $2\pi/L$. Then, according to Eq. (8.12), the phase constant previously designated as β_{-1} will increase by $2\pi/L$ to become the previous β_0. In general, each β_n will change by $2\pi/L$ so that it will equal the previous β_{n+1}, thereby changing the designation of each old β_n but leaving the whole sets of β_n's unchanged. This renumbering process clearly cannot change the physical situation in the slightest, for we must find the same coefficient for each of the new β's as that for the old β, which had the same numerical value, and the frequency as determined from the field must also be the same.

Third, the *group velocity v_g, being equal to $d\omega/d\beta_0$, must be zero at $\beta_0 = n\pi/L$.* This follows from the fact that waves reflected from the equally spaced disks of Fig. 8.1 add in phase when the cell length corresponds to an integral multiple of half of a guide wavelength, that is, when $L = n\lambda_g/2$ or equivalently $\beta_0 = n\pi/L$.

Using the above statements, we can almost sketch the general features of the ω–β_0 diagram. Before we do so, however, it would be expedient for us to deduce the ω–β plot for two limiting cases: (1) when $b \to a$, that is, when there are no disks; and (2) when $b \to 0$, that is, when there are no holes in the disks at all.

When $b \to a$, the fields in the structure would be expected to approach those for an empty guide. Consequently, the ω–β diagram is governed by the relation (3.5)

$$\omega^2 \mu\varepsilon = \beta_0^2 + k_c^2, \qquad (8.15)$$

where we have changed γ_0^2 to $-\beta_0^2$ since the guide is considered lossless. In accordance with Eq. (8.15), the ω–β_0 diagram is plotted in Fig. 8.2. Note that the ω–β_0 curve is a hyperbola intersecting the vertical axis at $\omega\sqrt{\mu\varepsilon} = k_c$; for convenience, we have plotted the normalized frequency, that is, ω divided by $1/\sqrt{\mu\varepsilon}$ along the vertical axis, where $1/\sqrt{\mu\varepsilon}$ is the velocity of light in an infinite medium of dielectric constant ε and permeability μ. Therefore, as $\omega \to \infty$, the curve approaches the 45° line, indicating that in this limit the phase velocity $v_p = \omega/\beta_0$

approaches the velocity of light in the medium. For simplicity, we have shown only the ω–β_0 curve for the dominant mode.

On the other hand, as $b \to 0$, the individual cells approach perfectly enclosed cavities. We studied the behavior of such cavities in some detail in Chapter 3. If $b = 0$, each cell will function as a completely enclosed cavity. We can easily determine the resonance frequencies ω_n of these cavities by requiring that L be equal to an integral multiple of half of a guide wavelength, that is, $L = n\lambda_g/2$. Correspondingly, we have

$$\beta_0 = 2\pi/\lambda_g = n\pi/L. \tag{8.16}$$

Substituting this expression into Eq. (8.15), we find for ω_n

$$\omega_n\sqrt{\mu\varepsilon} = \sqrt{k_c^2 + (n\pi/L)^2}. \tag{8.17}$$

Thus we see that the ω_n's are determined by the intersection of the ω–β_0 hyperbola of an empty guide, as given by Eq. (8.15) and plotted in Fig. 8.2, and the vertical line $\beta_0 = n\pi/L$. These points are marked by crosses in Fig. 8.2.

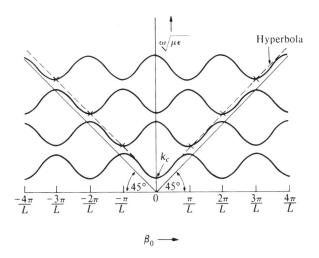

Fig. 8.2. ω–β diagram of a periodically loaded waveguide.

After considering the two limiting cases above, the reciprocal and periodic nature of β_0, and the fact that $d\omega/d\beta_0$ must be zero at values of β_0 equal to $n\pi/L$, we can sketch the ω–β_0 diagram of a periodically loaded guide as shown in Fig. 8.2. It should be noted here that when $\beta_0 = n\pi/L$, L is equal to an integral multiple of $\lambda_g/2$. At a frequency corresponding to the β_0's, the cells are integral multiples of half guide wavelengths long, so that the reflections from the various disks interfere constructively. Indeed, the reflection proves to be complete in this case and no power can be transmitted. Since power flow is proportional to the group velocity $v_g = d\omega/d\beta_0$, the ω–β_0 curves must have zero slope at $\beta_0 = n\pi/L$, as stated above.

When $0 < b < a$ and $\beta_0 = n\pi/L$, it is reasonable to assume on the basis of Eq. (8.11) that the field in each cell can be represented by a linear combination of symmetric and antisymmetric functions, i.e., sines and cosines, with the relative amount of each dependent on the ratio b/a. For example, if $b = 0$, there would be only sinelike components as the tangential electric field must be zero at the disks. In the intermediate case where $0 < b < a$, the ω corresponding to $\beta_0 = n\pi/L$ for sinelike waves will change very little since nodes exist at the disk planes. Thus the presence of the disks should have little influence on the frequency corresponding to $\beta_0 = n\pi/L$. On the other hand, the presence of the disks should have considerably more influence on the cosinelike waves since antinodes now exist at the disk planes.

Therefore, starting with the cutoff frequency in Fig. 8.2 and moving toward higher values of ω, we would expect the ω–β_0 curve to follow closely the ω–β_0 hyperbola of an empty guide until $\beta_0 \to \pi/L$. Otherwise, the reflected waves from the disks, being not equal to $n(2\pi)$ difference in phase, will not interfere constructively. However, at $\beta_0 = \pi/L$ the reflection is complete and the ω–β_0 curves must have a zero slope at this value of β_0. Since the sinelike and cosinelike waves are affected differently by the presence of the disks, as discussed in the last paragraph, there should be two frequencies corresponding to this value of β_0. For the sinelike wave, ω is not much affected; thus it is determined by the intersection of the ω–β_0 hyperbola of the empty guide and the straight line $\beta_0 = \pi/L$. Note that this frequency is also a resonance frequency of the completely enclosed cavity as given by Eq. (8.17), since its waves are also sinelike in distribution. We conclude from the above that there is a range of frequencies within which no β_0 can be found and the fields in the structure decay exponentially with distance.

When we proceed to an even higher value of ω, we again follow closely the ω–β_0 hyperbola until we reach $\beta_0 = 2\pi/L$. Here, as before, $d\omega/d\beta_0 \to 0$ and there are two frequencies corresponding to this value of β_0. In other words, various pass bands and stop bands are now introduced into the ω–β_0 diagram because of its periodic physical attributes.

Although the rest of Fig. 8.2 is based on the reciprocal and periodic nature of β_0, a word of clarification is nevertheless helpful. As we observed from Fig. 8.2, $v_p = \omega/\beta$ decreases with increasing n, giving rise to the possibility of obtaining a v_p less than c to synchronize with an electron beam. Note that v_p can be either positive or negative depending on whether β_0 is positive or negative respectively. Thus, waves can propagate in both directions of the periodic structure. On the other hand, $v_g = d\omega/d\beta$ can also take on positive or negative values dependent upon the value of β_0. In fact, at a given frequency, we can find waves whose phase and group velocities are in the same direction and waves whose phase and group velocities are oppositely directed.

The ω–β_0 information given by Fig. 8.2 must be supplemented by a consideration of the amplitudes of the spatial Fourier components. When $b \simeq a$, the reflections are small so that the forward wave, corresponding to a phase constant of β_0', say, will have by far the largest amplitude, except in the vicinity of $\beta_0 = n\pi/L$, where total reflections occur. However, as b decreases the amplitude of the reflected waves corresponding to $\beta_0' + n\pi/L$ increases. If b is sufficiently large, a component with β_0 equal to $\beta_0' + m\pi/L$, appropriate for synchronization with an electron beam, may have sufficiently high amplitude to be of practical significance in active microwave devices. Unfortunately, in order for this to occur b may have to be reasonably small. In this case, waves with $n \neq m$ may also have significant amplitudes. These waves carry power and thereby increase dissipation in the guide.

When $b \simeq a$, we would expect the stop bands to be narrow in frequency, while for b small we would expect the pass band to be small. Indeed, for $b \to a$ the stop bands shrink to zero and the amplitude of the reflected waves also approaches zero. On the other hand, as $b \to 0$ the passband approaches zero and the ω–β_0 diagram approaches the horizontal straight lines obtainable from Eq. (8.17).

Before closing our discussion, we note from Fig. 8.2 that v_g can be positive or negative for a given ω in the passband. This fact, coupled with the possibility of negative as well as positive v_p's, shows that forward as well as backward devices can employ periodic structures. Note also that we have shown the case of only one mode in Fig. 8.2. If this mode is the lowest in the guide and if the operating frequencies are lower than the cutoff frequency of the next higher-order mode, this simplification is entirely justified. However, since the lowest-order modes in both the rectangular and cylindrical guides are TE modes, they cannot be used in active microwave devices because they have no longitudinal electric field to axially accelerate or decelerate the beam. However, if we use a cylindrically symmetric higher-order mode, it alone can be excited since circular disks do not perturb the cylindrical symmetry of the field and thus excite other modes. Indeed, the linear electron accelerator uses the TM_{01} mode. If it did not, the pass and stop bands of the various excited modes would overlap and create a very complicated situation indeed.

The disk-loaded waveguide of Fig. 8.1 is, of course, only one of a number of possible periodic waveguide configurations. It is used in the linear electron accelerator, which operates at a single frequency. From an observation of Fig. 8.1 we see that due to this single-frequency requirement the hole on the disk can be reasonably small, presumably just large enough to allow the electron beam to traverse with ease. Thus the pass bands are narrow, since under these circumstances the cells act merely as coupled cavities. However, because of the large perturbation of the waveguide fields by the disks, the Fourier components with $v_p < c$, which permits field-beam synchronization, will have appreciable amplitudes for interaction. If we desire broad frequency band operation, the disk holes must be larger and will therefore have the accompanying disadvantage of decreasing the amplitude of the desired higher-order Fourier component. In wave amplifiers (TWT). The disk-on-rod structure of Fig. 8.3(a), a variation of the disk-loaded structure, has low-pass rather than band-pass properties.[2]

The most commonly used periodic structure for the TWT is the helix shown in Fig. 8.3(b). In this structure, the wave travels at approximately the velocity of light along the helix wire while the electron beam travels down the axis of the helix. Thus, for synchronization to occur the electron beam velocity v_0 must be approximately equal to the velocity of light multiplied by the ratio of the pitch p to the circumference $2\pi a$:

$$v_0 \simeq c \frac{p}{2\pi a}. \tag{8.18}$$

For example, if $p/2\pi a$ is 0.1, $v_0 \simeq 0.1c$.

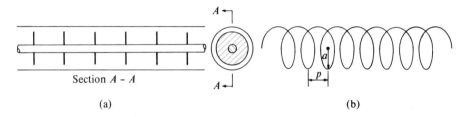

Fig. 8.3. (a) Disk-on-rod structure. (b) Helix.

One of the advantages of the helix structure over the disk-loaded or disk-on-rod structure is that its axial velocity is nearly independent of frequency if the frequency is not too low. This feature, of course, permits wideband operation at a fixed v_0. In contrast, the disk-loaded and disk-on-rod structures are dispersive, i.e., their phase velocities vary with frequency. However, the helix structure, enclosed in an evacuated glass tube, carries little power compared to the all-metal structures of Figs. 8.1 and 8.3(a).

The common structures used for the backward wave oscillator are the folded line structure of Fig. 8.4(a) and the interdigital structure of Fig. 8.4(b). In both cases, we may think of the wave as traveling in a serpentine fashion along the folded transmission line so that its axial velocity is lower than the velocity of light.

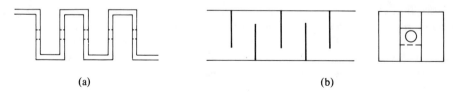

Fig. 8.4. (a) Folded line structure. (b) Interdigital structure.

---

[2] Lester M. Field, *Proc. Inst. Radio Engrs.* **37,** 34 (1949)

2. ACTIVE MICROWAVE DEVICES

In the last section we studied the behavior of passive periodic structures. We shall now examine the behavior of electromagnetic waves in these structures in the presence of an electron beam. We may surmise from the outset that the reciprocal nature of the field-electron interaction causes power to flow from the wave to the beam as well as from the beam to the wave. If there is a net power flow from the beam to the wave, the wave is amplified and the device functions as an amplifier and, under appropriate circumstances, as an oscillator. On the other hand, if the direction of net power flow is reversed, the electrons in the beam will be accelerated and the device becomes an accelerator.

In principle, at least, we can determine the field distribution in an active microwave device by a simultaneous solution of Maxwell's equations (4.1) through (4.4) and the combined Lorentz force law and Newton's second law (4.19). In practice, however, this task is difficult. Thus we shall first examine a simplified but reasonably realistic model of an active device to obtain some physical idea of its operational characteristics.

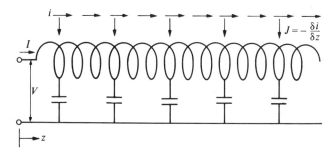

Fig. 8.5. Distributed circuit excited by electron beam.

Let us represent the periodic structure by a distributed semi-infinite transmission line, as shown in Fig. 8.5. Because of the presence of the electron beam and the signal applied to the circuit at $z = 0$, there will be a voltage V and current I on the line, both functions of z. Assume that an electron beam carrying a convection current $\mathbf{i}$ flows alongside the transmission line or circuit. If the electron beam is very narrow and flows very close to the circuit, the displacement current along the stream will be negligible compared with that from the stream to the circuit. In other words, under these circumstances nearly all the electric field lines from the beam will be coupled to the circuit. Since, according to the law of charge conservation, the sum of the displacement and convection currents into any infinitesimal volume of the electron beam must be zero, the displacement current $\mathbf{J}$ per unit length in amperes per meter impressed on the circuit must equal $-\delta \mathbf{i}/\delta z$.

If **i** and **J** are sinusoidal with time, the equations for the excited transmission line of Fig. 8.5 are

$$\frac{\delta I}{\delta z} = -jBV + J, \qquad (8.19)$$

$$\frac{\delta V}{\delta z} = -jXI, \qquad (8.20)$$

where B and X are respectively the shunt susceptance and series reactance per unit length of the transmission line. In the derivation of Eqs. (8.19) and (8.20), we have implicitly assumed that **i** and **J** are sinusoidal functions of time. If all quantities vary as $e^{-\gamma z}$, Eqs. (8.19) and (8.20) become

$$-\gamma I = -jBV + \gamma i, \qquad (8.21)$$

$$-\gamma V = -jXI, \qquad (8.22)$$

where we have made use of the relation $J = -\delta i/\delta z$. Eliminating I from the above equations, we find

$$V(\gamma^2 + BX) = -j\gamma X i. \qquad (8.23)$$

In the absence of the beam, $i = 0$. The above equation then reduces to that for a passive transmission line, and the propagation constant γ_1 and characteristic impedance Z_0 are given by

$$\gamma_1 = j\sqrt{BX}, \qquad (8.24)$$

$$Z_0 = \sqrt{X/B}. \qquad (8.25)$$

Eliminating B from Eqs. (8.24) and (8.25), we can express X in terms of γ_1 and Z_0 as

$$X = -jZ_0\gamma_1. \qquad (8.26)$$

Substituting Eqs. (8.24) and (8.26) into Eq. (8.23), we find V in terms of i as

$$V = \frac{-\gamma\gamma_1 Z_0 i}{\gamma^2 - \gamma_1^2}. \qquad (8.27)$$

Equation (8.27), known as the circuit equation, gives the circuit voltage for a given convection current in the electron beam. Keeping in mind that what we seek is the expression for γ so that we can find how the wave amplitude on the line varies with z, we obviously need another expression which gives the convection beam current for a given circuit voltage. From the Lorentz-Newton force law (4.19), we find

$$\frac{d(u_0 + v)}{dt} = \eta \frac{\delta V}{\delta z}, \qquad (8.28)$$

where u_0 and v are respectively the average and r.f. velocities of the electrons in the beam and $\eta = e/m$ is the charge to mass ratio of the electron. Noting that u_0 is independent of time and

$$dv/dt = \delta v/\delta t + (\delta v/\delta z)(dz/dt)$$

we find that Eq. (8.28) becomes

$$\frac{\delta v}{\delta t} + \frac{\delta v}{\delta z}(u_0 + v) = \eta \frac{dV}{\delta z}. \quad (8.29)$$

For small signals, $v \ll u_0$. Therefore, we can neglect second-order terms in the r.f. quantity v. In this case Eq. (8.29) is linear. It follows that v, V will be proportional to $e^{j\omega t - \gamma z}$, and Eq. (8.28) becomes

$$v = -\eta \gamma V/u_0(j\beta_e - \gamma), \quad (8.30)$$

where $\beta_e = \omega/u_0$ can be thought of as the phase constant of a wave traveling with a phase velocity equal to u_0.

In order to find i in terms of V from Eq. (8.30), we must find the relationship between v and i. The total current is the total velocity times the total charge density, or

$$-I_0 + i = (u_0 + v)(\rho_0 + \rho), \quad (8.31)$$

where $-I_0 = u_0 \rho_0$ is the dc beam current and ρ_0 and ρ are dc and r.f. charge densities. Neglecting second-order terms in the r.f. quantities, Eq. (8.31) reduces to

$$i = \rho_0 v + u_0 \rho. \quad (8.32)$$

We now proceed to find ρ in terms of i by invoking the equation of continuity, (4.25). Thus we have

$$\rho = -j\gamma i/\omega, \quad (8.33)$$

again using the fact that i and ρ are proportional to $e^{j\omega t - \gamma z}$. Combining Eqs. (8.30), (8.32), and (8.33), we finally obtain the *electronics equation*, that is, i in terms of V, as

$$i = \frac{jI_0 \beta_e \gamma V}{2V_0(j\beta_e - \gamma)^2} \quad (8.34)$$

The circuit equation (8.27) expresses V in terms of the convection current i, while the electronics equation (8.34) expresses i in terms of V. If the two expressions are to be consistent, the ratio V/i of Eq. (8.27) and that of Eq. (8.34) must be equal. Accordingly, we find

$$\frac{jZ_0 I_0 \beta_e \gamma^2 \gamma_1}{2V_0(\gamma_1^2 - \gamma^2)(j\beta_e - \gamma)^2} = 1. \quad (8.35)$$

For the TWT, we are particularly interested in a wave that propagates in the direction of electron flow at nearly the electron speed and which will account for the observed gain. Making the reasonable assumption that the electron speed is equal to that of the circuit wave in the absence of the beam, we have $-\gamma_1 = -j\beta_e$.

If we further assume that γ differs from β_e by a small quantity ξ, we have

$$-\gamma = -j\beta_e + \xi. \tag{8.36}$$

Substituting Eq. (8.36) into Eq. (8.35), we find

$$1 = \frac{-Z_0 I_0 \beta_e^2(-\beta_e^2 - 2j\beta_e \xi + \xi^2)}{2V_0(2j\beta_e \xi - \xi^2)\xi^2}. \tag{8.37}$$

In accord with the assumption that $\xi \ll \beta_e$, we may neglect the first- and second-order terms in ξ in the numerator and the fourth-order term in ξ in the denominator. In this case, Eq. (8.37) simply reduces to

$$\xi^3 = -j\beta_e^3 C^3, \tag{8.38}$$

where

$$C^3 = Z_0 I_0 / 4V_0. \tag{8.39}$$

Here C is known as the gain parameter of the tube. Because of the functional relationship between ξ, β_e, and C as expressed in Eq. (8.38), it is expedient to perform a change of variable from ξ to δ via the relation

$$\xi = \beta_e C \delta. \tag{8.40}$$

In terms of δ, Eq. (8.38) becomes simply

$$\delta = (-j)^{1/3} = \left[e^{j(2n-1/2)\pi} \right]^{1/3}, \tag{8.41}$$

where $n = 0, 1, 2$. The three roots of this equation are

$$\delta_1 = \frac{\sqrt{3}}{2} - j\frac{1}{2},$$

$$\delta_2 = -\frac{\sqrt{3}}{2} - j\frac{1}{2}, \tag{8.42}$$

$$\delta_3 = j.$$

Recalling that the r.f. quantities are proportional to $e^{j\omega t - \gamma z}$, we find with the help of Eqs. (8.36) and (8.40)

$$e^{j\omega t - \gamma_n z} = e^{\beta_e C \delta_n z} e^{j(\omega t - \beta_e z)}.$$

Accordingly, we find that the propagation factors for the three waves are

$$e^{j\omega t - \gamma_1 z} = e^{\beta_e C(\sqrt{3}/2)z} e^{j(\omega t - \beta'_e z)},$$

$$e^{j\omega t - \gamma_2 z} = e^{-\beta_e C(\sqrt{3}/2)z} e^{j(\omega t - \beta'_e z)}, \tag{8.43}$$

$$e^{j\omega t - \gamma_3 z} = e^{j(\omega t - \beta''_e z)},$$

where

$$\beta'_e = \beta_e \left(1 + \frac{C}{2}\right) = \frac{\omega}{v_{p1}} = \frac{\omega}{v_{p2}},$$

$$\beta''_e = \beta_e(1 - C) = \frac{\omega}{v_{p3}}. \tag{8.44}$$

We therefore conclude that the wave corresponding to δ_1 grows with increasing z and propagates at a phase velocity slightly lower than the beam velocity u_0. Likewise, the wave corresponding to δ_2 attenuates with increasing z and propagates at the same phase velocity as that of δ_1. In contrast, the wave corresponding to δ_3 will neither grow nor decay with distance and travels at a velocity slightly higher than the beam velocity. In other words, energy flows from the beam to the first wave, and conversely, from the second wave to the beam, but no net energy exchange occurs between the beam and the third wave.

We have now obtained only three solutions to Eq. (8.37), although it is of the fourth order in ξ. The missing solution clearly must be inconsistent with the assumption leading to Eq. (8.36). To obtain this solution, recall that Eq. (8.36) presupposes the near equality of the wave and electron velocities. In other words, the waves corresponding to Eq. (8.36) travel in the direction of the electron stream and with a velocity nearly equal to it. But there must also be a wave traveling at about the same speed as the electron stream except in the opposite direction since, in the absence of the beam, there are forward and backward traveling waves on the transmission line. Accordingly, in place of Eq. (8.36) we should have

$$-\gamma = j\beta_e + \xi \tag{8.45}$$

for the missing wave. Substituting Eq. (8.45) into Eq. (8.37) and neglecting appropriate higher-order terms in ξ, we easily find the missing solution:

$$\delta_4 = -j\frac{C^2}{4}. \tag{8.46}$$

Inasmuch as the typical value of C is about 0.02, δ_4 is much smaller than δ_1, δ_2, or δ_3. Thus the velocity of the wave traveling in the $-z$-direction is only minutely higher than that in the absence of the beam. Furthermore, the presence of the electron stream doubles the number of traveling waves.

We must emphasize that the "backward wave" corresponding to Eq. (8.46) is not the wave used in the backward oscillator (BWO). If there is to be net energy exchanged between the wave and the beam, the phase velocity of the wave and the velocity of the beam must be nearly equal. Otherwise, the slipping by of the wave with time will cause the electrons in the beam to alternately accelerate and decelerate as they encounter accelerating and retarding electric fields. Thus, since the wave corresponding to Eq. (8.46) travels in a direction opposite to the electron stream, it cannot be the wave connected with oscillations in the BWO.

From the above discussion, it is easy to surmise that in the BWO, as in the TWT, the wave responsible for the operation must have a phase velocity whose magnitude and direction are nearly equal to those of the electron velocity. However, whereas the wave grows from left to right in for example, the TWT, it grows from right to left in the BWO. In other words, the phase and group velocities are in the same direction for the TWT but in opposite directions for the BWO. In this connection, note from Fig. 8.2 that it is entirely possible to select a wave that has phase and group velocities in opposite directions for interaction with the electron beam.

3. TRAVELING WAVE TUBE

A. Gain of TWT

In the last section we demonstrated that the electron beam interaction with a distributed circuit produces four waves in all, three of which travel in the direction of the electron stream. Of particular interest in the study of the TWT is the fact that one of these waves, namely that corresponding to δ_1 of Eq. (8.42), grows with increasing z. To determine the amplification of the tube, we clearly need to know the amplitude of this growing wave at the input ($z = 0$) and output ($z = L$) points. Since energy is being exchanged between the wave and the electron beam all along the tube, we cannot determine the functional dependence of the growing-wave amplitude in isolation from that of the other waves.

To begin with, let us assume that V, v, and i are respectively the total circuit voltage, total r.f. velocity, and total r.f. current of the electron stream at the input, that is, at $z = 0$. It follows that

$$V_1 + V_2 + V_3 = V, \tag{8.47}$$

where V_1, V_2, and V_3 are forward traveling waves corresponding to the δ's of Eq. (8.42). We have implicitly assumed here that the structure is perfectly matched so that there is no backward wave. Therefore, $V_4 = 0$. We must now obtain similar boundary conditions for v and i in terms of the voltages. Combining Eqs. (8.30), (8.34), (8.36), and (8.40) and noting that typically $C\delta \ll 1$, we easily find

$$v = \frac{\eta}{ju_0 C}\left(\frac{V_1}{\delta_1} + \frac{V_2}{\delta_2} + \frac{V_3}{\delta_3}\right) \tag{8.48}$$

and

$$i = -\frac{I_0}{2V_0 C^2}\left(\frac{V_1}{\delta_1^2} + \frac{V_2}{\delta_2^2} + \frac{V_3}{\delta_3^2}\right). \tag{8.49}$$

Simultaneously solving Eqs. (8.47), (8.48), and (8.49), we have

$$V_1 = \frac{V - \dfrac{u_0 C}{\eta}\left(\dfrac{1}{2} + j\dfrac{\sqrt{3}}{2}\right)v - \dfrac{2V_0 C^2}{I_0}\left(\dfrac{1}{2} - j\dfrac{\sqrt{3}}{2}\right)i}{3},$$

---

[3] J. R. Pierce, *Travelling-Wave Tubes*, D. Van Nostrand Co., New York, 1950; pp. 10 and 106.

$$V_2 = \frac{V - \frac{u_0 C}{\eta}\left(\frac{1}{2} - j\frac{\sqrt{3}}{2}\right)v - \frac{2V_0 C^2}{I_0}\left(\frac{1}{2} + j\frac{\sqrt{3}}{2}\right)i}{3},$$

$$V_3 = \frac{V + \frac{u_0 C}{\eta}v + \frac{2V_0 C^2}{I_0}i}{3}. \tag{8.50}$$

Thus we see that if the beam enters the propagating region unmodulated, that is, if $v = i = 0$, then $V_1 = V_2 = V_3 = V/3$, a result which can only be true if the impedances of the three waves are nearly equal. Indeed, we can easily show from Eqs. (8.26) and (8.36) that

$$Z_{0n} \simeq Z_0(1 - jC\delta_n), \tag{8.51}$$

where we have recalled that $\gamma_1 = j\beta_e$. Since $C\delta_n \ll 1$, neglecting this term compared to unity will give $Z_{01} \simeq Z_{02} \simeq Z_{03} \simeq Z_0$.

For the TWT, the beam is usually unmodulated at the input end[4] so that $V_1 \simeq V_2 \simeq V_3 \simeq V/3$. Thus, combining the three waves of Eq. (8.43) with equal amplitude ($V/3$), we find the voltage $V(z)$:

$$V(z) = \left(\frac{V}{1} e^{\beta_e C(\sqrt{3}/2)z} e^{-j\beta_e(1+C/2)z} + e^{-\beta_e C(\sqrt{3}/2)z} e^{-j\beta_e(1+C/2)z} + e^{-j\beta_e(1-C)z}\right). \tag{8.52}$$

The gain A in decibels of the TWT follows immediately as

$$A = 10 \log_{10}\left|\frac{V(z)}{V}\right|^2. \tag{8.53}$$

We see from Eq. (8.52) that the growing wave predominates for sufficiently large values of z, as it should. In this case the gain, as obtained from Eq. (8.52) and (8.53), is simply

$$A\Big|_{z \to \infty} \simeq 10 \log_{10}\left(\frac{1}{9} e^{\sqrt{3}\beta_e Cz}\right) = 47.3\, CN, \tag{8.54}$$

where N is the length of tube L divided by the wave length λ corresponding to a phase constant β_e, that is, $\lambda = 2\pi/\beta_e$. A plot of Eqs. (8.53) and (8.54) as a function of CN shows that the former curve approaches the latter one asymptotically as $CN \to \infty$, while the latter curve, a straight line, intercepts the gain axis at -9.54 db. Thus for sufficiently large CN, the gain A is nearly given by the expression

$$A = -9.54 + 47.3 CN. \tag{8.55}$$

---

[4] This statement neglects the effect of the r.f. fringing field at the input on the beam. At most, however, this is roughly equivalent to moving the input ahead by a small amount.

As expected, for small values of $CN(< 0.2)$, this expression departs radically from that given by Eq. (8.53).

Recall that the foregoing analysis assumes that the input and output terminals of the propagating structure are perfectly matched to the source and load respectively. This is usually not the case, of course, and some reflections at these terminals are unavoidable. Reflections can cause undesirable oscillations in the tube if proper measures are not taken to avoid them.

Assume first that the output is slightly mismatched. When the forward wave arrives at the output, part of it will be reflected, thereby setting up a wave traveling unattenuated in the reverse direction, from output to input. If the input is also slightly mismatched, part of this backward traveling wave will in turn be reflected and set up three waves that travel from input to output. As the analysis above indicates, one of these waves will be amplified with increasing z. When the waves arrive at the output, the whole sequence of events described above will recur. Given enough such round trips, with amplification occurring in each forward trip, the feedback voltage—i.e., the amplitude of the backward wave traveling—will soon be sufficient to cause oscillations within the tube.[5] Since we can trace the oscillations to the backward traveling wave set up by reflections at the output, an obvious way to eliminate them is to greatly reduce the backward traveling wave by introducing attenuation in the tube. But an attenuator will also reduce the forward amplifying wave. Other things being constant, then, the gain of the tube will be reduced. We can recover this loss gain, however, by increasing the tube length. The question of optimum location of the lossy material is left as a problem for the reader.

On the other hand, if an anisotropic material such as a ferrite sample of proper shape is placed in a region of circular polarization of the r.f. field and appropriately magnetized, we can obtain unilateral attenuation. In this way, a backward wave can be greatly attenuated due to magnetic resonance absorption, while the growing forward wave is relatively unaffected.[6,7] However, due to the added complexity in this case, ferrites are not usually employed to eliminate oscillations in TWTs.

B. The Helix

The most commonly used TWT propagating structure is the helix. Having just examined the general properties of periodic structures, we shall now inquire further into the details of field distribution in a helix. We shall solve the associated boundary value problem by idealizing the real helix by a fictitious sheath helix.

The actual helix, composed of discrete wire wound in the form of a coil, is very difficult to deal with analytically. For this reason, we shall idealize the actual

---

[5] Since phase conditions must also be satisfied, these oscillations can occur only over small bands of frequencies giving rise to irregular output over the bandwidth of the tube. This is clearly undesirable for an amplifier.

[6] J. S. Cook, R. Kompfner, and H. Suhl, *Proc. Inst. Radio Engrs.* **42,** 1188 (1954).

[7] J. A. Rich and S. E. Weber, *Proc. Inst. Radio Engrs.* **43,** 100 (1955).

helix by a helically conducting cylindrical sheet of the same mean radius, as shown in Fig. 8.6. Let us assume that the cylinder is perfectly conducting in a helical direction that makes an angle ψ, the pitch angle, with a plane normal to the axis, but is perfectly nonconducting in a helical direction normal to this direction, called the ψ-direction.

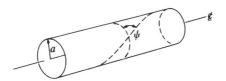

Fig. 8.6. A helically conducting cylindrical sheet.

To solve the boundary value problem, we must first find the solution to Maxwell's equations inside as well as outside the helix and then impose the proper electromagnetic boundary conditions. These conditions are: (a) the electric field components in the ψ-direction, inside and outside the helix, are zero; (b) the electric field components in the direction normal to the ψ-direction, inside and outside the helix, are equal; and (c) the components of the magnetic field in the ψ-direction, inside and outside the helix, must be equal, since there can be no current flow in the direction normal to the ψ-direction. That the current density is zero in the direction normal to the ψ-direction follows directly from Eq. (2.49).

For a plane wave having circular symmetry and propagating in the z-direction with phase velocity $v_p = \omega/\beta$, the wave equation (3.14) reduces to

$$\frac{\delta^2 E_z}{\delta r^2} + \frac{1}{r}\frac{\delta E_z}{\delta r} + k_c^2 E_z = 0, \tag{8.56}$$

where $k_c^2 = (\omega/c)^2 - \beta^2$. Since v_p must be less than c, it follows that if the wave is synchronized with the electron beam, k_c is imaginary. It then follows from Eq. (3.20) that the solution for Eq. (8.56) is

$$E_z = A J_0(j|k_c|r) + B_1 H_0^{(1)}(j|k_c|r). \tag{8.57}$$

Noting that

$$I_0(|k_c|r) = J_0(j|k_c|r) \quad \text{and} \quad K_0(k_c r) = j(\pi/2) H_0^{(1)}(j|k_c|r),$$

we can rewrite Eq. (8.57) to read

$$E_z = A J_0(|k_c|r) + B K_0(|k_c|r), \tag{8.58}$$

with the factor $e^{j(\omega t - \beta z)}$ understood. Since $K_0(|k_c|r) \to \infty$ as $r \to 0$ and $I_0(|k_c|r) \to \infty$ as $r \to \infty$, we conclude that *inside* the helix,

$$H_{z1} = B_1 I_0(|k_c|r),$$
$$E_{z3} = B_3 I_0(|k_c|r) \tag{8.59}$$

while *outside* the helix,

$$H_{z2} = B_2 K_0(|k_c|r)$$
$$E_{z4} = B_4 K_0(|k_c|r). \tag{8.60}$$

The transverse components of **E** and **H** follow directly from Eqs. (3.28) and (3.29) and the longitudinal components above. Inside the helix, we find

$$\begin{aligned} H_{\phi 3} &= B_3 \frac{j\omega\varepsilon}{|k_c|} I_1(|k_c|r), \\ H_{r1} &= B_1 \frac{j\beta}{|k_c|} I_1(|k_c|r), \\ E_{\phi 1} &= -B_1 \frac{j\omega\mu}{|k_c|} I_1(|k_c|r), \\ E_{r3} &= B_3 \frac{j\beta}{|k_c|} I_1(|k_c|r). \end{aligned} \tag{8.61}$$

Outside the helix, we have

$$\begin{aligned} H_{\phi 4} &= -B_4 \frac{j\omega\varepsilon}{|k_c|} K_1(|k_c|r), \\ H_{r2} &= -B_2 \frac{j\beta}{|k_c|} K_1(|k_c|r), \\ E_{\phi 2} &= B_2 \frac{j\omega\mu}{|k_c|} K_1(|k_c|r), \\ E_{\phi 4} &= -B_4 \frac{j\beta}{|k_c|} K_1(|k_c|r). \end{aligned} \tag{8.62}$$

In the derivation of Eqs. (8.61) and (8.62), we have also made use of the constitutive equations (2.5) and (2.6).

In accordance with the boundary conditions previously enumerated, we conclude that at the cylindrical surface, where $r = a$,

$$E_{z3} \cos \psi = E_{z4} \cos \psi,$$
$$E_{\phi 1} \sin \psi = E_{\phi 2} \sin \psi, \tag{8.63}$$

and

$$H_{z1} \sin \psi + H_{\phi 3} \cos \psi = H_{z2} \sin \psi + H_{\phi 4} \cos \psi. \tag{8.64}$$

Substituting Eqs. (8.59), (8.60), (8.61), and (8.62) into Eqs. (8.63) and (8.64), we obtain a set of three simultaneous algebraic equations containing the unknown constants B_1, B_2, B_3, B_4. Setting the determinant of these equations equal to zero, we find the secular equation for $|k_c|$:

$$(|k_c|a)^2 \frac{I_0(|k_c|a)K_0(|k_c|a)}{I_1(|k_c|a)K_1(|k_c|a)} = (\beta_0 a \cot \psi)^2, \qquad (8.65)$$

where $\beta_0 = \omega/c$. If we evaluate the unknown constants B_1, B_2, and B_4 in terms of B_3, we find that inside the helix

$$E_{zi} = B_3 I_0(|k_c|r),$$

$$E_{ri} = jB_3 \frac{\beta}{|k_c|} I_1(|k_c|r),$$

$$E_{\phi i} = -B_3 \frac{I_0(|k_c|a)}{I_1(|k_c|a)} \frac{1}{\cot \psi} I_1(|k_c|r),$$

$$H_{zi} = -j \frac{B_3}{Z_0} \frac{|k_c|}{\beta_0} \frac{I_0(|k_c|a)}{I_1(|k_c|a)} \frac{1}{\cot \psi} I_0(|k_c|r), \qquad (8.66)$$

$$H_{ri} = \frac{B_3}{Z_0} \frac{\beta}{\beta_0} \frac{I_0(|k_c|a)}{I_1(|k_c|a)} \frac{1}{\cot \psi} I_1(|k_c|r),$$

$$H_{\phi i} = j \frac{B_3}{Z_0} \frac{\beta_0}{|k_c|} I_1(|k_c|r),$$

and outside the helix,

$$E_{zo} = B_3 \frac{I_0(|k_c|a)}{K_0(|k_c|a)} K_0(|k_c|r),$$

$$E_{ro} = -jB_3 \frac{\beta}{|k_c|} \frac{I_0(|k_c|a)}{K_0(|k_c|a)} K_1(|k_c|r),$$

$$E_{\phi o} = -B_3 \frac{I_0(|k_c|a)}{K_1(|k_c|a)} \frac{1}{\cot \psi} K_1(|k_c|r), \qquad (8.67)$$

$$H_z = j \frac{B_3}{Z_0} \frac{k_c}{\beta_0} \frac{I_0(|k_c|a)}{K_1(|k_c|a)} \frac{1}{\cot \psi} K_0(|k_c|r),$$

$$H_r = \frac{\beta}{\beta_0} \frac{B_3}{Z_0} \frac{I_0(|k_c|a)}{K_1(|k_c|a)} \frac{1}{\cot \psi} K_1(|k_c|r),$$

$$H_\phi = -j \frac{B_3}{Z_0} \frac{\beta_0}{|k_c|} \frac{I_0(|k_c|a)}{K_0(|k_c|a)} K_1(|k_c|r),$$

where $Z_0 = \sqrt{\mu/\varepsilon}$.

To recapitulate, in this subsection we have derived the electromagnetic field expressions of a helix in terms of, among other quantities, $|k_c|$ as determined by the transcendental secular equation (8.65) for a given set of ω, a, and ψ. Returning to this equation and taking the square root of both sides, we find

$$(\omega/c)a \cot \psi = |k_c|a\sqrt{\frac{I_0(|k_c|a)K_0(|k_c|a)}{I_1(|k_c|a)K_1(|k_c|a)}}, \tag{8.68}$$

using the relation $\beta_0 = \omega/c$. The expression $(\omega/c) a \cot \psi$ is clearly proportional to ω for given values of a and ψ. It is instructive to plot some quantity that is proportional to the phase velocity v_p against $(\omega/c) a \cot \psi$ in order to see how v_p varies with frequency.[8] In this way, we can see over what range of frequencies the electron beam can travel in near synchronism with the forward wave.[9] In view of the form of Eq. (8.68), it appears expedient to plot v/c, instead of v, against $(\omega/c) a \cot \psi$. From the expression $k_c^2 = (\omega/c)^2 - \beta^2$, we can easily show that

$$\frac{v_p}{c} = \frac{\omega}{c\beta} = \frac{1}{\sqrt{1 + (|k_c|c/\omega)^2}}. \tag{8.69}$$

For large values of $(\omega/c) a \cot \psi$, we can show from Eq. (8.68) that $(k_c c/\omega) \to \cot \psi$. In this case, v_p/c of Eq. (8.69) approaches $\sin \psi$. We see from Fig. 8.6 that at high frequencies the wave travels along the ψ-direction with velocity of light c as we would intuitively expect. Accordingly, the significant quantity to be plotted against $(\omega/c) a \cot \psi$ is

$$\frac{v_p}{c} - \sin \psi = \frac{1}{\sqrt{1 + (|k_c|c/\omega)^2}} - \sin \psi,$$

which represents the deviation of the normalized v_p from its high frequency value. This plot is shown in Fig. 8.7, with $\tan \psi$ as a parameter. We note that if $(v_p/c) - \sin \psi$ is less than, say, 0.01 for reasonable values of ψ, $\beta_0 a \cot \psi$ must be larger than about 3.

Note that we can also write the abscissa variable as $(2\pi a/\lambda) \cot \psi$, where $\lambda = 2\pi c/\omega$ is the free-space wavelength. Thus for a given frequency increasing the radius/wavelength ratio and decreasing ψ will increase $(\omega/c) a \cot \psi$, thereby decreasing the normalized phase velocity deviation $(v_p/c) - \sin \psi$. Increasing a and decreasing ψ, however, modifies field-electron beam interaction and changes the accelerating voltage requirement.

In order to build a TWT or BWO, we must somehow match into the helix from a coax or waveguide. Unfortunately, as in the waveguide case [see Eqs. (3.83) to (3.85)], no impedance assigned to the helically conducting sheet can give full information for this matching. As we found for transducers between a coax

---

[8] Note that once ω, a, and ψ are fixed, $|k_c|$ can be found from Eq. (8.68). The quantity $(\omega/c) a \cot \psi$ contains all these independent variables which completely characterize a helix operating at a given frequency.

[9] Note that here we are speaking about waves in a "cold helix," i.e., a helix in the absence of an electron beam.

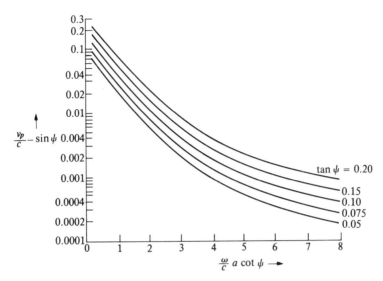

Fig. 8.7. Deviation of normalized phase velocity from its high frequency value vs. normalized frequency. (After J. R. Pierce, [3].)

and a waveguide or between waveguides of different cross sections, discontinuity effects as well as the impedance are important. Although we can make helix impedance calculations using various bases (e.g., a voltage-power basis) to get some approximate value for the impedance, matching into or out off the helix is usually done experimentally.

4. BACKWARD WAVE OSCILLATORS

From Fig. 8.2 and Eq. (8.13), we note that for a given ω in the passband there can be waves whose phase and group velocities are in the same direction, and waves whose phase and group velocities are oppositely directed. This is equivalent to saying that there can be a pair of waves, one (forward wave) with phase and group velocities in the $+z$-direction and the other (backward wave) with phase velocity in the $+z$-direction and group velocity in the $-z$-direction. Even in the presence of an electron beam, a backward wave may well have almost the same phase velocity as associated beam. Under these circumstances there can be cumulative interaction between the electron beam and the same backward wave if the electron velocity and phase velocity are nearly synchronized. This implies, of course, that energy transfer can occur between the beam and the wave. However, although the energy of the forward wave travels in the direction of the electron beam, that of the backward wave travels in the direction opposite to the beam.

A little reflection will show that whereas a forward wave will propagate on a transmission line with positive series reactance and positive shunt susceptance, a backward wave will propagate on a transmission line with negative series reactance and negative shunt susceptance. It follows that the sign of X in Eq. (8.26) should be changed. The circuit equation (8.27) thus read

$$V = \frac{\gamma \gamma_1 Z_0 i}{\gamma^2 - \gamma_1^2}. \tag{8.70}$$

The electronics equation (8.34), on the other hand, remains the same for the backward wave. Consequently, Eq. (8.37) becomes

$$1 = \frac{Z_0 I_0 \beta_e^2 (-\beta_e^2 - 2j\beta_e \xi + \xi^2)}{2V_0 (2j\beta_e \xi - \xi^2)\xi^2}. \tag{8.71}$$

Corresponding to Eq. (8.42) for the forward wave, we have for the backward wave

$$\delta_1' = \frac{\sqrt{3}}{2} + j\frac{1}{2},$$

$$\delta_2' = -\frac{\sqrt{3}}{2} + j\frac{1}{2}, \tag{8.72}$$

$$\delta_3' = -j.$$

Note that the δ_n''s here differ from those of Eq. (8.42) in that their imaginary parts are opposite in sign. Thus the wave related to δ_1' grows with increasing z and propagates at a phase velocity slightly higher than the beam velocity. The wave related to δ_2' attenuates with increasing z and propagates at the same phase velocity as that of δ_1'. The wave related to δ_3' neither grows nor decays with distance and travels at a velocity slightly less than the beam velocity.

The fourth solution which corresponds to expression (8.46) for the forward wave is

$$\delta_4' = j\frac{C^2}{4}, \tag{8.73}$$

which is again quite small compared to the other δ_n''s. As before, we shall neglect this "backward-traveling component" of the backward wave and emphasize again that it has little to do with the operation of the BWO.

To find the oscillation condition for the BWO, we must derive an equation for the voltage as a function of z similar to Eq. (8.52) for the TWT. We realize of course, that for oscillations to occur, the feedback signal must have the proper magnitude and phase.[10] Since energy in the BWO travels in the $-z$-direction, feedback must be accomplished via the electron beam. Since the phase velocity of this backward wave is nearly equal to the velocity of the electron beam traveling in the $+z$-direction, the wave can cumulatively modulate the beam, which in turn will carry the modulation and serve as the feedback medium. From an examination of the form of Eq. (8.35), we would expect the characteristic values of γ to depend sensitively on $(\gamma_1 - \gamma)$, particularly for the BWO because of the

---

[10] See discussion in Section 4, Chapter 7.

stringent magnitude and phase requirements for oscillation. We may recall in connection with Eqs. (8.36) and (8.71) that we assumed that $-\gamma_1 = -j\beta_e$, that is, that the phase velocity in the absence of the beam is exactly equal to the beam velocity. Whereas this synchronism assumption may be acceptable as far as the general behavior of the TWT is concerned, it may not be at all permissible for the BWO, again because of the sensitive magnitude and phase requirements for oscillation. For this reason, we shall write in place of $-\gamma_1 = -j\beta_e$:

$$-\gamma_1 = -j\beta_e - j\beta_e Cb. \tag{8.74}$$

Substituting Eqs. (8.36) and (8.74) into Eq. (8.71), we find an equation for δ'':

$$\delta''^2(\delta'' + jb) = +j, \tag{8.75}$$

where $\xi = \beta_e C \delta''$ as defined by Eq. (8.40). In the derivation of this equation, we assume that $|\delta''|$ is of the order unity, $|b|$ ranges from zero to unity, and $C \ll 1$. In the case of synchronous velocity, $b = 0$, the δ_n'''s reduce to the δ_n''s of Eq. (8.72).

Designating the solution of Eq. (8.75) by δ_1'', δ_2'', and δ_3'', we can write the total voltage $V(z)$ at z in terms of the voltage V at $z = 0$:

$$V(z) = V\left\{\left[\left(1 - \frac{\delta_2''}{\delta_1''}\right)\left(1 - \frac{\delta_3''}{\delta_1''}\right)\right]^{-1} e^{-j\beta_e(1 + jC\delta_1'')z}\right.$$

$$+ \left[\left(1 - \frac{\delta_3''}{\delta_2''}\right)\left(1 - \frac{\delta_1''}{\delta_2''}\right)\right]^{-1} e^{-j\beta_e(1 + jC\delta_2'')z}$$

$$\left. + \left[\left(1 - \frac{\delta_1''}{\delta_3''}\right)\left(1 - \frac{\delta_2''}{\delta_3''}\right)\right]^{-1} e^{-j\beta_e(1 + jC\delta_3'')z}\right\}. \tag{8.76}$$

For a given value of b, we can determine the δ_n'''s from Eq. (8.75) and substitute them into Eq. (8.76) to obtain $V(z)/V$. Thus we can plot this quantity against b with $\beta_e Cz$ constant. Figure 8.8 is a plot of $|V(z)/V|$ as a function of b for $\beta_e Cz = 2$.

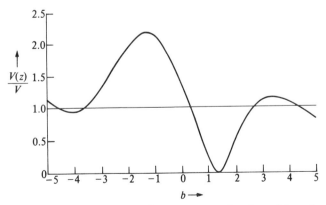

Fig. 8.8. Normalized voltage at z of a BWO structure as a function of the velocity deviation parameter.

Here z represents the length of the tube L. We see that for several ranges of b, $|V(z)/V| < 1$, or the input voltage $|V(z)|$ is less than the output voltage V. In other words, "backward gain" can occur for values of b within these ranges. Of course it follows that a backward wave amplifier (BWA) can be constructed if we choose the proper combination of b, C, β_0, L, etc. Furthermore, at $b \simeq 1.4$, $V(z)/V = 0$, that is, we obtain an output without an input. This is, of course, just the condition for the occurrence of oscillation. Thus a backward wave oscillator (BWO) is realizable.

In the relation (8.74), as well as the simpler relation $-\gamma_1 = -j\beta_e$ used in connection with Eq. (8.36), we implicitly assume that the propagating circuit is lossless. If this assumption is removed, Eq. (8.74) must be generalized to read

$$-\gamma_1 = -j\beta_e - j\beta_e Cb - \beta_e Cd. \tag{8.77}$$

Substituting Eqs. (8.36) and (8.77) into Eq. (8.37) and recalling that $\xi = \beta_e C\delta$, we find

$$\delta^2 = \frac{1}{-b + jd + j\delta}, \tag{8.78}$$

where we have assumed that $|\delta|$ is of the order unity, $|b|$ and $|d|$ range from zero to unity or a little larger, and $C \ll 1$. Equation (8.78) is applicable to the TWT. For the BWO, we must substitute Eq. (8.77) into Eq. (8.71) and then solve the resultant equation for δ. The latter turns out to be the same as Eq. (8.78), except that the left-hand side is multiplied by a minus sign.

In the foregoing analysis, we also implicitly assumed that there is only one mode of propagation. Actually, it is more reasonable to assume that γ is nearly equal to the propagation constant γ_1 of one active mode, is not near to the propagation constant of any other mode, and varies over a small fractional range only. Then we can regard the sum of terms due to all other possible modes of propagation as a constant over the range of γ under consideration. Detailed analysis shows that in this case the circuit equation (8.27) must be augmented by another term to read[11]

$$V = \left(\frac{-\gamma \gamma_1 Z_0}{\gamma^2 - \gamma_1^2} - \frac{\gamma}{j\omega C_1} \right) i, \tag{8.79}$$

where C_1 turn out to have the dimension of a capacitance per unit length. The first term represents the excitation of the γ_1 mode while the second term represents the excitation of passive and "nonsynchronous" active modes.[12] The latter also gives the field produced by the electrons in the absence of a wave propagating on the circuit, or the field due to the "space charge" of the bunched electron stream.

---

[11] J. R. Pierce, *op. cit.*, p. 111. Note that this equation is for the TWT. For the BWO, the sign of the first term should be changed.

[12] If γ_n is real, the mode is passive, while if γ_n is imaginary, the mode is active.

The second term can further be represented by distributed capacitances C_1 between the transmission circuit of Fig. 8.5 and the electron beam.

Equating V/i of circuit equation (8.79) to that of electronic equation (8.34), we find with the help of Eqs. (8.36) and (8.77)

$$\delta^2 = \frac{1}{(-b + jd + j\delta)} - 4QC \tag{8.80}$$

and

$$Q = \frac{\beta_e}{2Z_0 \omega C_1}, \tag{8.81}$$

where we have again recalled that $\xi = \beta_e C \delta$.

The backward wave oscillator can operate over an extremely wide frequency band (one waveguide band or more). We can change its oscillation frequency merely by changing the dc beam voltage. For this reason, it is the most versatile tunable oscillator and is widely used for wideband applications.

5. LINEAR ACCELERATORS

In the preceding two sections we were mainly concerned with the behavior of active microwave devices such as the TWT or BWO. For these devices, there is net energy flow from the electrons to the waves. As a consequence, electrons in the beam were slowed down and deliver part of their kinetic energy to the waves. There is nothing in our theory, however, that prohibits in principle the reverse net flow of energy, from the waves to the beam. In the latter case, electrons in the beam will be accelerated as they proceed from the gun end toward the collector. A device in which electrons are accelerated is known as a *linear accelerator*; the collector in this case is usually a target whose nuclear properties are to be examined. For example, we can study the charge and mass distribution of protons in a target by bombarding it with highly energetic electrons and examining the associated scattering of electrons by protons. In order to penetrate a nucleus, electrons in the beam must overcome Coulomb repulsion forces between them and the electrons surrounding positively charged nuclei. Therefore, it is desirable to accelerate the electrons to as high a velocity as possible. In fact, they are usually accelerated by microwave waves to relativistic speeds.

Since electrons are continuously accelerated and eventually reach relativistic speeds, the simple analysis of a transmission line by an electron beam that we gave previously for the TWT and BWO is not applicable to an accelerator for two reasons. First, the analysis is for small signal levels. Therefore, the total electron velocity, dc plus r.f., differs very little from the dc velocity itself. Second, the transmission-line model may not be applicable if the beam velocity is relativistic.[13]

A detailed analysis of relativistic electron motion in a linear accelerator is

---

[13] J. R. Pierce, *op. cit.*

clearly beyond the scope of this discussion. Thus we shall examine only some of the general features of relativistic electronic motion, using our knowledge of relativity theory (see Chapter 5).

According to Eq. (5.34), the force of repulsion, F_z, as seen by a stationary observer between two electrons on the z-axis separated by a distance Δz and moving with a velocity $\mathbf{u}_0$ in the positive z-direction is given by

$$F_z = \frac{1}{\kappa^2} \frac{e}{4\pi\varepsilon_0 (\Delta z)^2}, \qquad (8.82)$$

where $1/\kappa^2 = (1 - u_0^2/c^2)$. Similarly, if the electrons are located on a line perpendicular to the direction of motion and separated by a distance Δy, say, Eq. (5.36) shows that the force of repulsion F_y as seen by a stationary observer is given by

$$F_y = \frac{1}{\kappa} \frac{e^2}{4\pi\varepsilon_0 (\Delta y)^2}. \qquad (8.83)$$

It is interesting to note that both F_z and F_y approach zero as v approaches the velocity of light c. This implies that the longitudinal and transverse forces of repulsion between electrons in an electron beam traveling at relativistic speeds approach zero as $v \to c$, as far as the laboratory observer is concerned. Therefore, except for the first few feet of the accelerator within which the electrons are accelerated, the electron beam in a high-energy microwave electron accelerator—unlike that in the TWT or BWO—needs no electrostatic or magnetostatic focusing. Indeed, to keep the beam from spreading because of the earth's magnetic field, a coil with only a small number of ampere-turns is wrapped around the accelerator tube.

In the first few feet of an accelerator, electrons are accelerated to higher and higher velocities by the microwaves fed in and propagated along the periodic structure. Beyond this section, the increase in electronic energy is evident mainly in an increase in the electron mass as can be seen from the relation

$$m = \frac{m_0}{\sqrt{1 - (u_0/c)^2}}. \qquad (8.84)$$

As $u_0 \to c$, we see that very little additional increase in u_0 will drastically increase m. Since u_0 changes from the initial value determined by the acceleration potential of the electron gun (75kV, say) to essentially the velocity of light c in the first few feet, the periodic length of the propagating structure may have to be varied in this section to keep the electrons in near synchronism with the wave.

Figure 8.9 shows the elements of a linear accelerator. Microwave energy from Klystrons is fed in at regular intervals (10 ft, say) along the accelerator tube. All the Klystron amplifiers are driven by a signal from a magnetron so that their outputs can be in phase. If we introduce proper phase shifts between successive

Klystrons, a wave in the accelerator can in principle travel in near synchronism with the electron beam that travels axially down the tube through coaxial holes on the disks.

Before we begin a detailed analysis of the disk-loaded guide shown in Fig. 8.9, we should consider some of the desirable general characteristics of the accelerating wave. First, the desired wave must have an axial electric field so that it can cumulatively accelerate the electron beam. Second, the field along and near the axis must be as high as possible so as to maximize its interaction with the electron beam. To meet these requirements, we selectively excite the TM_{01} waveguide mode and make the holes on the disks just large enough to pass the beam. The various cells will then act almost like coupled cavities of high Q. This scheme, although capable of amplifying fields in the cell by a factor Q, is restricted in practice to a single frequency operation.[14] In the design of an accelerator, however, this restriction does not present any severe drawbacks.

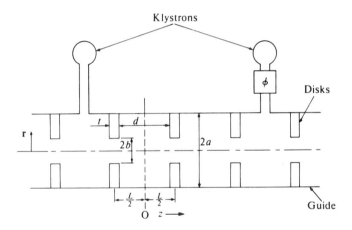

Fig. 8.9. Schematic diagram of a linear accelerator.

Let us divide the structure shown in Fig. 8.9 into two regions, region I for $r < b$ and region II for $b < r < a$. As usual, we first write down the solutions of Maxwell's equation with arbitrary coefficients for both regions. We then apply boundary conditions at $r = b$ and $r = a$ to determine the arbitrary coefficients.

For region I or $r < b$, the fields are given by an expression similar to that for TM modes inside a helix, i.e. by Eq. (8.59):

$$E_{z1} = \sum_{-\infty}^{\infty} A_n I_0(|k_{cn}|r). \tag{8.85}$$

---

[14] See the discussion connected with Fig. 8.2.

Substituting Eq. (8.85) into Eq. (3.29), we find

$$H_{\theta 1} = \sum_{-\infty}^{\infty} \frac{A_n \beta_0}{Z_0 |k_{cn}|} I_1(|k_{cn}|r), \qquad (8.86)$$

where $\beta_0 = \omega\sqrt{\mu\varepsilon}$ and $Z_0 = \sqrt{\mu/\varepsilon}$ as before.

To evaluate A_n, we must specify E_{z1} at $r = b$. Fortunately, the secular equation we seek here is not critically dependent upon the form of the field assumed. Therefore, let

$$E_{z1}\bigg|_{r=b} = \frac{E_0}{[1 - K(2z/d)^2]^{1/2}} \qquad (8.87)$$

for $|z| < d/2$. Here E_0 is a constant and $K = 0$ or 1. The $K = 0$ case corresponds to the situation in which $t \gg d$ so that region II looks like a thin radial transmission line. The expression for $K = 1$ is widely used in the theory of velocity modulation gaps to represent the field in which the gaps have rather thin sharp lips.[15] Using Eq. (8.87) in conjunction with Eq. (8.85) we can evaluate A_n as

$$A_n = \frac{E_0 d}{L I_0(|k_{cn}|b)} \begin{cases} (\pi/2) J_0(\beta_n d/2), & K = 1, \\ \sin(\beta_n d/2)/(\beta_n d/2), & K = 0, \end{cases} \qquad (8.88)$$

where β_n is the phase constant of the wave with amplitude A_n.

In region II, $E_{zII} = 0$ at $z = \pm d/2$ and $E_{zII} = 0$ at $r = a$. If we assume that there is no θ-variation, the expression for E_{zII} that satisfies the above boundary conditions is

$$E_{zII} = \sum_{-\infty}^{\infty} \left[n J_0(k'r) - \frac{J_0(k'a)}{N_0(k'a)} N_0(k'r) \right] \cos \frac{n\pi z}{d}, \qquad (8.89)$$

where $k' = [\beta_0^2 - (n\pi/d)^2]^{1/2}$. Usually, we use only a single term of this series. Equating E_{zII} at $r = b$ to expression (8.87) for E_{z1} at $r = b$, etc., we find the coefficient β_0:

$$\beta_0 = \frac{E_0}{J_0(k'b) - \dfrac{J_0(k'a)}{N_0(k'a)} N_0(k'b)} \begin{cases} \pi/2, & K = 1, \\ 1 & K = 0. \end{cases} \qquad (8.90)$$

Substituting Eq. (8.89) into Eq. (3.29), we can find $H_{\theta II}$. The tangential component of the magnetic field can be approximately matched at $r = b$ by equating the average values of $H_{\theta I}$ and $H_{\theta II}$:

$$\frac{1}{d}\int_{-d/2}^{d/2} H_{\theta 1}\, dz = \frac{1}{d}\int_{-d/2}^{d/2} H_{\theta II}\, dz = j\frac{B_0}{Z_0}\left[J_1(k'b) - \frac{J_0(k'a)}{N_0(k'a)} N_1(k'b) \right], \qquad (8.91)$$

which leads to the secular equation

---

[15] A. H. W. Beck, *Velocity Modulated Thermionic Tubes*, Cambridge University Press, Cambridge, 1948; pp. 64–65.

$$\frac{1}{k'}\left[\frac{J_1(k'b)N_0(k'a) - N_1(k'b)J_0(k'a)}{J_0(k'b)N_0(k'a) - N_0(k'b)J_0(k'a)}\right] = \frac{db}{L}\sum_{-\infty}^{\infty}\frac{I_1(|k_{cn}|b)}{|k_{cn}|I_0(|k_{cn}|b)}$$

$$\times \begin{cases} [\sin(\beta_n d/2)/(\beta_n d/2)]J_0(\beta_n d/2), & K = 1, \\ \sin^2(\beta_n d/2)/(\beta_n d/2)^2, & K = 0. \end{cases}$$

(8.92)

For a given set of values of ω, b, a, d, and L, we can plot the left-hand and right-hand sides of Eq. (8.92) against $|k_{cn}|$. The intersection of these two curves determines the permissible value of $|k_{cn}|$. Recalling that $k_{cn}^2 = (\omega/c)^2 - \beta_n^2$, we see that the ω–β diagram tells us how the phase velocity v_{pn} varies with frequency. Furthermore, we can find the axial electric field available for acceleration of the beam by means of Eq. (8.85). Let us merely say here that detailed analysis of these results shows that the electric field on the axis is very small for low phase velocities ($0.1c$ to $0.2c$) unless b is made very small compared to a wavelength. Thus only an electron beam of small cross-sectional area can pass the disk holes. Furthermore, under these conditions the structure, acting like a cavity, is highly dispersive. Therefore in most cases only single frequency operation is practical. As expected, these results agree with the physical conjectures we made at the beginning of this section.

6. MAGNETRON OSCILLATORS

The magnetron oscillator is the only available source of large amounts of power at frequencies exceeding 1 Gc. Its efficiency is very high, ranging from 40 to 70%. Typical power outputs are hundreds of kilowatts peak and kilowatts continuous wave (c.w.). Although magnetrons can operate at lower frequencies, their principal usefulness is at frequencies above 500 Mc, where there are no available triodes to generate large amounts of power.

In a magnetron, as in other microwave tubes, part of the energy gain by the electron in falling through an accelerating dc electrical potential is converted into r.f. energy. However, whereas electrons are preaccelerated before they enter the interaction structure in the TWT and BWO, the dc and r.f. potentials act on the electrons simultaneously in the interaction space of a magnetron. Furthermore, whereas a longitudinal magnetic field is applied to confine the beam in the TWT or BWO, a magnetic field in a direction normal to the electric field is essential for the operation of a magnetron.

Figure 8.10 is a schematic diagram of a typical magnetron oscillator. We see that the resonant system of a magnetron consists of a number of individual cavities, all coupled together via the interaction space between cathode and anode. In the absence of the magnetic field H_0, electrons emitted from the cathode will travel radially toward the anode under action of the electric force $-e\mathbf{E}_0$. When the magnetic field is present, however, the electron path will be bent by the magnetic force $-e\mathbf{v} \times \mathbf{B}_0$, according to Eq. (4.17), and the electron will acquire a

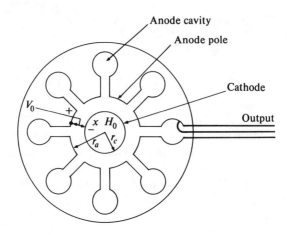

Fig. 8.10 Schematic diagram of cavity magnetron.

tangential as well as a radial component of velocity. Whether an electronic path is bent enough so that the electron heads back toward the cathode clearly depends on the relative magnitudes of V_0 and B_0. Application of Eq. (4.19) and the law of conservation of energy easily shows that the electron would just graze the anode if V_0 and B_0 were related as follows:

$$B_{0c} = \frac{\sqrt{8V_0/(e/m)}}{r_a[1 - (r_c^2/r_a^2)]}, \tag{8.93}$$

where e/m is the charge to mass ratio of the electron. Here B_{0c}, expressed in terms of V_0, is known as the cutoff magnetic field. If $B_0 > B_{0c}$ for a given V_0, the electrons will not reach the anode.

Since the cavities are coupled together, we would expect the number of possible resonant modes in a magnetron to equal the number of cavities. In other words, the mean length LS of the periodic structure must be a whole number of half guide wavelengths:

$$LS = n\frac{\lambda_g}{2}, \tag{8.94}$$

where n is an integer greater than or equal to S. Here S is the number of cavities and L is the mean separation between cavities. Since $\beta_0 \lambda_g = 2\pi$, we have

$$\beta_0 L = \frac{n\pi}{S}. \tag{8.95}$$

The most common mode of operation of a magnetron is the π-mode. For this mode, $n = S$, and the phase shift between adjacent poles is π rad.

In order to understand the mechanism by which oscillations in a magnetron are generated, we shall first study the behavior of emitted electrons under the simultaneous influence of a dc anode voltage and an axial magnetic field. For simplicity and without any appreciable loss in insight we shall consider the linear version of the cylindrical magnetron shown in Fig. 8.10.

Referring to Fig. 8.11, we can easily show by means of the Lorentz-Newton force equation (4.19) that the trajectories of the electrons starting from rest at the cathode at $t = 0$ are given by

$$x = -\frac{E_0 m}{eB_0^2}\sin\frac{eB_0}{m}t + \frac{E_0}{B_0}t,$$
$$y = -\frac{E_0 m}{eB_0^2}\sin\frac{eB_0}{m}t + \frac{E_0 m}{eB_0^2}. \tag{8.96}$$

Thus the electron motion is a combination of circular motion and uniform translation, which is equivalent to the motion of a marker on the spoke (or its extension) of a rolling wheel.

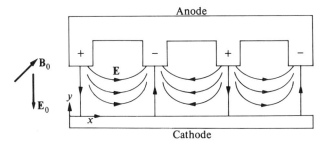

Fig. 8.11. Electric and magnetic field distributions in a linear magnetron structure.

Assume now that π-mode oscillations exist in the cavity structure. Lines of r.f. electric field **E** at some instant of time will then form as shown in Fig. 8.11. A half-cycle later the entire r.f. field pattern will shift one segment to the right, i.e., all field lines will be reversed in direction. We have already seen that in the absence of the r.f. field the path of an electron is the common cycloid given by Eq. (8.96). The presence of the r.f. field modifies this path. However, since the average drift velocity of all cycloids is the same and equal to E_0/B_0, the phase velocity of the wave propagating along the structure must also be nearly equal to this value if cumulative interaction is desired.

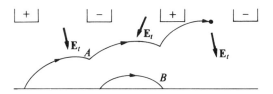

Fig. 8.12. Representative trajectories of the linear magnetron. Note that the plus and minus signs as well as the directions of E_t are for the instant of electron emission. A half-cycle later, the plus signs will have changed to minus signs and vice versa.

Consider the two typical trajectories shown in Fig. 8.12. The starting time of the electrons is such that they reach the gaps when the magnitude of the r.f. field is maximum. Electron B experiences a net accelerating electric force and gains kinetic energy from the r.f. field, while electron A experiences a net retarding electric force and loses kinetic energy to the field. If electron A travels to the right at about the same speed as the r.f. electric field wave does, it will again experience a retarding force when it reaches the next cavity and will lose more energy to the r.f. field. Since the average total electric field $\mathbf{E}_t$ felt by electron A is not vertical but instead tilts to the right, the direction of the drift velocity (which is perpendicular to $\mathbf{E}_t$ and $\mathbf{B}_0$) is tilted upward. For this reason, electron A will drift toward the anode following the path sketched in Fig. 8.12. Since the average drift velocity is E_0/B_0, a constant, and electron A is, by and large, moving towards regions of higher dc potential, the dc kinetic energy of electron A must be continuously converted into r.f. energy of the oscillation as it drifts toward the anode. On the other hand, the average electric field felt by electron B tilts to the left, so that electron B shifts slightly downward and strikes the cathode, giving rise to "back heating" and secondary emission, and is quickly removed from the interaction space.

In actuality, the detailed cavity structure of a magnetron does not greatly modify its operation. However, certain types of structures may increase the frequency separation between the π and adjacent modes, thus enhancing the stability of operation. Magnetrons can be tuned by introducing foreign metallic objects which interfere with the electric or magnetic fields of the cavities.

PROBLEMS

8.1 a) For most cases of practical interest, wavelike Eqs. (8.4) and (8.5) are very difficult to solve. However, the essential features of the ω–β diagram obtained from the solution of these equations are similar to those obtained from the Schrödinger equation (1.13) for an electron traveling through a periodic potential. Consider, therefore, the case in which the potential energy U of an electron is in the form of a periodic array of square wells, as shown in Fig. 8.13. Find the solution for the wave function ψ by assuming that $\psi \sim e^{ikx}u_k(x)$, where k is the wave number. Note that $u_k(x)$ must be continuous at $x = 0$ and at $x = -b$ and that, due to the periodic nature of U, $\partial u_k(x)/\partial x$ evaluated at $x = -b$ is equal to $\partial u_k(x)/\partial x$ evaluated at $x = a$.

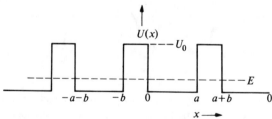

Fig. 8.13. Periodic potential.

b) Plot E vs. k to exhibit allowed and forbidden energy bands. Note that $E = \hbar\omega$ and k is equivalent to β in electromagnetic theory.

8.2 a) Find the field distribution of the lowest-order mode in a rectangular guide of height b and width a partially loaded by a dielectric slab of dielectric constant ε and thickness δ against the left sidewall of the guide. Assume that the dielectric slab is of height b.
b) Sketch the electric field distribution of the lowest-order mode for (a) $\varepsilon \gg \varepsilon_0$, and (b) $\varepsilon \ll \varepsilon_0$.
c) Would the phase and group velocities in the guide increase, decrease, or remain the same after the dielectric slab was introduced?

8.3 To prevent load reflections of a TWT from reaching the input and possibly leading to oscillations, loss has to be introduced into a short section of the propagating structure. As the loss is increased more and more, the gain must approach a constant value. Thus the circuit will in effect be severed as far as the electromagnetic wave is concerned, and any excitation in the output will be due to the ac velocity and convection current of the electron stream which crosses the lossy section. For simplicity, assume that the helix is severed and terminated looking in each direction, so that the helix voltage falls off to zero in zero distance while the ac beam voltage and current remain unchanged in traversing the lossy section. Find the optimum location of the lossy section to yield maximum gain for a given TWT.

8.4 How are oscillations in a BWO initiated? Describe the buildup process of these oscillations.

8.5 a) Utilizing the expression (8.84) for the relativistic mass, derive the Einstein mass-energy relation $E = (m - m_0)c^2$.
b) Show that the velocity of the electron u after being accelerated through a potential V_0 is

$$u = \left[\frac{2eV_0}{m_0} \frac{1 + \dfrac{e}{m}\dfrac{V_0}{2c^2}}{\left(1 + \dfrac{e}{m}\dfrac{V_0}{c^2}\right)^2}\right]^{\frac{1}{2}}.$$

8.6 In a traveling wave tube, net energy flows from the electron beam to the microwave, giving rise to wave amplification. In a linear accelerator, net energy flows from the wave to the beam, giving rise to electron acceleration. Yet the physical structure of these devices is rather similar. Delineate the parameters in the design of these devices that determine the direction of net power flow.

8.7 Show that the π-mode operation of a magnetron is equivalent to that of a polyphase motor with some given number of poles.

CHAPTER 9

BEAM FORMATION, FOCUSING, AND PLASMA OSCILLATIONS

In the last two chapters, we studied the behavior of microwave devices with cavity or periodic structures. We implicitly assumed that electron beams are somehow produced and focused into a pencil-like geometry. In this chapter, we shall deal explicitly with the problems of forming and focusing an electron beam. The discussion will include thermionic emission, the Pierce gun, and uniform and periodic magnetic focusing. We shall also examine the plasma oscillations generated by Coulomb repulsion forces between charges. These oscillations create the so-called space-charge waves in electron beams.

1. THERMIONIC EMISSION

Electrons may be ejected from a metallic cathode by various means, including (1) photo emission, whereby the metallic surface is exposed to photons in a light beam; (2) release of secondary electrons by bombarding the surface with primary electrons or ions; (3) thermionic emission by heating the cathode material to high temperatures; and (4) high-field emission. If a weak or modest electric field is applied to the cathode surface, the thermionic emission current can be increased because of the lowering of the work function. If the electric field strength is sufficiently strong, quantum-mechanical tunneling can occur as a result of the reduction in the width of the surface potential barrier. In this case, no heating of the cathode is necessary and the phenomenon is referred to as cold emission.

Of the methods used for electron liberation previously mentioned, thermionic emission is by far the most common; therefore, we shall discuss it in some detail. First, we note that when an electron approaches the surface of a metal from within, it experiences a force directed back into the interior of the metal because of the attraction of unbalanced ionic charges at or near the emission surface. This force decreases the velocity component in the direction of the surface normal. Unless this component exceeds a certain critical value, the electrons will clearly not be able to escape from the metal. If we heat the cathode material to sufficiently high temperatures, however, the electrons can acquire a velocity component exceeding the critical value. Therefore, the object of our search here is a relation between emission current density and cathode temperature. Of course, this current density must also be a function of the strength of the retarding force of the unbalanced surface ionic charges, i.e., a function of the potential barrier at the surface of the metal, commonly known as the *work function* of the material.

An exact treatment of thermionic emission from metals is prohibitively complicated due to the extremely large number of atoms in the crystal. We shall therefore ignore the periodic nature of the potential within the crystal and use the box model of a metal with finite walls, as shown in Fig. 9.1. In this model, we assume that a potential energy difference of $(E_F + W)$ exists between the electrons that just emerge from the surface and those just beneath the surface where E_F and W are respectively the Fermi level and work function of the material.

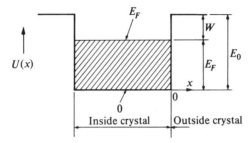

Fig. 9.1. Box model of potential energy of a metal.

Since electrons are indistinguishable and must obey the Pauli exclusion principle inside the metal, they satisfy Fermi-Dirac statistics. If the probability that a state at energy E will be occupied in an ideal electron gas at thermal equilibrium temperature T is given by P_{FD}, then,[1]

$$P_{FD}(E) = \frac{1}{e^{(E-E_F)/kT} + 1}, \tag{9.1}$$

where k is Boltzmann's constant.

Now, if E_0 is the energy of the vacuum level, an inspection of Fig. 9.1 shows that

$$W = E_0 - E_F. \tag{9.2}$$

Usually, W has a value of several electron volts which is at least one or two orders of magnitude larger than kT. This value, along with the fact that $E > E_0$ for any emitted electron, indicates that $e^{(E-E_F)/kT} \gg 1$ in Eq. (9.1). Consequently, the expression for P_{FD} simplifies to

$$P_{FD}(E) \simeq e^{(E_F-E)/kT}. \tag{9.3}$$

---

[1] For the derivation of this expression, see, for example, C. L. Hemenway, R. W. Henry, and M. Coulton, *Physical Electronics*, John Wiley and Sons, New York, 1963; p. 45.

Since an electron within the metal must expend an amount of energy E_0 in order to overcome the potential barrier at the surface of the metal, the quantity $(E - E_0)$ is equal to the kinetic energy E_1 of the emitted electrons. Therefore, Eq. (9.3) can be rewritten as

$$P_{FD}(E_1) \simeq e^{-W/kT} e^{-E_1/kT}, \tag{9.4}$$

where we have also made use of the relation (9.2). We see from expression (9.4) that as far as emitted electrons are concerned, the distribution of E_1 obeys classical Maxwell-Boltzmann statistics. Experimentally, the velocity distribution of electrons thermionically emitted has been found to be Maxwellian, in agreement with Eq. (9.4).

We shall now calculate the emission current density. First, due to energy conservation, we have

$$v_{x(\min)} = \sqrt{2(E_F + W)/m}, \tag{9.5}$$

where $v_{x(\min)}$ is the minimum value of the x-component of electron velocity and m is the electron mass. The x-directed force decreases v_x, and unless v_x exceeds $v_{x(\min)}$, the electron will be "reflected" back into the crystal. Of course, the x-directed force cannot influence the y- and z-components of the electron velocity.

Next, the number of electrons with velocities in the differential range v_x to $v_x + dv_x$, v_y to $v_y + dv_y$, v_z to $v_z + dv_z$ reaching an area A of the surface in time dt is simply the number having velocities in that range within a distance $v_x\, dt$ of the surface, or the number within a physical volume $V = v_x A\, dt$. The number of allowed electronic states corresponding to velocities within a differential range may be found as follows. Since $p_x = n_x h/2L_x$, etc. and $dp_x = h/2L_x$, etc., the related "volume" in momentum space is $h^3/8L_x L_y L_z = h^3/8V$. It is important to note, however, that this expression is applicable to standing but not traveling waves. Since n_x, n_y, and n_z are all positive integers, all permissible values of p appear in one octant of momentum space. Each point in such a diagram represents the summation of six traveling waves, two in each of the three direction, which interfere to produce the total standing wave. In other words, n_x, n_y, n_z in the traveling wave representation can take on negative as well as positive values. Therefore, permissible points will now appear in all eight octants formed by the cartesian axes p_x, p_y, p_z. Since the number of quantum states within an energy interval must be independent of the method of describing the states, the density of points plotted in all eight octants is one-eighth that plotted in a single octant. Therefore, if we are to represent a wave function in terms of traveling rather than standing waves, $dp_x dp_y dp_z = h^3/V$. It follows that the density of states dS or number of electronic states within the differential momentum volume $dp_x\, dp_y\, dp_z$ is given by

$$dS = \frac{2V}{h^3} dp_x\, dp_y\, dp_z, \tag{9.6}$$

where the factor of 2 is introduced to account for the fact that an electron spin can have a projection of $+\frac{1}{2}$ or $-\frac{1}{2}$ along a given axis.

Equation (9.6) gives us the number of states within a given momentum range. To determine the number of electrons dN within the differential velocity range that reach area A in time dt, we must multiply expression (9.6) by the probability that a given state is occupied, that is, by P_{FD} of Eq. (9.1). Thus we have

$$dN = \frac{2v_x A m^3 \, dt}{h^3} \frac{1}{1 + e^{(E - E_F)/kT}} \, dv_x \, dv_y \, dv_z, \tag{9.7}$$

using the relations $p_x = mv_x$, etc. and $V = v_x A \, dt$. Since the potential energy inside the crystal is taken to be zero, the energy E appearing in P_{FD} is just

$$E = \tfrac{1}{2} m(v_x^2 + v_y^2 + v_z^2). \tag{9.8}$$

Combining Eqs. (9.7) and (9.8), we find that the total number of electrons per unit area per unit time crossing the surface is

$$\iiint \frac{dN}{A \, dt} = \frac{2m^3}{h^3} \int_{v_{x(\min)}}^{\infty} \int_{-\infty}^{\infty} \int_{-\infty}^{\infty} \frac{v_x \, dv_x \, dv_y \, dv_z}{1 + e^{-E_F/kT} e^{m(v_x^2 + v_y^2 + v_z^2)/2kT}}. \tag{9.9}$$

We have already shown in connection with Eq. (9.3) that the exponential term in the denominator of Eq. (9.9) is much larger than unity for the emitted electrons. Thus integral (9.9) can be readily evaluated to yield

$$\iiint \frac{dN}{A \, dt} = \frac{4m\pi k^2}{h^3} T^2 e^{-W/kT}, \tag{9.10}$$

where we have used Eq. (9.5). It now follows that the emitted current density J_{th} is given by

$$J_{\text{th}} = \left(\frac{4me\pi k^2}{h^3}\right) T^2 e^{-W/kT}. \tag{9.11}$$

This expression is known as the Richardson-Dushman equation. It agrees reasonably well with experiment as far as its temperature dependence is concerned, but the experimentally observed value of the quantity in parantheses is only about half as much as predicted by Eq. (9.11). There are several explanations for this discrepancy. First, partial reflections of electronic wave functions occur at the crystal surface even for those electrons with enough x-directed velocity to escape. For such electrons, there is a finite probability, quantum-mechanically, that they will escape upon reaching the surface. Second, since the emitting surface is polycrystalline, the surface of the crystal is composed of different crystal faces with significantly different work functions. Due to the strong exponential dependence of J_{th} on W, the faces with the lowest W contribute most of the electrons while the remaining portions of the surface are relatively poor emitters.

We see from the above discussion that electrons will emerge from heated cathode surfaces with a Maxwellian distribution of velocity in all three cartesian coordinate directions. In other words, electrons will emerge from the surface with, in general, nonzero velocity and at various angles to the surface normal. This behavior has two important implications. First, the random emission of electrons on the microscopic scale gives rise to the shot noise found in microwave tubes. Second, the resulting crossing of electron trajectories actually dictates the minimum diameter that a convergent beam can attain. These thermal effects on the behavior of electron beams have been studied by a number of workers.[2]

We shall now conclude our discussion of thermionic emission by examining the prope ties of various cathode materials used in microwave tubes. First, to obtain appreciable emission cathodes must be heated to above 1000°C. It follows that only materials with melting temperatures substantially above this value can be used as emitting surfaces. These materials can be divided into three general types: pure metal cathodes, oxide-coated cathodes, and composite cathodes.

Examples of pure metals used as cathodes for microwave tubes are tungsten, with a melting temperature of 3370°C and a work function of 4.5 V[3] and tantalum, with a melting temperature of 3000°C and a work function of 4.1 V. Because of the relatively high work function of tungsten, emission is low unless the tungsten cathode is operated at very high temperatures. However, tungsten has good mechanical stability at high temperatures and can withstand high-energy positive ion bombardment when immersed in an imperfect vacuum. On the other hand, tantalum, with its lower work function, can easily yield ten times the emission current density of tungsten over, admittedly, the smaller operating temperature range due to its substantially lower melting point. However, a tantalum cathode's emission ability can be more easily impaired by oxide formation resulting from attacks by molecules of residual gases.

Because of the exponential dependence of J_{th} on W given by Eq. (9.11), the emission current is rather sensitively dependent upon the work function. We can obtain lower values of W (1 to 2 V) by using oxide-coated cathodes. Examples of such materials are the oxides of barium, strontium, calcium or a combination thereof. These cathodes are usually indirectly heated by a tungsten filament and can produce very large emission currents over short intervals. This high rate of emission cannot be maintained, however, since the cathode must have time to recover before the next pulse. With a suitable duty cycle, we can obtain pulsed currents 100 or more times greater than those for C.W. operation. Unfortunately oxide-coated cathodes are easily poisoned by various gases and vapors and may disintegrate as a result of positive-ion bombardment.

Thoriated tungsten and the L-cathode are examples of composite cathodes. Thorium evaporates from tungsten at a lower rate than from itself, and a thin layer

---

[2] See, for example, J. R. Pierce and L. R. Walker, *Jour. Appl. Phys.* **24,** 1328 (1953); C. C. Cutler and M. E. Hines, *Proc. Inst. Radio Engrs.* **43,** 307 (1955).

[3] The work function W in Eq. (9.11) has the dimensions of energy and is therefore in, say, electron volts. However, values of work function are frequently quoted in volts or in units of W/e, where e is the electronic charge.

of thorium ($W = 3.4$ V) forms at the surface of the tungsten. The thorium forms a dipole layer, partially neutralizing the electrostatic forces on the tungsten surface with a consequent reduction of W to 2.6 V for the composite structure. In such cathodes, the evaporated thorium is continuously replenished by diffusion from within. In an L-cathode, a porous tungsten or porous nickel pellet is impregnated with an alkaline-earth metal compound.[4] The layer that forms on the surface again reduces W and gives rise to very high C.W. emission at moderate temperatures of 1000 to 1300°C.

2. CURRENT-VOLTAGE RELATION

In Eq. (9.11) the thermal emission current density J_{th} is given as a function of temperature T. Multiplying J_{th} by the area of the emitting surface gives the total emission current from a cathode. We must emphasize, however, that this product is the *maximum* or *saturation* current that can be drawn from the cathode at a given temperature when the accelerating voltage V_a is sufficiently high to draw electrons away from the cathode as fast as they are emitted. Were this not the case, an electron emerging from the cathode might find a pool of other electrons in the vicinity of the cathode exerting a retarding force and tending to turn it back toward the cathode. We refer to the latter mode of emission as space-charge limited.

On the basis of the preceding discussion, we would expect the current density J to increase with voltage up to the maximum value J_{th} given by Eq. (9.11). Once this value of current is reached, further increase in the voltage will result in no further increase in current. We refer to the range in which J varies with V as the range of *space-charge-limited operation*. In this case, J is independent of temperature. On the other hand, under *temperature-limited operation*, J ($= J_{th}$) is independent of voltage but is a function of temperature. The general J–V_a behavior with T as a constant parameter is sketched in Fig. 9.2. The transition between the space-charge-limited and temperature-limited domains is usually not perfectly sharp, since not all parts of the emitting surface reach saturation simultaneously because of the variation in work function mentioned previously.

We must now find the relationship between J and V. According to Maxwell's equation (2.3) and the constitutive equation (2.6), $\nabla \cdot \mathbf{E} = \rho/\varepsilon_0$. For the static case, $\nabla \times \mathbf{E} = 0$ according to Eq. (2.1). Therefore, we can let $\mathbf{E} = -\nabla V$, where V is a scalar potential. Thus the V–ρ relation is given by Poisson's equation:

$$\nabla^2 V = -\rho/\varepsilon_0. \tag{9.12}$$

For the parallel-plane electrode configuration shown in Fig. 9.3(a), the voltage $V(z)$ with no emission, that is, at $\rho = 0$, is simply

$$V = \frac{V_a}{d} z, \tag{9.13}$$

---

[4] R. Levi, *Jour. Appl. Phys.* **24**, 233 (1953).

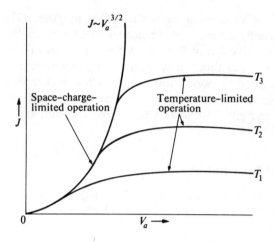

Fig. 9.2. Current-voltage relation at constant cathode temperatures. Note that $T_3 > T_2 > T_1$.

obtained by integrating Eq. (9.12). Equation (9.13) is plotted in Fig. 9.3(b), in which we observe that V varies linearly with z or the electric field E is independent of z. If we increase the cathode temperature until significant current flows, a finite negative charge density exists between the plates. When ρ is negative, integration of Eq. (9.12) gives

$$\frac{dV}{dz} = \frac{V_a}{d} - C_1 + \int_0^z \frac{|\rho(z)|}{\varepsilon_0}\, dz, \qquad (9.14)$$

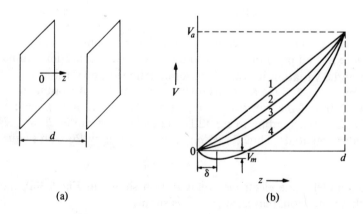

Fig. 9.3. (a) Parallel plane electrode configuration. (b) Potential distribution between planes. For curve 1, $\rho = 0$; curves 2, 3, 4 correspond to $\rho \neq 0$ and dV/dz at $z = 0$ and are positive, zero, and negative respectively.

where the constant C_1 is given by

$$C_1 = \frac{1}{d}\left[\int_0^z \left[\int_0^z \frac{\rho|z|}{\varepsilon_0} dz\right] dz\right]_{z=d}. \tag{9.15}$$

An examination of Eq. (9.14) shows that dV/dz, in contrast to the $\rho = 0$ case, is now a function of z. Furthermore, we note that at the cathode, $dV/dz < V_a/d$, the value corresponding to $\rho = 0$, for all $|\rho(z)|$. Indeed, depending on the emission current density, dV/dz at $z = 0$ can be positive, zero, or negative, corresponding to V_a/d being larger than, equal to, or less than C_1. In all three cases, dV/dz must increase with z. Thus, if dV/dz is negative at $z = 0$, dV/dz must go through zero at some plane between the cathode and anode, since V must be equal to V_a at $z = d$. Between this plane and the cathode, V must be negative. Figure 9.3(b) shows the potential distribution for three representative cases where $dV/dz|_0$ is positive, zero, or negative.

Most microwave tubes are operated in the space-charge-limited mode. In this mode of operation, as we can see from Fig. 9.2, the current is determined by the accelerating voltage and is rather insensitive to small temperature changes. In Fig. 9.3(b), this mode of operation is represented by the sketch with $dV/dz < 0$ at $z = 0$. Note that the potential just outside the cathode is depressed below the cathode value. Therefore, the energy in electron volts of an electron emitted from the cathode must be greater than the maximum potential depression at $z = \delta$ for the electron to have enough energy to reach the anode. Thermionically emitted electrons have a Maxwellian distribution of velocities, as discussed in the last section, and those electrons unable to pass the potential minimum are returned to the cathode by the retarding field in the region between $z = 0$ and δ. It turns out that the normally directed thermal energy is about 0.1 eV at typical operating temperatures, while, depending on the rate of emission, a potential depression of up to several tenths of a volt can be expected. Consequently, a pool of space charge may indeed form in the vicinity of the cathode if the emission current density is sufficiently high.

Having obtained some physical insight into the space-charge phenomenon, we must now return to our task of obtaining the functional relationship between J and V from a solution of Poisson's equation (9.12). Although the space-charge-limited operation usually corresponds to the case depicted in Fig. 9.3(b), where $dV/dz < 0$ at $z = 0$, it is analytically expedient to solve Eq. (9.12) for the transition case, where $dV/dz = 0$ at $z = 0$. This is actually quite permissible since typically $\delta \ll d$ and $V_m \ll V_a$. We shall make these assumptions in what follows.

Combining Eq. (9.12) with the energy conservation relation,[5]

$$eV = \tfrac{1}{2} mv^2, \tag{9.16}$$

---

[5] We implicitly neglect the thermal velocity of emitting electrons.

where e, m, and v are respectively the mass, charge, and velocity of the electron, we find

$$\frac{d^2V}{dz^2} = -\frac{J}{\varepsilon_0\sqrt{2(e/m)V}}, \qquad (9.17)$$

also using Eq. (4.20).[6] Integrating Eq. (9.17) and imposing boundary conditions $V = 0$, $dV/dz = 0$ at $z = 0$, we obtain the sought-after Langmuir-Child law:

$$J = -\frac{4\sqrt{2}}{9}\varepsilon_0\sqrt{\frac{e}{m}}\frac{V^{3/2}}{z^2} = -2.33 \times 10^{-6}\frac{V^{3/2}}{z^2}, \qquad (9.18)$$

in A/m².

Electrodes in microwave tubes are often made of concentric cylinders or concentric spherical caps. For example, in convergent-beam tubes, the cathode corresponds to the outer spherical cap. Since the deviation of the J–V_a relation is similar to that used for the parallel planes we leave it as an excercise for the reader. Suffice to say, however, that the $V^{3/2}$ dependence of J applies quite generally to any arbitrary cathode-anode configuration. This conclusion leads naturally to the definition of *perveance*, a quantity which is extremely important in electron gun design and which we shall now discuss.

3. ELECTRON GUN DESIGN

As mentioned previously, most microwave tubes operate under the space-charge-limited condition to minimize current fluctuation resulting from changes in temperature. Under this condition, we found in the last section that the total current is proportional to $\frac{3}{2}$ power of voltage. It seems reasonable, therefore, that a figure of merit describing the performance of an electron gun may be defined as

$$K = I/V_a^{3/2}, \qquad (9.19)$$

where K is expressed in A/V $^{3/2}$ and is known as the *perveance* of the gun. For example, if an electron gun yields 1 mA at 100 V, it is said to have a perveance of 1×10^{-6}, or simply 1 with the factor 10^{-6} understood.

In the last section, we considered the electron flow between infinite planes. The flow between such planes, or between infinitely long complete cylinders or spheres, must be rectilinear, since space-charge repulsion forces are all balanced to zero at every point. In other words, every point on a plane z in Fig. 9.3(a), for example, is equivalent to all other points in the plane so that there is no tendency for an electron to move in the y- or x-direction, thus ensuring streamline flow in the z-direction.

The actual cathode configuration used in microwave tubes is seldom completely symmetrical or rectilinear in nature. For example, it is frequently desirable to

---

[6] Here J is assumed to be independent of z; that is, there is no continuing pile-up or loss of electrons to walls of the tube.

obtain a beam of circular cross section by means of a disk-shaped cathode. In this case, because of the uncompensated space-charge repulsion forces, there is a tendency for the beam to spread as it traverses the cathode-anode region. We obtain rectilinear flow, however, by supplying compensating forces along the beam edge by placing electrodes of certain shapes and potentials just outside the beam. This method is due to J. R. Pierce, and an electron gun designed in this way is known as a Pierce gun.[7]

Let consider specifically a rectangular beam originating from a cathode which extends infinitely in the z-direction, as shown in Fig. 9.4. To ensure rectilinear flow between the electrodes, it is clear that we must establish a potential which satisfies Laplace's equation in the charge-free region outside the beam, reduces to the correct value along the beam edge given by Eq. (9.18), and gives zero electric field or potential gradient normal to the boundary.

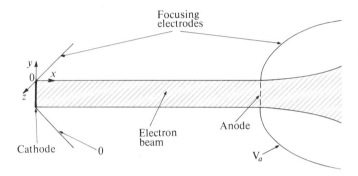

Fig. 9.4. Elements of a Pierce gun.

The potential along the beam edge $y = 0$ is given by Eq. (9.18) as

$$V(x,0) = [-(9J/4\varepsilon_0)\sqrt{m/2e}]^{2/3} x^{4/3}. \qquad (9.20)$$

Outside the beam, the potential must satisfy Laplace's rather than Poisson's equation, since the charge density there is zero. Pierce has shown that the potential distribution,

$$V = [-(9J/4\varepsilon_0)\sqrt{m/2e}]^{2/3} [\text{Re}(x + jy)^{4/3}], \qquad (9.21)$$

satisfies the Laplace equation $\nabla^2 V = 0$ outside the beam, reduces to (9.20) at $y = 0$, and has a zero potential gradient at the beam edge, that is, $\partial V/\partial y = 0$ at $x = 0$. We can easily obtain the equipotential surfaces from Eq. (9.21) as

$$(x^2 + y^2)^{2/3} \cos(\tfrac{4}{3}\theta) = \frac{V}{[(-9J/4\varepsilon_0)\sqrt{m/2e}]^{2/3}}, \qquad (9.22)$$

where $\theta = \tan^{-1}(y/x)$.

---

[7] J. R. Pierce, *Jour. Appl. Phys.* **11**, 548 (1940)

Thus, to obtain rectilinear flow between planes finite in the y-direction we need only enclose the region by metal surfaces shaped according to Eq. (9.22). For convenience, the two surfaces chosen are usually those corresponding to cathode and anode potentials, as indicated in Fig. 9.4. Note that the zero equipotential is a plane whose intersect with the xy plane is given by $y = z \tan 67.5°$. It is usually separated from the cathode by a small gap in order to thermally insulate it from the cathode.

In principle, the electrodes shaped according to Eq. (9.22) must be infinite in extent in the x- and y- as well as z-directions. In practice, however, we can use finite compensating electrodes to obtain rectilinear flow between finite or incomplete cathode and anode electrode configurations. In these cases, it is clearly not feasible to obtain analytical solutions for the equipotentials, as was done previously. Pierce has devised a convenient experimental approach based on the tilted electrolytic-tank technique originally suggested by Bowman-Manifold and Nicoll.[8] In this approach, a thin wedge of the beam is represented by a tilted electrolytic tank. The edge of the electrolyte represents the axis of symmetry, while a strip of insulating material represents the edge of the beam. A number of metal probes are inserted into the strip, and the potential distribution along the beam edge is measured. Thin metal strips represent compensating electrodes and their shapes are adjusted until the prescribed distribution along the beam edge is obtained.

The additional electrodes compensate for space-charge repulsion forces only in the cathode-anode region. Once beyond the anode, the beam tends to spread because of uncompensated Coulomb forces. Furthermore, the anode aperture acts as a divergent electrostatic lens. To obtain converging or parallel-flow beams in the region beyond the anode, we must usually use disk or spherical-cap cathodes to yield an initially converging beam. Equipotentials obtained for disk[9] and spherical-cap cathodes[10] have been determined using the electrolytic tank. It has been found that the prescribed voltage distribution along the beam edge can be obtained by a large variety of electrode shapes, including some that can be easily fabricated. Presumably, this "nonuniqueness" of equipotential shapes results from the use of finite compensating electrodes.

4. BEAM FOCUSING

If the electron beam is still convergent after emerging from the anode aperture, it will travel some distance beyond the anode before Coulomb repulsion forces cause it to diverge. Thus, if the drift tube diameter is larger than the beam diameter at the anode aperture, the beam can be transmitted through the tube for a small distance without appreciable loss of electrons to the walls of the drift tube. This

---

[8] M. Bowman-Manifold and F. H. Nicoll, *Nature* **142**, 39 (1938)

[9] J. R. Pierce, *op. cit.*

[10] R. Helm, K. R. Spangenberg, and L. M. Field, *Elec. Comm.* **24**, 101 (1947).

is the case with the reflex Klystron, in which there is no beam focusing beyond the anode aperture since the beam current density is relatively small. However, if the beam travels over an extended distance beyond the anode aperture (such as in the TWT) and/or if the current density is sufficiently high, some form of beam focusing beyond the anode is required. The focusing may be electrostatic or magnetostatic, but magnetic focusing is by far the most commonly used.

A. Brillouin flow

In our discussion of Pierce-gun design, we implicitly assumed that there is no magnetic field in the gun region. It follows that the axial magnetic field used to confine the electron beam beyond the anode aperture must originate to the right of the anode aperture, as illustrated in Fig. 9.5(a). We can see from this figure that electrons in the beam must cross regions where the magnetic field has a radial as well as an axial component. The interaction of the axial electron velocity with the radial component of the magnetic field gives rise to a tangential velocity, which in turn interacts with the field in the axial region to produce an inward focusing force. This force counteracts the coulomb repulsion force and centrifugal force of circular motion. This method of focusing is known as *magnetic* focusing employing Brillouin flow.

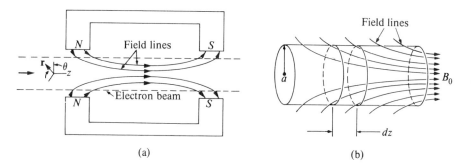

Fig. 9.5. (a) Magnetic flux distribution in Brillouin flow. Magnetic flux entering an electron beam.

To begin our quantitative discussion of Brillouin flow, let us calculate the coulomb repulsion force in a cylindrical electron beam. For simplicity, consider the case of an outermost electron. From Maxwell's equation (2.3) and Gauss's theorem (2.36), we easily find the radial electric field in terms of linear charge density λ as

$$\mathbf{E}_r = -\frac{\hat{r}}{r}\frac{\lambda}{2\pi\varepsilon_0 a}, \qquad (9.23)$$

where a is the radius of the beam and $\hat{r}/r$ is the outward-directed unit radial vector. We can express λ in terms of dc beam current I_0 and dc beam velocity v_0

by combining Eqs. (4.20) through (4.22) to yield

$$\lambda = I_0/v_0, \tag{9.24}$$

where we have also used the relation $\rho = \lambda/A$. If we invoke the law of conservation of energy, $\frac{1}{2} mv_0^2 = eV_0$, where V_0 is the dc beam acceleration voltage, Eq. (9.24) becomes

$$\lambda = \frac{I_0}{\sqrt{2(e/m) V_0}}. \tag{9.25}$$

Substituting this expression into Eq. (9.23) for λ, we find the coulomb repulsion force acting on an outermost electron:

$$\mathbf{F}_{r1} = (-e)\mathbf{E}_r = \frac{\hat{r}}{r} \frac{eI_0}{2\pi\varepsilon_0 a \sqrt{2(e/m) V_0}}. \tag{9.26}$$

The force $\mathbf{F}_{r1}$ is directed outward from the center of the beam and tends to cause the beam to diverge; the focusing force $\mathbf{F}_\theta$ required to counteract this divergent force originates in the angular momentum acquired by the electrons in their passage through the region where there is an appreciable radial component of magnetic flux density B_r. From Fig. 9.5(a), we have

$$\mathbf{F}_\theta = \hat{\theta} ev_0 B_r = \hat{\theta} e \frac{dz}{dt} B_r, \tag{9.27}$$

where $\hat{\theta}$ is a unit vector in the θ-direction.

To complete our analysis, we must calculate v_θ of the electron acquired in crossing radial field lines. Equating the torque about the z-axis to the time rate of change of angular momentum, we have

$$\frac{d}{dt}(mrv_\theta) = \frac{d}{dt}(mr^2\dot{\theta}) = rF_\theta. \tag{9.28}$$

If the beam is perfectly confined, the distance of an outer electron from the axis is constant, so that $r = a$ in Eq. (9.28). Substituting the expression (9.27) for F_θ into Eq. (9.28) and multiplying both sides of the resulting equation by $2\pi\, dt$, we find

$$2\pi ma^2\, d\dot{\theta} = e 2\pi a B_r\, dz. \tag{9.29}$$

If the angular velocity is zero before the electron enters the region of finite B_r, this equation can be readily integrated to yield

$$\int_0^{\dot{\theta}} 2\pi ma^2\, d\dot{\theta} = 2\pi ma^2 \dot{\theta} = e \int 2\pi a B_r\, dz, \tag{9.30}$$

where $\dot{\theta}$ is the angular velocity acquired by the electron in crossing B_r. Now, according to Fig. 9.5(b), the magnetic flux $d\phi$ entering a length dz of the beam of radius a is equal to $2\pi a B_r\, dz$. It follows that we can find the total flux which enters the cylindrical surface of the beam by integrating this expression over the length of the cylinder. The totality of flux entering the cylindrical wall of the beam must of course emerge from the end of the cylinder, where the flux is axial and with intensity B_0. Thus we have

$$\int_{\substack{\text{cylindrical}\\\text{surface}}} 2\pi a B_r \, dz = \pi a^2 B_0. \tag{9.31}$$

Combining Eqs. (9.30) and (9.31), we finally find the expression for $\dot{\theta}$ in terms of B_0 as

$$\dot{\theta} = \frac{eB_0}{2m} = \frac{\omega_c}{2}, \tag{9.32}$$

where ω_c is the cyclotron frequency. Thus, the electrons rotate about the axis with an angular velocity $\dot{\theta}$.

Let us now consider the behavior of electrons in the region where the magnetic field is entirely axial. In practice, of course, B_r is finite everywhere except in the transverse plane situated halfway between the pole pieces.[11] For mathematical convenience, however, we shall assume that the axial extent of the region within which B_r is appreciable is sufficiently small that it takes only a short time interval Δt for an electron to pass through it. Thus an electron passing this region will receive a short impulse $\mathbf{F}_\theta \Delta t$. When this electron reaches the region of axial field, it has acquired a tangential velocity $\mathbf{v}_\theta = \hat{\theta} v_\theta$ perpendicular to the axial field $\mathbf{B}_z = \hat{z} B_0$. The interaction of $\mathbf{v}_\theta$ and $\mathbf{B}_z$ clearly gives rise to the focusing field $\mathbf{F}_{r2}$ that we are seeking, that is,

$$\mathbf{F}_{r2} = -\frac{\hat{r}}{r} e v_\theta B_0. \tag{9.33}$$

The magnetic force $\mathbf{F}_{r2}$ acting on an outermost electron must be large enough not only to counteract the outward radial electric force resulting from coulomb repulsion $\mathbf{F}_{r1}$ but also to give it sufficient centripetal acceleration to keep it moving in a helical path about the beam axis, thus confining the beam to the radius a. Applying Newton's second law to the radial direction, we equate $|\mathbf{F}_{r2}|$ to $|\mathbf{F}_{r1}|$ plus the centrifugal force $m\dot{\theta}^2 a$. Thus we have

$$ea\dot{\theta}B_0 = \frac{eI_0}{2\pi\varepsilon_0 a \sqrt{2(e/m)V_0}} + m\dot{\theta}^2 a, \tag{9.34}$$

where we have used Eqs. (9.26) and (9.33) and the relation $v_\theta = a\dot{\theta}$. Substituting the expression (9.32) for $\dot{\theta}$ into Eq. (9.34) and solving for B_0^2, we find

$$B_0^2 = \frac{\sqrt{2}\, i}{\varepsilon_0 (e/m)^{3/2} V_0^{1/2}}, \tag{9.35}$$

where $i = I_0/\pi a^2$ is the current density in the beam. The flux density B_0 given by Eq. (9.35) is known as the *Brillouin field* and is the minimum value of flux

---

[11] This plane corresponds to the right-hand cross section of the cylinder in Fig. 9.5(b).

density required to confine a beam of current density i accelerated by a voltage V_0.[12]

We can show that Eq. (9.35) also applies to electrons inside the beam provided the current density is independent of radius. Therefore, as the beam travels in the region of uniform axial field all of its electrons rotate about the beam axis with angular velocity $\dot{\theta} = \omega_c/2$, according to Eq. (9.32). In other words, the entire beam of electrons twists about the axis as if it were a rigid body. Note that in a practical beam, not only may the current density vary with r within the beam, but the initial electron velocity on entering the field need not be along the z-axis. Thus the outer edge of the beam may be poorly defined. Furthermore, as exhibited by Eq. (9.34), true Brillouin flow depends on a delicate balance of forces. Deviation of quantities such as I_0 or B_0 with z will clearly disturb this balance and is likely to lead to variation of beam diameter with z.

B. Confined flow

Another way of focusing an electron beam is to have the electrons "born" into an axial magnetic field. To accomplish this, the permanent magnet or solenoid must be appropriately longer than the cathode-collector distance so that in the entire region of electron flow the field is axially directed. Because of the radially directed coulomb force $\mathbf{F}_{r1}$ given by Eq. (9.26), the electron beam tends to spread. This gives rise to a radial component of velocity $\dot{r}$ and a θ-directed force,

$$\mathbf{F}'_\theta = \hat{\theta} e \dot{r} B, \tag{9.36}$$

where B is the axial or z-directed flux density. Equating rF'_θ, the torque about the z-axis, to the time rate of change of angular momentum, we have

$$er\dot{r}B = \frac{d}{dt}(mr^2\dot{\theta}). \tag{9.37}$$

Integrating with respect to t, we find

$$\dot{\theta} = \frac{\omega_c}{2} + \frac{C}{r^2}, \tag{9.38}$$

where $\omega_c = eB/m$ is the cyclotron frequency and C is a constant of integration. If electrons start at the cathode with zero angular velocity, then $C = -\omega_c r_c^2/2$, where r_c is the radius of the cathode. Thus Eq. (9.38) becomes

$$\dot{\theta} = \frac{\omega_c}{2}\left(1 - \frac{r_c^2}{r^2}\right). \tag{9.39}$$

---

[12] Actually, the effective voltage V_0 representing the axial velocity is usually less than the acclerating voltage, since some axial kinetic energy is converted into rotational energy. See J. R. Pierce, *Theory and Design of Electron Beams*, D. Van Nostrand Co., New York, 1954; p. 153.

This $\dot\theta$ gives rise to a tangential velocity $\mathbf{v}_\theta = \dot\theta r \hat\theta$ and a consequent restoring force,

$$\mathbf{F}'_r = -\frac{\hat{r}}{r} e r \dot\theta B. \tag{9.40}$$

Combining Eqs. (9.39) and (9.40), we find

$$\mathbf{F}'_r = -\frac{\hat{r}}{r} \frac{e r(e/m) B^2}{2}\left(1 - \frac{r_c^2}{r^2}\right). \tag{9.41}$$

We see from this equation that if the beam is well focused so that r is kept close to r_c, B must be very large to provide the necessary restoring force. This is the disadvantage of using confined rather than Brillouin flow. However, the former usually results in a better defined constant-diameter beam than does the latter.

C. Periodic Focusing

From Eq. (9.41), we see that a very large magnetic field is required for good focusing when confined flow is used. For Brillouin flow, we can obtain perfect focusing, according to Eq. (9.35), by using relatively large but still finite fields. In practice, a Brillouin field of several hundred oersteds is required for focusing typical electron beams used in traveling wave tubes. This field must be supplied by permanent magnets or solenoids. In either case, the magnet structure is inconveniently heavy, typically weighing about 40 lb for Brillouin flow. If a solenoid is used, a considerable amount of power must be supplied in addition to meet the electrical losses.

Clearly, then, what we seek is a focusing method that can substantially reduce the field requirement below the Brillouin flow value. To this end, we find that if we use an axially periodic field of alternating polarity the required weight of the magnet will be drastically reduced. Indeed, under ideal conditions we can expect a reduction of $1/N^2$, where N is the number of magnets used in the periodic case.[13] We can appreciate this reduction from a consideration of Fig. 9.6. Figure 9.6(a) shows one large magnet supplying a field over a length $10L$, while Fig. 9.6(b) shows 10 magnets supplying a field of alternating polarity each over a period L. At a point p sufficiently far away from the magnets, the field resulting from the 10 small magnets will be considerably smaller than that resulting from the large magnet, since the fields that result from the adjacent north and south poles nearly cancel in the former case because of their proximity. Therefore, other factors being constant, the energy stored outside the 10-magnet assembly is reduced over that of the one-magnet case. Consequently, less magnetic material is required to provide the same axial field when period focusing is used. If the magnets are assembled in the N–S, N–S, ... instead of the N–S, S–N, ... configuration shown in Fig. 9.6(b), fields of similar polarity will result. This configuration has no particular

---

[13] J. T. Mendel, C. F. Quate, and W. H. Yocom, *Proc. Inst. Radio Engrs.* **42,** 800 (1954).

advantage over the one-magnet arrangement, since reasoning similar to that given above shows that no external field reduction can be expected.

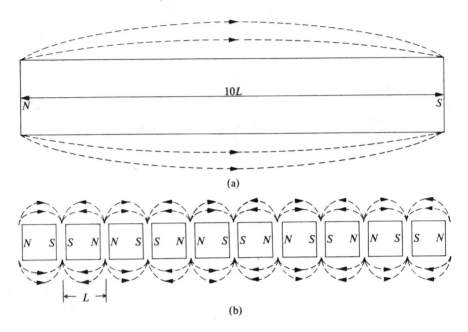

Fig. 9.6. (a) Focusing field supplied by one large magnet. (b) Periodic focusing field supplied by 10 small magnets.

Before we examine periodic focusing analytically, it may be instructive to make a forecast of what we can expect from it based on our knowledge of Brillouin flow. We see from Fig. 9.5(a) that when an electron enters the field axially, an angular velocity $\hat{\theta}\dot{\theta}$ is imparted to it and uniform flow is obtained in the region of axial field. When the electron emerges from the magnet on the right-hand side of the figure, it must cross field lines whose radial component is directed away rather than toward the axis. Thus an additional angular velocity $-\hat{\theta}\dot{\theta}$ will be imparted to the electron; the electron will "unwind" itself in such a way that it emerges from the right-hand side of the magnet z-directed and with no net angular velocity. If this electron is now allowed to enter a second magnet with field lines oppositely directed compared to the first, an angular velocity $-\hat{\theta}\dot{\theta}$ will be supplied to it. When it emerges from this magnet it will again have zero angular velocity, having received an additional $\hat{\theta}\dot{\theta}$ in crossing radially directed field lines on the right-hand side. If this electron is allowed to pass through a third magnet polarized like the first, then a fourth magnet polarized like the second, and so on, the process described above will be repeated again and again.

It thus follows that if the magnets are placed in close contact with each other, we can in principle obtain perfect focusing. In practice, however, some separation between magnets is unavoidable, since pole pieces must be used to concentrate the field in the region of the beam. Therefore, unavoidable beam divergence will occur in the intervening region between magnets. If the combination of values of magnetic flux density B_0, accelerating voltage V_0, and cell length L is such that this divergence is not cumulative with increasing z, then we would expect only small oscillations of the beam boundary with z; focusing would be nearly perfect and 99% beam transmission could result. However, if conditions are such that the divergent effect is cumulative, most of the electrons in the beam could be lost to the walls of the tube and little collector current would result. We designate this region of small beam transmission as the stop band and the region of high transmission as the pass band. We shall now make a simplified calculation to illustrate and confirm these expectations.

For mathematical simplicity, we shall assume that [14]

$$B_z = B_0 \cos \frac{2\pi z}{L}, \tag{9.42}$$

and that the electric field due to space charge acts only in the radial direction. For this problem, it is convenient to work in the Lagrangian formulation. The Langrangian $\mathscr{L}$ for an electron in an electric and magnetic field is given by [15]

$$\mathscr{L} = \frac{m}{2}(\dot{r}^2 + r^2\dot{\theta} + \dot{z}^2) + e(\mathbf{v} \cdot \mathbf{A}) - eV, \tag{9.43}$$

where A is the vector magnetic potential and V the scalar electric potential Since $\mathbf{H} = \nabla \times A$ and $\mathbf{H} = \hat{z}H_z$, we have the component equations

$$\frac{1}{r}\frac{\partial A_z}{\partial \theta} - \frac{\partial A_\theta}{\partial z} = 0,$$

$$\frac{\partial A_r}{\partial z} - \frac{\partial A_z}{\partial r} = 0,$$

$$\frac{1}{r}\frac{\partial (rA_\theta)}{\partial r} - \frac{1}{r}\frac{\partial A_r}{\partial \theta} = 0. \tag{9.44}$$

These equations can be satisfied if we let

$$A_r = A_z = 0,$$
$$A_\theta = \frac{rH_z}{2}. \tag{9.45}$$

---

[14] In practical periodic magnetic structures this is still the dominant term being perhaps 5 or 6 times larger than the next term, which represents the third harmonic component.

[15] N. W. MacLachlan, *Theory and Application of Mathieu Functions*, Oxford University Press, New York, 1947.

For a shielded cathode, as in the case of Brillouin flow, we have

$$\dot{\theta} = eB_z/2m \tag{9.46}$$

according to Eq. (9.32). On the other hand, for an electron at the edge of the beam, Eq. (2.3) leads to an expression for V:

$$V = r_0^2 \rho / 2\varepsilon_0 r, \tag{9.47}$$

where r_0 is the beam radius at the entrance of the focusing structure and ρ is the volume charge density. If we substitute Eqs. (9.45), (9.46), and (9.47) into Eq. (9.43), the Lagrangian becomes

$$\mathscr{L} = \frac{m}{2}\left(\dot{r}^2 - \frac{e^2 B_z^2}{4m^2} r^2 + \dot{z}^2\right) + \frac{e r_0^2 \rho}{2\varepsilon_0 r}. \tag{9.48}$$

The equation of motion obtained from the Lagrangian equations is

$$\frac{\partial \mathscr{L}}{\partial q_k} - \frac{d}{dt}\left(\frac{\partial \mathscr{L}}{\partial \dot{q}_k}\right) = 0, \tag{9.49}$$

where $q_k = r, \theta$, or z. Substituting Eq. (9.48) into Eq. (9.49), we find that $\dot{z} = \text{const} = u_0$, where u_0 is the dc beam velocity and the equation in r is

$$\ddot{r} + \frac{(e/m)^2 B_z^2}{4} r - \frac{(e/m) r_0^2 \rho}{2\varepsilon_0 r} = 0. \tag{9.50}$$

By an appropriate change of variables, we can convert Eq. (9.50) to

$$\ddot{\sigma} + \alpha(1 + \cos 2T)\sigma - \frac{\beta}{\sigma} = 0, \tag{9.51}$$

where $\ddot{\sigma} = d^2\sigma/dT^2$ and

$$\sigma = \frac{r}{r_0}, \quad T = \omega t, \quad \alpha = \tfrac{1}{2}(\omega_L/\omega)^2, \quad \beta = \tfrac{1}{2}(\omega_p/\omega)^2,$$

$$\omega_L = \tfrac{1}{2}(e/m) B_0, \quad \omega = \frac{2\pi u_0}{L}, \quad \omega_p^2 = \frac{\rho(e/m) r_0}{\varepsilon_0}. \tag{9.52}$$

The nonlinear equation (9.51) has been solved with the aid of a computer.[13] For convenience, we have assumed that the electrons are injected with no radial velocity at $T = 0$ [or $z = 0$, since $z = u_0 t$ has been assumed in the derivation of Eq. (9.51)]. In other words, they are injected at the point where B_z is a maximum [see Eq. (9.42)], a condition which is attainable by proper positioning of the electron gun. Figure 9.7 depicts typical beam contours encountered with a change in magnetic field parameter α and a fixed value of β. As might be expected, for a given set of beam parameters, optimum focusing or minimum beam ripple is obtained when the rms magnetic field strength equals the corresponding Brillouin field. From a large set of these curves representing wide-ranging values of β, we can conclude that optimum focusing occurs when $\alpha = \beta$ (curve b). From Eq. (9.52), this equality can be written as

$$I_0 = \frac{\pi \varepsilon_0}{\sqrt{2}} \left(\frac{e}{m}\right)^{3/2} \left(\frac{B_0}{\sqrt{2}}\right)^2 r_0^2 V^{1/2}. \tag{9.53}$$

This relationship is the same as that for Brillouin flow provided the rms value of the flux density, $B_0/\sqrt{2}$, and the average beam diameter are associated with the axial flux density and beam diameter of Brillouin flow. Equations (9.35) and (9.53) are strictly applicable only if electrons enter the focusing structure with no transverse velocities. But such is never the case in a physical system, and in practice the required field for both Brillouin flow and optimum periodic focusing is roughly twice the theoretical value.

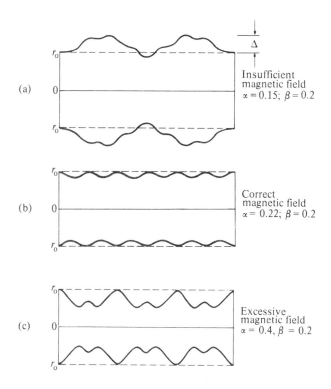

Fig. 9.7. Typical beam contours for various values of magnetic field parameter α and space charge parameter β. (After Mendel, Quate, and Yocom, [13].)

We would expect, and calculations confirm, that the percentage ripple Δ/r_0 increases with the space charge parameter β when optimum focusing is used.

To determine the location of the stop bands, we must find the conditions under which Eq. (9.51) yields divergent solutions for σ. It turns out that as far as the location of the stop bands is concerned, the space-charge term, i.e., the last term, in Eq. (9.51) is not important. Neglect of this term changes it into Mathieu's equation with arbitrary constants specified. For certain ranges of α, $\sigma(T)$ or $\sigma(z)$ is known to be periodic or quasi-periodic and bounded, while for other values of α, σ increases without limit, indicating an unstable flow.[16]

For the sake of completeness, we might mention that Hahn and Metcalf[17] have studied periodic fields of axial symmetry produced by electrostatic lenses. However, this focusing system is rarely used in microwave tubes.

5. SPACE-CHARGE EFFECTS IN ELECTRON BEAMS

In our study of microwave tubes so far, we have implicitly assumed that electronic charges in the beam are unneutralized. Actually, this is not necessarily the case. Since perfect vacuum cannot be realized, there must be residual gas molecules (about $10^{10}/cm^3$ for a pressure of 10^{-8} torr) inside the tube envelope. Positive ions are therefore produced at a constant rate as a result of the collision of residual gas molecules with electrons in the beam. In normal vacuum tubes, a sufficient number of ions may be produced to neutralize the beam with the excess escaping to the walls of the tube and recombining there. Such neutralization will greatly reduce the dc potential within the beam. However, since positive ions are orders of magnitude more massive than electrons, they cannot easily follow rapidly changing electric fields at microwave frequencies. Consequently, whereas positive ion neutralization can facilitate dc electronic motion, the ions have little influence on r.f. motion of the electrons. Accordingly, as electrons congregate to form bunches in a drift tube, large coulomb repulsion forces come into play.

Consider the distance-time relationship of three typical electrons; electron A, which crosses the gap when the voltage is maximum retarding; electron B, which crosses the r.f. gap voltage is zero, and electron C, which crosses when the gap voltage is maximum accelerating. According to Eq. (7.4), the velocity of these electrons in the drift space is determined by the time t_1 at which they cross the gap, that is,

$$v_1 \simeq v_0 \left(1 + \frac{V_1}{2V_0} \sin \omega t_1\right). \tag{7.4}$$

Since the drift tube is a field-free region of space, this velocity will remain constant at values given by Eq. (7.4). It further follows that the arrival time t_2 of the electrons at the catcher is given by

$$t_2 \simeq t_1 + \frac{L}{v_0}\left(1 - \frac{V_1}{2V_0} \sin \omega t_1\right). \tag{7.10}$$

---

[16] N. W. MacLachlan, op. cit.
[17] W. C. Hahn and G. F. Metcalf, Proc. Inst. Radio Engrs. **27**, 106 (1939).

Accordingly, crossing of electron paths will occur when $dt_2/dt_1 = 0$, yielding

$$t_1 = \frac{1}{\omega} \cos^{-1}\left(\frac{2\omega v_0 V_0}{LV_1}\right) \qquad (9.54)$$

This situation is depicted by dashed curves in the distance-time diagram of Fig. 9.8. At the values of t_1 given by Eq. (9.54), many electrons are crossing the collector gap simultaneously. Note, however, that this is possible only if the electrons are considered mass points of zero extent and if coulomb forces are neglected. In practice, as electrons congregate to form bunches, intense coulomb forces come into play (these forces vary as the inverse square of the distance between particles). Therefore, electrons are repelled by fellow electrons as they approach each other. It follows that crossing of electron trajectories will not occur; instead, spacing between electrons will be minimum at time t_1 given by Eq. (9.54). Beyond this value of t_1, electronic paths will diverge as depicted in Fig. 9.8. Since this divergence implies that dt_2/dt_1 can never be zero, the singularities in i_2 given by Eq. (7.11) will no longer occur, because $i_2 = I_0 \, dt_1/dt_2$.

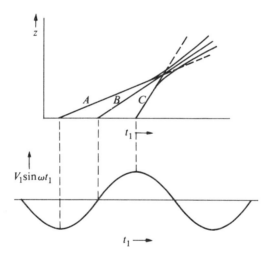

Fig. 9.8. Distance-time diagram of electrons in a drift tube.

If we assume that the dc electron beam is neutralized, then each electron in the bunch must have an equilibrium position about which it can oscillate when subjected to the coulomb repulsion forces of other electrons. If the deviation of an electron's position from its equilibrium point is small, we may expect the restoring force, to a first approximation, to be proportional to the deviation. It follows that an electron executes simple harmonic motion about its equilibrium position in the beam. Since the beam as a whole is moving in the axial direction, the equilibrium position also moves in the axial direction with dc beam velocity. As far as the laboratory observer is concerned, each electron oscillates slightly back and forth in the z-direction as it moves toward the catcher. The collective motion of the electrons gives rise to what are known as space-charge waves. From what has been said above, we would expect the velocity of these waves to be nearly equal to the dc beam velocity as seen by a laboratory observer.

In our preceding discussion, we implicitly assumed that only longitudinal debunching is possible. This is the case if the electrons are restrained from moving in the radial direction by a focusing magnetic field, as in the case of Brillouin flow. Therefore, the three-dimensional problem of electron motion reduces to that of one dimension.[18]

6. PLASMA OSCILLATIONS

There is no static electric field in a completely neutralized beam when the electrons are in their equilibrium positions; the field lines merely extend from an electron to the adjacent ions. If the electrons are displaced from equilibrium while the positive ions remain fixed, however, there will be a net negative charge where they are bunched and a net positive charge where they are depleted. The associated field lines extend between positive charge concentrations and adjacent negative charge concentrations. This field provides a restoring force returning the electrons to their equilibrium positions. Since these electrons are likely to possess finite velocity when they reach their equilibrium positions, they may be expected to overshoot, thus resulting in a restoring force in the opposite direction that tends to return them again to equilibrium locations. Clearly, charge oscillations will occur about the electrons' average positions. If the amplitude of the oscillations is sufficiently small, they will be simple-harmonic with a characteristic frequency known as the *plasma frequency*.

We shall first determine the restoring force on a displaced electron. If we let $+\rho_0$ and $-\rho_0$ be the average charge density of ions and electrons respectively, it follows that

$$\rho_0 = n_0 e, \tag{9.55}$$

where n_0 is the average number of electrons per unit volume. Let us consider a constant differential volume $A\,dz$ located at plane z. Originally there is not net charge density inside this volume element; however, due to some disturbance accompanied by unbalanced inflow and outflow of charges from this volume, a net charge density

$$q = \rho A\,dz \tag{9.56}$$

appears inside $A\,dz$. Here ρ is the net charge density inside the volume. At the left surface of volume $A\,dz$ located at z, the electrons are assumed to be displaced by a distance s. Because of coulomb repulsion, a displacement of $(s + ds)$, with

---

[18] Since the beam is cylindrically symmetric, its coherent rotation as a rigid body about the beam axis does not invalidate conclusions based on our one-dimensional model.

ds negative, occurs for electrons at the right surface located at $z + dz$. Thus the net charge q inside $A\,dz$ after the disturbance is

$$q = -\rho_0 A s - \left[-\rho_0 A \left(s + \frac{\partial s}{\partial z} dz \right) \right], \tag{9.57}$$

where we have implicitly assumed that $\rho \ll \rho_0$ and that the displacement of all electrons are considered at one given time. Combining Eqs. (9.56) and (9.57), we have

$$\rho = \rho_0 \frac{\partial s}{\partial z}. \tag{9.58}$$

From Maxwell's equation (2.3), we find

$$\frac{\partial E}{\partial z} = \frac{\rho_0}{\varepsilon_0} \frac{\partial s}{\partial z}, \tag{9.59}$$

also using Eq. (9.58). Integrating Eq. (9.59), we have

$$E = \frac{\rho_0}{\varepsilon_0} s, \tag{9.60}$$

since $E = 0$ at $s = 0$. Here E can be interpreted as the electric field seen by a displaced electron due to the organized bunching or collective motion of the rest of the electrons. Since the force on an electron displaced a distance s from equilibrium is $-eE$, Newton's second law (4.19) for the electron becomes

$$-\frac{e\rho_0}{\varepsilon_0} s = m\ddot{s}. \tag{9.61}$$

The electron therefore executes simply harmonic motion about its equilibrium position with an angular frequency given by

$$\omega_p = \sqrt{e\rho_0/m\varepsilon_0}, \tag{9.62}$$

where $\rho_0 = n_0 e$ according to Eq. (9.55). For electron beams having electron densities encountered in microwave tubes, f_p is of the order of a few hundred megacycles.

In the preceding one-dimensional problem, we implicitly assumed that the beam has an infinite cross section. An actual beam not only has a finite cross section but may also be enclosed by a metal tube. Thus **E** is no longer entirely axial; the axial restoring force and hence the plasma frequency are reduced as compared to the infinite beam case. Because of the three-dimensional nature of the general problem, infinitely many modes or frequencies of oscillation are possible. In simple cases, the mode of oscillation associated with the fundamental frequency (the reduced plasma frequency) is the only one which is excited with significant amplitude.

In the foregoing discussion, we have implicitly assumed that we are in a coordinate system which moves in the axial direction with the dc beam velocity, i.e., the equilibrium position of the electrons moves in the axial direction with the average beam velocity.

7. SPACE-CHARGE WAVES

We note from the above discussion that electrons oscillate at the natural resonant frequency of the neutralized electron beam or plasma. In other words, in a stationary plasma electromechanical waves can propagate only with the frequency ω_p characteristic of the medium. Although electrons oscillate with simple harmonic motion at the same frequency ω_p about their equilibrium positions, they do so with different amplitudes and phases. In the laboratory frame, their collective motions give rise to the phenomenon known as space-charge waves.

The equation of motion of an electron as seen by an observer moving axially with the average beam velocity v_0 is given by Eq. (9.61), rewritten here as

$$\ddot{s} + \omega_p^2 s = 0, \tag{9.61}$$

where ω_p is the plasma frequency defined by Eq. (9.62). The general solution of this equation is

$$s = A e^{j\omega_p t} + B e^{-j\omega_p t}, \tag{9.63}$$

where A and B are constants.

Consider the velocity and density modulation phenomena in a Klystron amplifier. If an electron is assumed to cross an infinitely thin buncher gap located at $z = 0$ at $t = t_1$, it will receive a velocity kick v_1 given by

$$v_1 = v_0 \left(1 + \frac{V_1}{2V_0} \sin \omega t_1\right), \tag{7.4}$$

where we recall that v_0 is the dc beam velocity, V_1, the amplitude of the modulation voltage, and V_0, the beam acceleration voltage. It follows that the initial conditions are

$$s \bigg|_{t=t_1} = 0, \tag{9.64}$$

$$\dot{s} \bigg|_{t=t_1} = -j \frac{v_0 V_1}{2V_0} e^{j\omega t_1}. \tag{9.65}$$

Note that $\dot{s}$ follows from Eq. (7.4) and is written in complex notation for mathematical convenience. Substituting Eq. (9.63) and its derivative into Eqs. (9.64) and (9.65), we find that arbitrary coefficients are

$$A = -\frac{v_0 V_1}{2V_0 \omega_p} e^{j(\omega - \omega_p)t_1}, \tag{9.66}$$

$$B = \frac{v_0 V_1}{2V_0 \omega_p} e^{j(\omega + \omega_p)t_1}. \tag{9.67}$$

Substituting Eqs. (9.66) and (9.67) into Eq. (9.63), we obtain

$$s = -\frac{v_0 V_1}{2V_0 \omega_p} e^{j(\omega - \omega_p)t_1} e^{j\omega_p t} + \frac{v_0 V_1}{2V_0 \omega_p} e^{j(\omega + \omega_p)t_1} e^{-j\omega_p t}. \tag{9.68}$$

To obtain the displacement in terms of z (separation between equilibrium position and buncher gap), we note that

$$z = v_0(t - t_1) \tag{9.69}$$

or

$$t_1 = t - \frac{z}{v_0}. \tag{9.70}$$

Substituting Eq. (9.70) into Eq. (9.68), we finally get

$$s = -\frac{v_0 V_1}{2V_0 \omega_p} e^{-j(\beta_f)z} e^{j\omega t} + \frac{v_0 V_1}{2V_0 \omega_p} e^{-j(\beta_s)z} e^{j\omega t}, \tag{9.71}$$

where

$$\beta_f = \frac{\omega - \omega_p}{v_0}, \tag{9.72}$$

$$\beta_s = \frac{\omega + \omega_p}{v_0}. \tag{9.73}$$

Thus, s can clearly be expressed in the form of two traveling waves with velocities

$$v_f = \frac{\omega}{\beta_f} = \frac{v_0}{1 - (\omega_p/\omega)}, \tag{9.74}$$

$$v_s = \frac{\omega}{\beta_s} = \frac{v_0}{1 + (\omega_p/\omega)}. \tag{9.75}$$

Therefore, if $\omega_p < \omega$, as is usually the case in microwave tubes, both v_f and v_s are positive, indicating that the corresponding waves both travel in the $+z$-direction. Furthermore, under this condition, $v_f > v_0$ and $v_s < v_0$; thus the associated waves are known respectively as fast and slow waves. Because we have introduced relation (9.69), the velocities of these waves are referred to the stationary laboratory reference frame.

We can now derive the expressions for r.f. charge density ρ, r.f. velocity v, and current density i_2. From Eq. (9.58), we have

$$\rho = \rho_0 \frac{\partial s}{\partial z}. \tag{9.58}$$

We now differentiate Eq. (9.71) and substitute into Eq. (9.58) to find

$$\rho = j\frac{\rho_0 V_1}{4V_0} \frac{\omega - \omega_p}{\omega_p} e^{-j[(\omega-\omega_p)/v_0]z + j\omega t}$$

$$- j\frac{\rho_0 V_1}{4V_0} \frac{\omega + \omega_p}{\omega_p} e^{-j[(\omega+\omega_p)/v_0]z + j\omega t} \tag{9.76}$$

Similarly, since

$$v = \frac{\partial s}{\partial z}\frac{dz}{dt} + \frac{\partial s}{\partial t} = \frac{\partial s}{\partial z}v_0 + \frac{\partial s}{\partial t},$$

we have

$$v = -j\frac{v_0 V_1}{4V_0} e^{-j[(\omega-\omega_p)/v_0]z + j\omega t} - j\frac{v_0 V_1}{4V_0} e^{-j[(\omega+\omega_p)/v_0]z + j\omega t}, \tag{9.77}$$

again using Eq. (9.71). Recalling from Eq. (8.32) that

$$i_2 \simeq I_0 - \rho_0 v + v_0 \rho, \tag{9.78}$$

we find

$$i_2 = I_0 \left[1 + z\omega \frac{V_1}{2V_0 v_0} \left(\frac{\sin(\omega_p z/v_0)}{\omega_p z/v_0} \right) e^{-j(\omega/v_0)z + j\omega t} \right], \tag{9.79}$$

using Eqs. (9.76) and (9.77). Comparing Eq. (9.79) with Eq. (7.11), we see that coulomb repulsion leads in effect to a reduction in modulation voltage V_1 as far as i_2 is concerned. The reduction factor is $[\sin(\omega_p z/v_0)]/(\omega_p z/v_0)$.

When we deal with actual microwave electronics structures, the presence of such things as conducting walls complicates the preceding simplified analysis. In general, we must combine the equation of continuity (4.25), Maxwell's equations (2.1) through (2.4), and the equation of motion (4.19) to obtain a wavelike equation involving driving sources:

$$\nabla^2 \mathbf{E} + \frac{\omega^2}{c^2}\mathbf{E} = \frac{\omega_p^2 \beta}{j(\omega - v_0\beta)^2}\nabla E_z + \hat{z}\frac{\omega^2}{c^2}\frac{\omega_p^2}{(\omega - v_0\beta)^2}E_z\hat{z}, \tag{9.80}$$

where we have assumed that the field quantities vary as $e^{j\omega t - \beta z}$. We can evaluate the arbitrary constants in the solution of this equation by imposing the proper boundary conditions for a given structure; e.g., at the wall of a cylindrical conducting drift tube, $E_z = 0$. Such detailed analysis, although necessary from an engineering standpoint, does not yield results that invalidate those found for the simple case. Therefore, we will not give its details here but will leave the derivation of Eq. (9.80) as an exercise for the reader.

We conclude this discussion by noting that plasma oscillations and space-charge waves are fundamental phenomena common to several types of microwave devices. For example, cyclotron waves in a magnetically confined electron beam can be utilized to build an Adler-type parametric amplifier, which we will discuss

PROBLEMS

9.1 a) A dc voltage V_0 is applied between two parallel-plane electrodes separated by a distance d with electrons emitted thermally from the cathode at temperature T. Find the average velocity of an electron at plane A situated at a distance z from the cathode. Given: $T = 1000°C$, $d = 1$ cm, and $V_0 = 1$ volt.
   b) Referring to Fig. 9.1, find the maximum kinetic energy of the emitted electrons given that monochromatic light of frequency v strikes the metal surface, which is at $0°K$.

9.2 Evaluate integral (9.9) without assuming that the exponential term in the denominator is much less than unity.

9.3 Derive the current-voltage relation for electrodes made of concentric cylinders in terms of cathode radius r_k, anode radius r_a, and anode voltage V_a. Do the same for spherical caps which are portions of concentric spheres in terms of cathode radius r_k, anode radius r_a, and the half angle of the cone that forms the cap θ, as well as anode voltage V_a.

9.4 Show that the $\frac{3}{2}$ power law ($J \propto V^{3/2}$) is applicable to any arbitrary cathode-anode configuration.

9.5 Show that Eq. (9.35) is also applicable to electrons inside the beam if the current density is independent of radius.

9.6 a) Periodic electrostatic focusing is often impractical because the limiting perveances are, in general, lower than those with magnetic focusing fields. A more promising idea is to set the beam spinning before it enters the region of periodic electrostatic focusing fields. Devise a means of spinning the electron beam and creating a periodic electrostatic field.
   b) Show that equilibrium can be established between various forces acting on the beam electrons in a manner similar to that used for Brillouin flow.

9.7 Show that the discussion of the collective behavior of electrons following Eq. (9.60) has meaning only when s is greater than the Debye length of the neutralized electron beam.

9.8 a) For a plane wave propagating in an ionized gas, the real part of the dielectric constant is given by

$$\varepsilon' = \varepsilon_0 \left[1 - \frac{Ne^2}{\varepsilon_0 m(\gamma^2 + \omega^2)} \right],$$

where ε_0 is the dielectric constant of free space, N is the electron density, e is the electronic charge, m is the electronic mass, γ is the collision frequency, and ω is the frequency. Find the imaginary part of the dielectric constant ε'' by means of the Kronig-Kramers relation:

$$\varepsilon'(\omega) = \varepsilon_0 + \frac{2}{\pi} \int_0^\infty \frac{\omega' \varepsilon''(\omega')\, d\omega'}{\omega'^2 - \omega^2}$$

$$\varepsilon''(\omega) = -\frac{2\omega}{\pi} \int_0^\infty \frac{[\varepsilon'(\omega') - \varepsilon_0]\, d\omega'}{\omega'^2 - \omega^2}.$$

b) Neglecting collisions, find the impedance of the plane wave in the medium and the frequency at which the transition from propagation to attenuation occurs as a function of the electron density N.

9.9 a) In our discussion of space-charge waves, we mentioned that positive ions recombine at the walls of the tube. Why do they recombine at the walls instead of in the region of the beam?

b) Estimate the residual gas density in a vacuum of 10^{-6} torr. Compare your result with the typical electron density of an electron beam in a microwave tube.

CHAPTER 10

PARAMETRIC AMPLIFIERS

In the last chapter we began our study of charge oscillation in an electron beam. We found that in the presence of r.f. modulation electromechanical space-charge waves can propagate in a neutralized electron beam. We shall now examine the use of an electron beam as well as a semiconductor or a ferrite material as the time-varying or nonlinear coupling element in a parametric amplifier. The distinguishing feature of a parametric amplifier is that it utilizes an ac rather than a dc supply of power. In this respect, it is analogous to the quantum amplifier maser we will discuss in the next chapter.

1. PARAMETRIC AMPLIFICATION

In a superheterodyne receiver, an r.f. signal may be mixed with a signal from the local oscillator in a nonlinear circuit element, the mixer, to produce sum and difference frequencies. In a parametric amplifier, the nonlinear element may be replaced by either a time-varying inductor or a time-varying capacitor. Frequency mixing is again possible and consequently energy supplied to the systems at one frequency can be converted into that at another frequency. Thus, unlike Klystrons, TWT's, magnetrons, etc., a parametric amplifier like the maser utilizes an ac rather than a dc power source. It is described as *parametric* because energies at different frequencies are coupled via a time-varying circuit element.

In order to appreciate the mechanism of energy transfer in a parametric amplifier, let us consider the qualitative behavior of the series RLC circuit in Fig. 10.1. We shall show that mechanical work supplied to the capacitor to

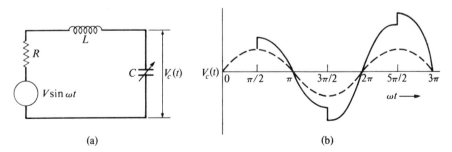

Fig. 10.1. Mechanically pumped parametric amplifier.

separate its plates appears in the form of electrical energy, as evidenced by an increase in the capacitor voltage V_c with time. If the capacitor plates are stationary, the sinusoidal source voltage V at frequency ω will give rise to a sinusoidal voltage V_c at frequency ω across the capacitor as indicated by the dashed curve of Fig. 10.1. At $\omega t = \pi/2$, when the dashed voltage is a maximum, the capacitor plates are suddenly pulled apart. Since unlike charges attract, mechanical work must be supplied to separate the capacitor plates. If the time interval involved in separating the plates is very small compared to the period of oscillation of the circuit,[1] little energy stored in the capacitance will be transferred to the inductance during this interval. It follows that the charge q residing on the capacitor plates is unchanged. However, as a result of the larger plate separation C decreases and $V_c = q/C$ accordingly increases, as indicated by the nearly vertical solid segment at $\omega t = \pi/2$. As shown in Fig. 10.1, this increased voltage across the capacitor continues its cycle and returns to zero at $\omega t = \pi$. If at this instant the plates are pushed back to their original position, no mechanical work is supplied or extracted from the capacitor since $V_c = 0$ and no charges reside on the plates. If the plates are successively pulled apart at $\omega t = \pi/2, 3\pi/2, 5\pi/2, 7\pi/2 \ldots$, and pushed together at $\omega t = \pi, 2\pi, 3\pi, 4\pi \ldots$, V_c will increase in amplitude with time until the energy dissipated per cycle in the circuit resistance R equals the mechanical energy supplied per cycle. In other words, the addition of mechanical energy by "pumping" the capacitor plates at twice the source or signal frequency leads to an increase in signal energy, producing amplification. Note further that maximum energy transfer requires C to be varied in the proper phase and at twice the signal frequency.[2]

At microwave frequencies, the capacitance (or inductance) variation is achieved by electronic means, since mechanical pumping is obviously impractical. The principle of operation, however, remains unchanged; i.e., energy at one frequency can be transferred to that at another if they are properly coupled by a time-varying storage element such as the junction capacitance of a semiconductor diode, an electron beam, or electron spins in a ferrimagnetic material.

Before we examine the general equivalent circuit of a parametric amplifier, let us perform a mathematical analysis of the circuit of Fig. 10.1. For mathematical simplicity, we shall assume that the capacitance variation, instead of being abrupt, is sinusoidal:

$$C = C_0 - C_1 \cos 2\omega t. \tag{10.1}$$

---

[1] To obtain maximum amplitude for V_c, the circuit should be tuned to resonance, that is, ω should be nearly equal to $1/\sqrt{LC}$, the circuit resonance frequency. In this and subsequent discussions, circuit Q's are assumed to be much larger than unity.

[2] Another sinusoidal voltage, of frequency ω but of opposite phase, can also be amplified. A device utilizing this degeneracy is called a *parametron*; the two states differing in phase by π rad can be employed to represent the bistable states of a binary memory.

In terms of the charge q flowing in the circuit, Kirchhoff's law gives[3]

$$\frac{d^2q}{dT^2} + \frac{\omega_0^2}{\omega^2}\left(1 + \frac{C_1}{C_0}\cos 2T\right)q + \frac{R}{\omega L}\frac{dq}{dT} = 0, \qquad (10.2)$$

where $T = \omega t$, $\omega_0 = 1/\sqrt{LC_0}$, and we have assumed that $C_1 \ll C_0$. For investigation of circuit stability, we can neglect the last term if the circuit Q is reasonably large. In this case, Eq. (10.2) reduces to

$$\frac{d^2q}{dT^2} + \frac{\omega_0^2}{\omega^2}\left(1 + \frac{C_1}{C_0}\cos 2T\right)q = 0. \qquad (10.3)$$

Note that (10.3) is a form of Mathieu's equation with arbitrary constants specified. The general Mathieu's equation reads

$$\frac{d^2q}{dT^2} + (a + 2b\cos 2T)q = 0, \qquad (10.4)$$

where a and b are variable parametrics. Comparing Eqs. (10.3) and (10.4), we find that

$$a = \frac{\omega_0^2}{\omega^2},$$

$$b = \left(\frac{\omega_0^2}{\omega^2}\right)\frac{C_1}{2C_0}. \qquad (10.5)$$

The nature of the solution for q as a function of T depends upon the ratios ω_0^2/ω^2 and C_1/C_0. For a certain range of ω_0^2/ω^2 and C_1/C_0, q is periodic or quasi-periodic and bounded, while for other values of ω_0^2/ω^2 and C_1/C_0, q increases without limit, thus indicating the onset of oscillation. The oscillations will grow until the energy supplied by the pump per signal cycle is just equal to the energy dissipated in the resistance R per cycle. Since $V_c = q/C$ and C is bounded, a limitless increase of q implies a limitless increase of V_c, consistent with the qualitative result given above.

We can easily find the unstable regions from the a vs. b stability plot of the general Mathieu equation (10.4) [12][4] and the straight line of Eq. (10.5). We see from Fig. 10.2 that as ω_0^2/ω^2 and C_1/C_0 increase along the straight line with slope given by Eq. (10.5), the solution for q passes through successive regions of stability and instability. Since $C_1 \ll C_0$, the slope of the straight line given by Eq. (10.5) is very large; in Fig. 10.2, C_1 is set equal to $0.1C_0$ for definiteness. We see that regions of instabilty occur only near $\omega_0/\omega = 1, 2, \ldots$. In other words, these regions are rather narrow and widely separated in frequency space.

---

[3] Here, we have set $V = 0$ and assumed that a signal of frequency ω originally exists in the circuit.
[4] Bracketed numbers refer to references listed at the end of this chapter.

166   Parametric amplification

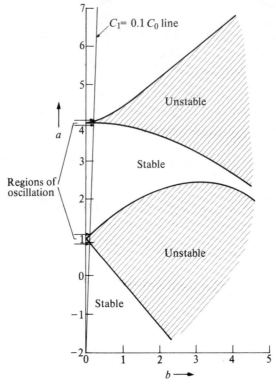

Fig. 10.2. Stability plot of Mathieu's equation.

For a discussion of the general parametric amplifier to follow, it is convenient to rewrite Eq. (10.2) as

$$L\frac{d^2q}{dt^2} + \frac{q}{C_0} + \frac{q}{C_0^2/C_1 \cos 2\omega t} + R\frac{dq}{dt} = 0. \tag{10.6}$$

Accordingly, we can redraw the circuit of Fig. 10.1 to have L, R, a constant capacitance C_0, and a time-varying capacitance $C' = C_0^2/C_1 \cos 2\omega t$ all connected in series as shown in Fig. 10.3(a). If $\omega = \omega_0 = 1/\sqrt{LC_0}$, we can say that Fig. 10.3(a) represents a high Q resonant circuit (composed of L, R, and C_0 in series) connected across a time-varying capacitor C'. It is now clear that the variation of C' can be accomplished electronically by connecting a pump source across C', again via a pump resonant circuit as shown in Fig. 10.3(b). Here L_s, R_s, C_s correspond respectively to L, R, C_0 of Fig. 10.3(a), while R_p, L_p, C_p forms the series resonant circuit of the pump.

Mixing of signal and pump frequencies will occur in the time-varying capacitor C'. Accordingly, the voltage of the sum and the difference frequency $\omega_p \pm \omega_s$, as well as ω_s and ω_p, will appear across C'. If loads are connected across appropriate terminals of the circuit, we can extract energy at frequencies ω_s, $\omega_p - \omega_s$, or

$\omega_p + \omega_s$. For example, to extract energy at the difference frequency $\omega_i = \omega_p - \omega_s$, the load circuit should be connected across C'. This circuit, which is totally passive unlike the pump and signal circuits that require external excitation, is appropriately called the idler circuit.

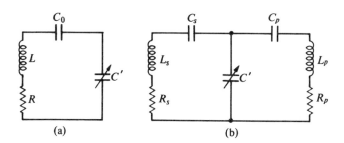

Fig. 10.3. (a) Series RLC circuit connected across time-varying capacitor. (b) Signal and pump circuits connected across time-varying capacitor.

In spite of its amazing simplicity, a parametric amplifier can perform a variety of amplification functions. A little reflection will show that it can be an amplifier at frequency ω_s, an up-converter at frequency $\omega_i > \omega_s$, and a down-converter at frequency $\omega_i < \omega_s$. Another major advantage of the parametric amplifier over a conventional amplifier is its low noise figure. Because the coupling element is either an inductor or a capacitor, it displays no Johnson noise.

Consider now the general parametric amplifier circuit shown in Fig. 10.4 in which three series resonant circuits, the signal, pump, and idler circuits, are connected across the time-varying capacitor C'. Whereas the signal and pump circuits are connected to external sources, the idler circuit is not. Note that here

$$\Omega_s = \frac{1}{\sqrt{L_s C_s}},$$

$$\Omega_p = \frac{1}{\sqrt{L_p C_p}},$$

$$\Omega_i = \frac{1}{\sqrt{L_i C_i}}, \tag{10.7}$$

where $\Omega_s, \Omega_p, \Omega_i$ are respectively the source, pump, and idler resonance frequencies. In subsequent discussions, we shall assume that the signal frequency ω_s is nearly equal to Ω_s, the pump frequency ω_p is nearly equal to Ω_p, and the idler frequency ω_i is nearly equal to Ω_i.

Consider a load resistance R_L connected across the idler output terminals i–i. If L_i and C_i are adjusted so that ω_i is equal to $\omega_p - \omega_s$, appreciable power develops across R_L. If the pump is much stronger than the signal, the voltage across R_L will be much larger than the signal voltage. Most of the load power thus comes from the r.f. pump source. If $\omega_i < \omega_s$, the device is known as a *down-frequency converter*. Conversely, if $\omega_i > \omega_s$ it is known as an *up-converter*. In either case, the signal at a given frequency is converted to an amplified output at a different frequency. Note that in contrast to the degenerate parametric amplifier of Fig. 10.1, ω_p need not be equal to $2\omega_s$. The phase requirement between the signal and pump needed for amplification is satisfied by the presence of the idler circuit. Of course, if $\omega_p = 2\omega_s$ and $\omega_i = \omega_s$, we obtain amplification at the signal frequency.

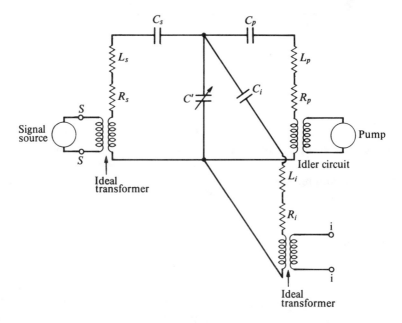

Fig. 10.4. General equivalent circuit for a parametric amplifier.

Alternatively, we can obtain signal amplification by having ω_i equal to $\omega_p - \omega_s$ and load R_L connected across the source terminals S–S. Now, the current at frequency $\omega_p - \omega_s$ which flows in the idler circuit beats with the pump current at frequency ω_p in the time-varying capacitor C' to produce a voltage across C' at the signal frequency ω_s. This voltage excites the signal circuit, producing a current that adds coherently to the original signal current. This amplified current results in power delivered to the load resistance R_L. Since the input and output terminals are identical in this amplifier, somewhat complicated circuitry is required to separate the input and output power.

2. ANALYSIS OF PARAMETRIC AMPLIFIERS[5]

A. General Considerations

In Fig. 10.5 we have the simplified equivalent circuit of Fig. 10.4. Here R'_s and R'_i include the effect due to the load resistance R_L. We have not explicitly retained the pump circuit except insofar as the time variation of C' is presumed to be due to the pump, that is,[6]

$$C' = C'_m \sin(\omega_p t + \phi_p). \tag{10.8}$$

Note that in general C'_m and ϕ_p can take on arbitrary values and ω_p need not be equal to $2\omega_s$.

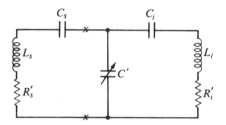

Fig. 10.5. Simplified equivalent circuit of a parametric amplifier. For simplicity, the ideal transformers of Fig. 10.4 have been assumed to have turn ratios of 1 : 1.

Omission of the pump circuit is justified, since we assume that both the signal and idler resonant circuits have sufficiently high Q's to act effectively as open circuits to currents at frequencies other than their respective resonant frequencies. Note also that

$$\omega_p = \Omega_s + \Omega_i, \tag{10.9}$$

where $\Omega_s = 1/\sqrt{L_s C_s}$ and $\Omega_i = 1/\sqrt{L_i C_i}$ are respectively the resonant frequencies of the signal and idler circuits. Of course, ω_s and ω_i are nearly equal to Ω_s and Ω_i respectively for effective parametric amplification.

Now let the signal and idler currents be

$$i_s(\omega_s) = I_s \sin(\omega_s t + \phi_s) \tag{10.10}$$

and

$$i_i(\omega_i) = I_i \sin(\omega_i t + \phi_i), \tag{10.11}$$

where

$$\omega_p = \omega_s + \omega_i, \tag{10.12}$$

---

[5] Our analysis here closely follows that of H. Heffner and G. Wade [9] except that in their case two parallel resonant circuits representing signal and idler circuits are connected in series with a coupling capacitor. However, our circuit of Fig. 10.5, composed of two series resonant circuits both connected across the coupling capacitor, is a more straightforward extension of the "mechanically pumped" single RLC circuit of Section 1.

[6] For the single equivalent circuit discussed in the last section, $C'_m = C_0^2/C_1$ and $\phi_p = \pi/2$.

Analysis of parametric amplifiers

since $\omega_s \simeq \Omega_s$ and $\omega_i = \Omega_i$ and because of the high-Q assumption. Accordingly, the capacitor current $i_{c'}$ is

$$i_{c'} = I_s \sin(\omega_s t + \phi_s) + I_i \sin(\omega_i t + \phi_i). \tag{10.13}$$

Since $v_{c'} = \int_0^t (i'_c/C')\,dt$, it is more convenient to write the relevant quantities in complex notion:

$$v_{c'} = \mathrm{Re}\left\{\int_0^t \frac{1}{C'_m(-j)e^{j(\omega_p t + \phi_p)}} \left[I_s \frac{e^{j(\omega_s t + \phi_s)} - e^{-j(\omega_s t + \phi_s)}}{2j}\right.\right.$$
$$\left.\left. + I_i \frac{e^{j(\omega_i t + \phi_i)} - e^{-j(\omega_i t + \phi_i)}}{2j}\right]dt\right\}. \tag{10.14}$$

Integrating and taking the real part, we find

$$v_{c'} = -\frac{1}{2C'_m}\left\{\frac{I_s}{\omega_i}\sin(\omega_i t + \phi_p - \phi_s) + \frac{I_s}{\omega_p + \omega_s}\sin[(\omega_p + \omega_s)t + \phi_p + \phi_s]\right.$$
$$+ \frac{I_s}{\omega_p}[\sin(\omega_p t + \phi_p - \phi_s) + \sin(\omega_p t + \phi_p + \phi_s)]$$
$$+ \frac{I_i}{\omega_s}\sin(\omega_s t + \phi_p - \phi_i) + \frac{I_i}{\omega_p + \omega_i}\sin[(\omega_p + \omega_i)t + \phi_p + \phi_i]$$
$$\left. + \frac{I_i}{\omega_p}[\sin(\omega_p t + \phi_p - \phi_i) + \sin(\omega_p t + \phi_p + \phi_i)]\right\}. \tag{10.15}$$

Note that the voltages of several frequencies—$\omega_i = \omega_p - \omega_s$, $\omega_p + \omega_s$, ω_p, $\omega_s = \omega_p - \omega_i$, and $\omega_p + \omega_i$—exist across the coupling condenser. However, because of the high-Q assumption, only the components with frequencies ω_s and ω_i can cause a significant current flow in the signal and idler circuits respectively. Therefore, for our purposes we need retain only these components in $v_{c'}$. Thus we have

$$v_{c'}(\omega_i) = -\frac{I_s}{2C'_m \omega_i}\sin(\omega_i t + \phi_p - \phi_s), \tag{10.16}$$

$$v_{c'}(\omega_s) = -\frac{I_i}{2C'_m \omega_s}\sin(\omega_s t + \phi_p - \phi_i). \tag{10.17}$$

From Eqs. (10.10) and (10.17) we can calculate the effective impedance presented to the signal resonant circuit, that is, at terminals x–x of Fig. 10.5.

Expressing $v_{c'}(\omega_s)$ of Eq. (10.17) and $i_s(\omega_s)$ of Eq. (10.10) in complex notation and taking their ratio, we obtain the effective impedance $Z(\omega_s)$ presented across the signal circuit at ω_s by the parallel combination of variable capacitor C' and the idler resonant circuit as

$$Z(\omega_s) = -\frac{1}{2C'_m \omega_s}\left(\frac{I_i}{I_s}\right)e^{j(\phi_p - \phi_i - \phi_s)}. \tag{10.18}$$

We see that $Z(\omega_s)$ is proportional to the ratio I_i/I_s. We can obtain the relationship between I_i and I_s by noting from Fig. 10.5 that

$$v_{C'}(\omega_i) = -i_i(\omega_i)\left[R_i' + j(\omega L_i - \frac{1}{\omega_i C_i})\right] = -i_i(\omega_i)Z_i. \qquad (10.19)$$

Expressing $v_{C'}(\omega_i)$ of Eq. (10.16) and $i_i(\omega_i)$ of Eq. (10.11) in complex notation, Eq. (10.19) becomes

$$\frac{I_i}{I_s} = \frac{1}{2C_m'\omega_i Z_i} e^{j(\phi_p - \phi_s - \phi_i)}. \qquad (10.20)$$

Since I_i and I_s are by definition current amplitudes, I_i/I_s or I_i^*/I_s^* must therefore both be real. Consequently, we can take the complex conjugate of Eq. (10.20) and substitute it into Eq. (10.18) to obtain

$$Z(\omega_s) = -\frac{1}{4\omega_s \omega_i C_m'^2 Z_i^*}. \qquad (10.21)$$

Since the deviation of Eq. (10.21) is not predicated on what gives rise to i_s and i_i in the signal and idler circuits, it is applicable to either an oscillator or an amplifier.

B. Oscillation

Let us briefly consider the oscillator first. In this case, the frequencies of the currents in the signal and idler resonant circuits are equal to the resonance frequencies of the respective circuits, that is, $\omega_s = \Omega_s$ and $\omega_i = \Omega_i$. It follows from Fig. 10.5 and Eq. (10.19) that Z_i^* is real and equal to R_i'. From Eq. (10.21), we then see that $Z(\omega_s)$ is real and negative representing a negative resistance. It is clear that $Z(\omega_s)$ must be equal in magnitude to R_s' to sustain oscillations. From Eq. (10.21), the required amplitude of variation of capacitance C' for sustained oscillation is

$$C_m' = \frac{1}{2\sqrt{\omega_s \omega_i R_s' R_i'}}. \qquad (10.22)$$

Thus the signal resonance circuit faces a resistance equal in magnitude and opposite in sign to its own series resistance. We can easily show that the same is true for the idler resonant circuit. In other words, variation of the coupling capacitance C' at the pump frequency results in substained oscillation in both the signal and idler circuits at their corresponding resonance frequencies. It is also interesting to note from Eq. (10.20) that

$$\phi_p = \phi_s + \phi_i \qquad (10.23)$$

for sustained oscillations since $Z_i = R_i'$ and I_i/I_s is real.

C. Amplification

Now consider the case of amplification. Assume that a voltage source v_g with

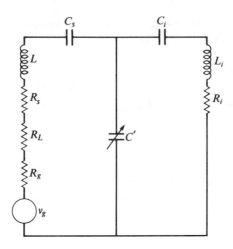

Fig. 10.6. Equivalent circuit of parametric amplifier showing signal voltage and load connected to the signal circuit.

internal resistance R_g as well as a load R_L is connected in series with the signal resonant circuit as depicted in Fig. 10.6. Let Q_s be the Q of the signal circuit, including the effects of R_g and R_L, and let Q_i be that of the idler circuit. Also let

$$\omega_s = \Omega_s + \Delta\omega \tag{10.24}$$

and

$$\delta = \frac{\Delta\omega}{\Omega_s}, \tag{10.25}$$

where $\Delta\omega$ and δ are respectively the actual and normalized deviation of the signal frequency from the signal circuit resonance frequency. It then follows from Eqs. (10.9), (10.12), and (10.24) that

$$\omega_i = \Omega_i - \Delta\omega. \tag{10.26}$$

Substituting Eq. (10.26) into the expression for Z_i, the impedance of the idling circuit given in Eq. (10.19), we find

$$Z_i = R_i[1 - j2\delta(\Omega_s/\Omega_i)Q_i], \tag{10.27}$$

where we have assumed that $\Delta\omega \ll \Omega_i$ and noted that $Q_i = 1/\Omega_i R_i C_i$. It follows that the impedance presented to the amplifying circuit due to the parallel combination of variable capacitance C' and the idler circuit is

$$Z(\omega_s) = -\frac{1}{4\omega_s\omega_i C_m'^2 R_i[1 + j2\delta(\Omega_s/\Omega_i)Q_i]}, \tag{10.28}$$

where we have used Eqs. (10.21) and (10.27). Note that $Z(\omega_s)$ is in general complex. However, at the center frequency, that is, when $\omega_s = \Omega_s$ or $\Delta\omega = \delta = 0$ according to Eqs. (10.24) and (10.25), Eq. (10.28) reduces to

$$Z(\omega_s) = -R = -\frac{1}{4\omega_s\omega_i C_m'^2 R_i'} \quad (10.29)$$

which is clearly a negative resistance representing the source of amplifying power.

The power gain of the amplifier is defined as the ratio of power dissipated in the load resistance divided by the available power from the generator. On this basis, we can show from Fig. 10.6 that to the order δ^2, power gain G is:

$$G = \frac{4R_g R_L}{\{R_{Ts} - R[1 - 4\delta^2(\Omega_s^2/\Omega_i^2)Q_i^2]\}^2 + 4\delta^2[R_{Ts}Q_s + R(\Omega_s/\Omega_i)Q_i]^2}, \quad (10.30)$$

where we have used Eqs. (10.24), (10.28), and (10.29) as well as

$$Q = \Omega_s L_s/R_{Ts} = 1/R_{Ts}\Omega_s C_s.$$

Here R_{Ts}, the total resistance of the signal circuit including generator resistance and load, is equal to $(R_g + R_L + R_s)$. The expression for the power gain at resonance is particularly simple:

$$\text{power gain at resonance} = \frac{4R_g R_L}{(R_{Ts} - R)^2}. \quad (10.31)$$

To find the half-power bandwidth of the parametric amplifier, we need merely equate (10.30) to Eq. (10.31) multiplied by $\frac{1}{2}$ and solve for δ. First, however, we can simplify our mathematics considerably by noting that if the gain is large, the second term in the denominator of Eq. (10.30), representing the off-resonance susceptance, increases more rapidly than the first as δ deviates from zero. To a good approximation, therefore, we need retain only this term in the bandwidth calculation. It follows that the normalized bandwidth is

$$2\delta = \frac{R_{Ts} - R}{Q_s[R_{Ts} + R(\Omega_s Q_i/\Omega_i Q_s)]}. \quad (10.32)$$

We see from this equation that the bandwidth decreases with increasing Q_s and increases with increasing ratio of idler to signal frequency. For a reasonable gain, the bandwidth of the parametric amplifier is rather narrow. For example, a parametric amplifier with a 20-db gain may have a bandwidth of only 0.01%.

D. Noise Figure

One of the important advantages of a parametric amplifier is its low noise level. The noise figure F of the amplifier is defined as

$$F = \left(\frac{S_i}{N_i}\right)\left(\frac{N_o}{S_o}\right) = \frac{1}{\text{power gain}} \frac{1}{kTB} N_o, \quad (10.33)$$

where

$\frac{S_i}{N_i}$ = available signal-to-noise ratio at the input,

$$\frac{N_o}{S_o} = \text{noise-to-signal ratio at the output,}$$

and k, T, and B are respectively the Boltzmann constant, the standard noise temperature (290°K), and the noise bandwidth of the amplifier. Significant noise sources which may contribute to N_0 are:

1. Thermal noise at ω_s in signal circuit.
2. Thermal noise at ω_i in idling circuit.
3. Noise current at ω_s emanating from C'.
4. Noise current at ω_i emanating from C'.
5. Noise fluctuations at ω_p in value of variable capacitance C'_m.
6. Noise fluctuations at $2\omega_s$ in value of variable capacitance C'_m.
7. Noise fluctuations at $2\omega_i$ in value of variable capacitance C'_m.
8. Noise fluctuations at $(\omega_s - \omega_i)$ in value of variable capacitance C'_m.

Since the total noise at the output, N_o, will be the summation of the noise due to each of the sources, we may write

$$F = \frac{1}{4kTB} \frac{(R_{Ts} - R)^2}{R_g R_L} \sum_{n=1}^{8} N_{on}.$$

For the derivation of the expressions for N_{on}, the reader is referred to Heffner and Wade [9]. Suffice to say here that one of the most important sources of noise is the thermal noise from the idling circuit. However, we can obtain a very low noise figure by choosing a large ratio of idling frequency to amplifying frequency or by artificially cooling the resonant circuits.

We can also use the signal-idling resonant circuit parametric system as a frequency up or down converter with large conversion gain by having the output taken from the idling rather than the signal circuit. The calculation of gain, bandwidth, and noise figure in these cases is similar to that for the amplifier above. For this reason, its analysis is left as an exercise for the reader.

3. COUPLING ELEMENT

A. General Considerations

Throughout the last two sections we assumed that the time-varying parameter was a capacitor. Actually, coupling can also be achieved via a time-varying inductance or any other energy-storage circuit element, including electron beams, junction capacitances of semiconductor diodes, and electron spins in ferrimagnetic materials or ferrites. In what follows, we shall briefly examine the characteristics of these coupling elements.

For electron beam parametric amplifiers and certain forms of bridged-diode amplifiers, the foregoing treatment of the coupling capacitance as a simple time-varying element is justified. For most other forms, however, the coupling capacitance should be treated as a nonlinear element whose capacitance is a function of its voltage. Thus there will be capacitance changes not only at the frequency of the pump voltage ω_p but also at the signal and idling frequencies ω_s and ω_i when these voltages are present. It turns out, however, that the expressions derived above for gain, bandwidth, and noise figure for the time-varying capacitance are correct for the case of nonlinear capacitance if effective capacitance change at ω_p is taken to be twice that produced by the pump alone [9].

B. Electronic Beams

Parametric amplifiers can be built by utilizing the space-charge of an electron beam [3,11]. According to the results of the last chapter, the velocities of these waves follow from Eqs. (9.74) and (9.75)

$$\left(v_f\right)_{sc} = \frac{v_0}{1 - \omega_p/\omega_s}, \tag{10.34}$$

$$\left(v_s\right)_{sc} = \frac{v_0}{1 + \omega_p/\omega_s}, \tag{10.35}$$

where v_0 is the dc beam velocity, ω_p the plasma frequency, and ω_s the signal frequency. Here $(v_f)_{sc}$ is the velocity of the fast wave while $(v_s)_{sc}$ is that of the slow wave. In a TWT, only the slow wave can interact with a slow wave structure in such a way as to give rise to tube gain, but either wave can bring about parametric amplification. However, there is a minimum noise figure obtainable in the slow-wave case since the noise present cannot be completely removed [7, 8]. Because noise elimination is in principle possible for the fast space-charge wave, our discussion here shall emphasize parametric amplification via the fast wave.

Consider a Klystron amplifier with the electron beam velocity-modulated at signal frequency ω_s at the buncher cavity. Ideally, slow and fast space-charge waves of equal amplitude are set up at the cavity gap which interfere to give the usual space-charge standing wave patterns as they propagate down the beam in the drift tube. If, however, a similar cavity at the input applies pump power at frequency $\omega_p = 2\omega_s$, this standing wave pattern will grow exponentially with distance as both the slow and fast waves at ω_s are parametrically amplified.

Actually, each of the normal space-charge waves on such a beam divides into one wave that grows and another that attenuates with distance. Either the fast or slow wave can be selectively amplified independently of each other. The character of the growing waves on the beam is such that optimum growth occurs when the phase velocity of the pumping wave and that of the signal wave are equal. As the velocities become unequal a greater depth of modulation is required for the onset of exponential gain.

If $\omega_s \neq \omega_p/2$, then, due to the nonlinear nature of the electron beam, an idler will be generated at frequency $\omega_i = \omega_p - \omega_s$, as well as at frequencies $\Omega_{lm} = l\omega_p + m\omega_s = -\Omega_{-l,-m}$ (where l, m are zero, negative, or positive integers). Ampli-

---

[7] Equations (9.74) and (9.75) were derived for a one-dimensional beam of infinite cross-section. For a cylindrical beam, ω_p must be multiplied by R, the plasma reduction factor [4].

tudes at these frequencies can grow with distance; thus the noise the frequencies carry will also be amplified. Since there is no simple way to suppress the additional Ω_{lm} idler frequencies, their additional noise contribution raises the noise figure of the parametric amplifier, even if the fast space charge is used. For this reason, apparently, parametric amplifiers using the longitudinal space charge waves have not had the expected low noise figure, although they have been successfully operated. On the other hand, the Adler tube [1,2] which utilizes the fast cyclotron rather than the space-charge wave of an electron beam in confined flow, yields a noise figure as low as 1.4.

If an electron beam premodulated by the signal and the pump is coupled to a slow-wave structure such as a helix, there are six instead of three forward waves, as in the case of the TWT. The nature of these waves depends on the pumping modulation, but in general there is more than one growing wave in the vicinity of both the point at which the fast space-charge wave becomes equal to the circuit wave and the point at which the slow space-charge wave is equal to the circuit wave.

To summarize, the energy required for building up the signal in an electron-beam parametric amplifier comes from the r.f. pump source rather than the dc beam. This feature allows the amplification of the fast space-charge wave, giving rise to the possibility of low-noise amplification. However, as mentioned previously, the presence of an infinite number of idler frequencies in the beam has made it impossible to obtain a low noise figure using the fast wave.

We shall now study the Adler tube, which attains a very low noise figure by utilizing the fast cyclotron wave for amplification.

In a manner similar to the derivation of Eq. (8.42) for the propagation constants of the TWT, we can show that the phase constant β of waves due to electrons traveling down the beam and oscillating transversely with cyclotron frequency $\omega_c = eB/m$ in a z-directed magnetic flux density B is given by

$$\beta = \frac{\omega_s}{v_0}\left(1 \pm \frac{\omega_c}{\omega}\right), \tag{10.36}$$

where ω_s is the signal frequency and v_0 the dc velocity of the beam [13]. It follows that the phase velocities of the fast and slow cyclotron waves $(v_f)_c$ and $(v_s)_c$ are given by

$$\left(v_f\right)_c = \frac{v_0}{1 - \omega_c/\omega_s}, \tag{10.37}$$

$$\left(v_s\right)_c = \frac{v_0}{1 + \omega_c/\omega_s}, \tag{10.38}$$

analogous to similar expressions for space-charge waves given by Eqs. (10.34) and (10.35).

As in the case of space-charge waves, noise elimination is in principle possible for the fast but not the slow cyclotron wave. Unfortunately, interaction with the fast wave does not produce gain in conventional devices. However, fast cyclotron waves can be excited by passing an electron beam through a set of parallel plates across which an r.f. signal voltage of frequency ω_s is applied, as shown in Fig. 10.7(a). The plates are part of a lumped resonant cavity called the Cuccia electron coupler [5]. The electron beam is assumed to be immersed in a homogeneous axial magnetic field whose intensity is so chosen that $\omega_c = \omega_s$ in order to maximize signal-beam interaction. A signal applied to the plates will cause electrons in the beam to acquire a transverse component of velocity. The interaction of this velocity component with the axial magnetic field B will in turn give rise to a force orthogonal to both. Thus a signal applied to the plates will cause the electrons to spiral outward and reach a maximum radius at the exit of the cavity at which point all the signal power supplied appears in the form of electron kinetic energy. This energy can be regained by passing the beam through a second Cuccia cavity in which the orbiting electrons will induce a current. If the cavity is properly loaded, the resulting field causes them to spiral inward and give up all their kinetic energy.

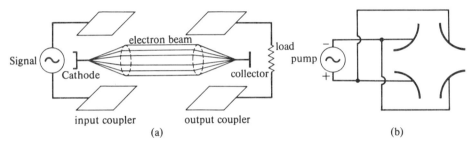

Fig. 10.7. (a) Cuccia couplers traversed by an electron beam. (b) Pump quadrupole structure.

However, the Cuccia coupler is not practical because it has neither gain nor any energy exchange between the transverse signal motion and the axial electron motion. All electrons in the beam throughout its journey orbit in phase at an infinite phase velocity in accordance with Eq. (10.37), with $\omega_s = \omega_c$. It is important to note, however, that whereas the output cavity absorbs all the available fast-wave signal power from the beam, the input cavity, if properly loaded, will absorb all the fast-wave noise power. In other words, the initial noise in the beam associated with transverse motions of electrons is damped out as the beam passes through the input coupler and receives its cyclotron wave modulation. Thus the beam is essentially noise free as it emerges from the input cavity, as has been experimentally demonstrated [1].

Since no amplification results from the two-coupler arrangement just described, each electron in the drift region follows a helical path of constant radius. Any mechanism that increases the radius of the electron path will result in amplification. For the amplification to be linear, the rate of growth of the radius must be proportional to the radius itself. We can increase the radius of the circular path by having the electron constantly experience a tangential force in the direction of its transverse motion. Furthermore, for the operation to be proper, the magnitude of such a tangential force on the electron must be a function of the position of the electron; as the radius of curvature of the electron path increases, the magnitude of the force encountered should also increase. Figure 10.7(b) shows an electrode structure ideally suited to supply the forces necessary to expand the radii of all electrons that enter with the proper phase. Note that one pair of opposite plates of the quadrupole structure are connected to one terminal of a pump source while the other pair of plates are connected to the other terminal [2]. For the Cuccia coupler, the phase velocity of the fast wave is infinite and that of the slow wave is one-half the axial electron velocity, according to Eqs. (10.37) and (10.38). Since the two velocities are so different it is easy to couple to one wave only. The quadrupole amplifier can operate over a range of signal frequencies close to the cyclotron frequency.

To summarize, as the electron beam traverses the input coupler the signal gives the electrons a spiraling cyclotron modulation at frequency ω_s ($\simeq \omega_c$). When the beam enters the quadrupole region, the cyclotron waves get additional energy from a tangential electric field oscillating at the pump frequency ω_p, providing that the quadrupole coupler is tuned to the beat or idler frequency $\omega_i = \omega_p - \omega_s$.[8] Of course, if $\omega_s = \omega_c$ and $\omega_p = 2\omega_c$, then $\omega_i = \omega_s = \omega_c$. In thise case, the beat or idler rotates in synchronism with the electrons and thereby transfers energy to them from the pump. The increased energy of the electrons at ω_s (or ω_i) is extracted at the output coupler, resulting in parametric amplification. It should be noted that the quadrupole amplifier is not inherently narrow-band, bilateral, or unstable, as is common in many cavity-type, solid-state versions. For example, a bandwidth of 10% for the Adler tube can be readily realized.

C. Semiconductor Diodes

We can use the incremental depletion-layer capacitance of a reverse-biased junction diode as the coupling element in a parametric amplifier. This capacitance is a function of the applied voltage V_c; in other words, it represents a nonlinear coupling reactance. Figure 10.8(a) shows some typical depletion layer capacitance C_D vs. V_c characteristics. Varactor diodes, which typically have low series resistance and low capacitance, are most successfully used in parametric applications.

If we introduce donor impurities into one side and acceptors into the other side of a single crystal semiconductor such as germanium or silicon, a p–n junction is formed, as shown in Fig. 10.8(b). Initially, there are nominally only p-type carriers to the left of the junction and n-type carriers to the right. Due to this density gradient across the junction, holes will diffuse across the junction to the right, and electrons to the left. As a result of the displacement of these charges, an electric field that opposes the flow of carriers will appear across the junction.

---

[8] Beating is due to the mixing of pump and signal frequencies in the electron beam, a nonlinear medium.

Diffusion will continue until this field becomes large enough to restrain the forces of diffusion. The positive holes that neutralized the acceptor ions near the junction in the p-type region have disappeared due to recombination with electrons which have diffused across the junction. Similarly, the neutralizing electrons in the n-type region have recombined with holes which have crossed the junction from the p-type region. Therefore, the unneutralized electric charges are confined to the neighborhood of the junction and consists of immobile ions.

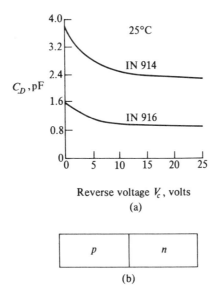

Fig. 10.8. (a) Typical depletion-layer capacitance vs. reverse voltage for silicon diodes. (b) A *p-n* junction.

A reverse bias (p-region negatively biased with respect to the n-region) causes majority carriers to move away from the junction, thereby uncovering more immobile charges. Hence the thickness of the space-charge or depletion layer d at the junction increases as the reverse-bias voltage increases. As might be expected, the depletion layer capacitance C_D is equivalent to that of a parallel-plate capacitor of area A and plate separation d containing a dielectric constant ε:

$$C_D = \varepsilon A/d. \tag{10.39}$$

Since d increases with V_c for the case of negative bias, C_D decreases as V_c gets increasingly negative, consistent with typical C_D-V_c characteristics depicted in Fig. 10.8(b).

An equivalent circuit for a varactor diode under reverse bias is shown in Fig. 10.9. Here R_s represents the body or ohmic series resistance of the diode.

Typical values of C_D and R_s are 20 pf and 85 Ω respectively at a reverse bias of 4 V. The reverse diode resistance R_r shunting C_D is large (> 1 MΩ), and hence usually neglected.

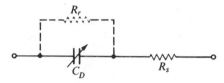

Fig. 10.9. Equivalent circuit of a varactor diode.

D. Ferromagnetic Material

A sample of ferromagnetic material can also be used as a coupling element in a parametric amplifier.[9] If it is immersed in a sufficiently strong, uniform dc magnetic field H_0, the magnetic moments due to electron spins in the material will precess at a frequency ω_r determined by H_0. This precession will die out quickly because of damping forces in the ferromagnet. However, if an r.f. field of frequency $\omega = \omega_r$ is applied to the sample in a direction orthogonal to the dc field, the precession can be maintained at a constant amplitude. In this case, all the magnetic moments in the material precess in unison. This mode of motion is termed, appropriately, the uniform precession. At low signal levels, the magnetic moments do indeed execute the uniform precession in a dc field. However, at high signal levels, another resonance, called the subsidiary resonance, can occur at a dc field smaller than that corresponding to the uniform precession at a given operating frequency. This subsidiary absorption peak at high signal levels involves the growth of disturbances or *magnetostatic* or nonuniform modes whose frequencies ω_1, ω_2 add up to the frequency ω of the uniform precession. Identifying ω_1 with the signal frequency ω_s, ω_2 with the idler frequency ω_i, and ω with the pump frequency ω_p, we may surmise that we can utilize this effect to construct a microwave parametric amplifier. Since we are dealing with magnetic quantities, the coupling element is an inductor rather than a capacitor. The equivalent circuit of the ferromagnetic amplifier is the same as that shown in Fig. 10.4 except that the variable capacitor $C'(t)$ is replaced by a time-varying inductor $L'(t)$ [14].

If we couple an input signal and an output load to the signal or idler loop and adjust the pumping power so that oscillations occur only without the extra load, stable power gain may result. However, there are an infinite number of magnetostatic mode pairs with frequencies that add up to approximately ω_p whose mutual

---

[9] Actually, ferromagnetic conductors such as nickel, iron, or cobalt have too much eddy current loss at microwave frequencies to be useful. Consequently, ferrimagnetic materials such as ferrites or garnets, materials whose conductivities are orders of magnitude lower, are used instead.

couplings have comparable strength and whose instability thresholds are therefore of similar magnitude. Thus selective amplification of the wanted mode pair may be difficult to attain, particularly because the presence of the output load raises the pair's threshold. As the pumping power or the depth of modulation of $L'(t)$ increases, unwanted mode pairs may become unstable before the wanted pair exhibits much, if any, gain. We can overcome this difficulty by separating the two functions performed by the sample, i.e., as a resonant system and as a coupling element.

It turns out that by adjusting H_0 or ω_p or both we can assure that there is no magnetostatic mode pair whose frequencies add up to ω_p. The sample in this case is placed into a triply resonant microwave structure with resonance frequencies ω_s, ω_i, and ω_p. If the r.f. field lines of the signal cavity at the sample are orthogonal to those of the idler cavity, power gain can be obtained as before. However, the stability threshold of the uniform precession now depends on the signal and idler cavity Q's rather than on the losses of the ferromagnetic sample. This type of operation is known as the *electromagnetic* operation and should have a threshold at least as low as that for the magnetostatic operation described above.

Another way of overcoming the difficulty encountered in the magnetostatic operation is to let the cavity supply one mode (the idler mode, say) and the sample the other. In this operation, known as the *semistatic* operation, the threshold depends on both cavity and sample losses.

Weiss [16] succeeded in building an amplifier using the electromagnetic mode. The most successful operation appears to be one that employs the magnetostatic mode with longitudinal pumping [6]. Thus far, the ferromagnetic amplifier has been found to be inferior to the maser using a paramagnetic salt in terms of pump power requirement or noise figure.[10]

In the analysis above, it appears necessary for the pump frequency ω_p to be larger than the signal frequency ω_s if amplification is to take place. However, it turns out [10] that amplification at a frequency larger than that of the pump is possible by the introduction of a fourth frequency ω'_i. A ferromagnetic traveling-wave amplifier of the parametric type has also been proposed [15]. In this case, amplification of signal power in the form of growing waves is obtained in a propagating structure which is partially or totally embedded in a ferromagnetic medium. The structure possesses two propagating modes, one excited by the signal and the other used as an idling circuit. A traveling wave excited by the pump provides a time-varying coupling between the two propagating modes via the motion of the magnetization (magnetic moment per unit volume).

Before we conclude our discussion of parametric amplifiers, let us summarize its principle of operation. *The pump and signal frequencies ω_p and ω_s are mixed in a nonlinear or time-varying medium to produce an idler frequency $\omega_i = \omega_p - \omega_s$,*

---

[10] In somewhat simplified terms, the operation of the ferromagnetic parametric amplifier relies on modulation of the real part of the susceptibility rather than on population inversion between two energy levels, as in the case of the maser (to be discussed in the next chapter).

which in turn excites the idling cavity into oscillation at an amplitude dependent upon the idling cavity Q, as well as the amplitude of the pump and signal fields. This reinforced idler[11] now reacts back on the coupling medium and, in conjunction with the pump, generates an additional signal frequency $\omega_s = \omega_p - \omega_i$ whose amplitude is clearly proportional to the magnitude of the pump and idler or, in other words, to the product of the initial signal field and the square of the pump field. This additional component at frequency ω_s tends to drive the signal cavity and, if it is in phase with the initial signal which initiated the process, energy is added to the signal cavity and the entire device behaves as a regenerative amplifier.

REFERENCES

1. R. Adler, *Proc. IRE* **46,** 1300 (1958).
2. R. Adler, G. Hrbeck, and G. Wade, *Proc. IRE* **47,** 1713 (1959).
3. A. Ashkin, *Jour. Appl. Phys.* **29,** 1646 (1958).
4. G. M. Branch and T. G. Mihran, *IRE Trans. Electron Devices* **ED-2,** 3 (1955).
5. C. L. Cuccia, *RCA Rev.* **10,** 270 (1949).
6. R. T. Denton, *Proc. IRE* **48,** 937 (1960).
7. H. A. Haus and D. L. Bobroff, *Jour. Appl. Phys.* **28,** 694 (1957).
8. H. A. Haus and R. N. H. Robinson, *Proc. IRE* **43,** 981 (1955).
9. H. Heffner and G. Wade, *Jour. Appl. Phys.* **29,** 1321 (1958).
10. C. L. Hogan, R. L. Jepsen, and P. H. Vartanian, *Jour. Appl. Phys.* **29,** 422 (1958).
11. W. H. Louisell and C. F. Quate, *Proc. IRE* **46,** 707 (1958).
12. N. W. MacLachlan, *Theory and Application of Mathieu Functions*, Oxford University Press, New York, 1947.
13. J. R. Pierce, *Traveling-Wave Tubes*, D. Van Nostrand Co., New York, 1950; p. 175.
14. H. Suhl, *Phys. Rev.* **106,** 384 (1957).
15. P. K. Tien and H. Suhl, *Proc. IRE* **46,** 700 (1958).
16. M. T. Weiss, *Phys. Rev.* **107,** 317 (1957).

PROBLEMS

10.1 Starting from Eq. (10.2), show that for oscillations to occur C_1/C_0 must be larger than or equal to $2/Q$, where Q is the circuit quality factor.

10.2 a) Sketch the waveform V_c of Fig. 10.1 as a function of ωt, assuming that the capacitance does not vary at exactly twice the frequency.

---

[11] Since the pump is assumed strong and idler cavity Q high, the reinforced idler can clearly be much stronger than the original signal.

b) In the up- or down-converter, the pump frequency ω_p need not be equal to twice the signal frequency ω_s. How is the phase requirement between the pump and signal frequencies used for amplification satisfied by the presence of the idler circuit?

10.3 Design a circuit that can separate the input and output power when it is desired that the load resistance R_L be connected across the source (or equivalent) terminals.

10.4 a) Can you think of another type of coupling element for a parametric amplifier aside from the three mentioned in the text?

b) Consider further the sentence: "... whereas the output cavity absorbs all the available fast-wave signal power from the beam, the input cavity, if properly loaded, will absorb all the fast-wave noise power" in Section 3B of this chapter. Can you justify this statement in more detail?

CHAPTER 11

MICROWAVE GENERATION AND AMPLIFICATION IN JUNCTION SEMICONDUCTORS

In Chapter 10, we showed that the pump source in a parametric amplifier can produce a negative resistance at the signal or idler terminals. It follows that amplification can be obtained since the negative resistance acts as a source of power at the signal frequency. Of course, any other source that has a negative resistance in a given range of operation can also be the source of power. Indeed, both junction and bulk semiconductor devices can function as microwave amplifiers.

If a p-n junction is heavily doped, quantum-mechanical tunneling can occur across the junction. Within a certain range of forward bias, the I–V characteristic of the diode exhibits a negative slope, i.e., a negative conductance. This phenomenon was first reported by Esaki; thus the diode is known as the *Esaki* or *tunnel* diode. Microwave emission can also occur in the so-called *Read* or *avalanche* diode due to avalanche breakdown in a junction semiconductor.

1. QUANTUM-MECHANICAL TUNNELING

Classically, a particle must have an energy equal to or greater than the height of a potential barrier if it is to move from one side of the barrier to the other. However, quantum mechanically, if the barrier is very thin, solution of the Schrödinger equation indicates that there is an appreciable probability that a particle will penetrate or tunnel through a potential barrier, even though it does not have enough kinetic energy in the classical sense to negotiate the same barrier. In an Esaki diode, the semiconductor is heavily doped to produce a very thin barrier, on the order of 100 Å or 10^{-6} cm. Thus quantum-mechanical tunneling is facilitated and, as we shall soon show, the result is a negative slope in the diode I–V characteristic.

Let us first examine how the junction barrier thickness depends on doping. Assume, for simplicity, that there is an abrupt change from acceptor ions on one side to donor ions on the other side of the junction, as depicted in Fig. 11.1.[1] Since the semiconductor is electrically neutral, we must require that the total acceptor

---

[1] See Section 3C, Chapter 10, for a more detailed discussion on the nature of this region.

charge be equal to the total donor charge, or

$$eN_A d_p = eN_D d_n, \tag{11.1}$$

where $(d_p + d_n) = d$ is the total width of the region occupied by uncompensated ions, e is the electronic charge, N_A is the acceptor ion concentration, and N_D is the donor ion concentration. The potential V and acceptor concentration N_A on the P-side are related by Poisson's equation:

$$\frac{d^2V}{dx^2} = \frac{eN_A}{\varepsilon}, \tag{11.2}$$

where ε is the dielectric constant of the semiconductor. Since electric field lines must start on the positive donor ions and terminate on the negative acceptor ions, there are no flux lines to the left of plane $x = 0$ in Fig. 11.1, or $dV/dx = 0$ at $x = 0$. Integrating Eq. (11.2) and applying this boundary condition, we find

$$\frac{dV}{dx} = \frac{eN_A}{\varepsilon} x. \tag{11.3}$$

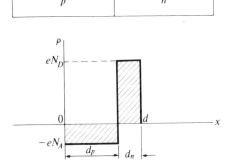

Fig. 11.1. *p-n* junction and its depletion layer charge distribution.

Also, since the zero of potential is arbitrary, we may choose V to be zero at $x = 0$. Thus, integrating Eq. (11.3), we have

$$V = \frac{eN_A}{2\varepsilon} x^2. \tag{11.4}$$

Similarly, on the *n*-side, the voltage V and donor concentration N_D are related by

$$\frac{d^2V}{dx^2} = -\frac{eN_D}{\varepsilon}. \tag{11.5}$$

Integrating, we find

$$\frac{dV}{dx} = -\frac{eN_D}{\varepsilon} x + C_1. \tag{11.6}$$

Integrating again, we have

$$V = -\frac{eN_D}{2\varepsilon}x^2 + C_1 x + C_2, \tag{11.7}$$

where

$$C_1 = \frac{e}{\varepsilon}d_p(N_A + N_D) \tag{11.8}$$

and

$$C_2 = -\frac{e}{2\varepsilon}d_p^2(N_A + N_D). \tag{11.9}$$

We evaluated constant C_1 by equating dV/dx at $x = d_p$, given by Eqs. (11.3) and (11.6). Similarly, we evaluated C_2 by equating the voltage V at $x = d_p$ given by Eqs. (11.4) and (11.7). If we further observe that $x = d$, V is equal to the barrier potential V_B, we find that the junction barrier width d is given by

$$d = \sqrt{2\varepsilon(N_A + N_D)V_B/eN_A N_D}. \tag{11.10}$$

The barrier potential V_B is equal to $V_0 - V$, where V_0 is the contact potential difference and V is the applied potential. For a forward bias, V is considered positive. Here we have assumed that the entire applied voltage appears across the depletion layer, i.e., there is no voltage drop across the contacts or the body of the semiconductor.

We can see from Eq. (11.10) that d varies inversely as $N_A N_D/(N_A + N_D)$ to the power $\frac{1}{2}$. Whereas the values N_A and N_D of an ordinary p–n junction are such that d is on the order of a few microns (10^{-4} cm), d of a tunnel diode is only about 100 Å (10^{-6} cm) because its doping is orders of magnitude higher.

We shall now show that very small barrier junction thickness is required if we are to obtain a significant transmission *through* the barrier. Consider for simplicity the one-dimensional problem depicted in Fig. 11.2. In this problem we wish to determine the probability that a particle of energy E ($<$ potential barrier energy U_0) will penetrate the barrier width d of region 2 and therefore go from region 1 into region 3.

According to classical mechanics, a particle moving in the positive x-direction with total energy E less than the potential barrier height U_0 will reverse the direction but not the magnitude of its velocity upon reaching the left potential boundary at $x = 0$. In other words, the potential wall acts like a barrier through which a particle with energy $E < U_0$ cannot penetrate. Were it possible for the particle to enter the barrier region, its kinetic energy, being equal to $E - U_0$, would be negative, a concept totally foreign to classical mechanics. On the other hand, if $E > U_0$, the particle will move right into the barrier with a reduction in its kinetic energy equal to the barrier height U_0. This loss in kinetic energy is recovered as the particle enters region 3.

In quantum mechanics, the wave function ψ of a particle in a given potential distribution is determined by the time-independent Schrödinger equation:

$$\frac{d^2\psi}{dx^2} + \frac{2m}{\hbar^2}[E - U(x)]\psi = 0, \qquad (11.11)$$

where m is the mass of the particle and $\hbar$ is Planck's constant divided by 2π. Again, E is the total energy while $U(x)$ is the x-dependent potential energy function. Specifically, the Schrödinger equations for the three regions of Fig. 11.2 are

$$\frac{d^2\psi_1}{dx^2} + \frac{2m}{\hbar^2}E\psi_1 = 0,$$

$$\frac{d^2\psi_2}{dx^2} + \frac{2m}{\hbar^2}(E - U_0)\psi_2 = 0, \qquad (11.12)$$

$$\frac{d^2\psi_3}{dx^2} + \frac{2m}{\hbar^2}E\psi_3 = 0.$$

Their solutions are

$$\psi_1 = Ae^{i\alpha x} + A'e^{-i\alpha x},$$
$$\psi_2 = Be^{\beta x} + B'e^{-\beta x}, \qquad (11.13)$$
$$\psi_3 = Ce^{i\alpha x} + C'e^{-i\alpha x},$$

where

$$\alpha^2 = \frac{2mE}{\hbar^2}, \qquad (11.14)$$

$$\beta^2 = \frac{2m(U_0 - E)}{\hbar^2}. \qquad (11.15)$$

Since $U_0 > E$, β and α are both real.

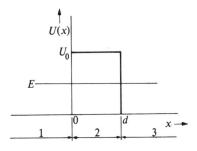

Fig. 11.2. Potential energy as a function of position for a one-dimensional barrier of finite width.

If we assume that region 3 extends to infinity, there will be no reflected wave traveling in the negative x-direction in region 3. Therefore, we can set C' equal to zero. By requiring that the wave function ψ and its first derivative $d\psi/dx$ be continuous at the barrier boundaries $x = 0$ and $x = d$, we can evaluate the constants A', B, B', and C in terms of A to yield

$$A' = A \frac{1 + \frac{\beta^2}{\alpha^2}(e^{-\beta d} - e^{\beta d})}{\Delta},$$

$$B = A \frac{2\left(1 + \frac{\beta}{i\alpha}\right)e^{-\beta d}}{\Delta}, \quad (11.16)$$

$$B' = A \frac{-2\left(1 - \frac{\beta}{i\alpha}\right)e^{\beta d}}{\Delta},$$

$$C = A \frac{4\left(\frac{\beta}{i\alpha}\right)e^{-i\alpha d}}{\Delta},$$

where

$$\Delta = \begin{vmatrix} 1 + \frac{\beta}{i\alpha} & 1 - \frac{\beta}{i\alpha} \\ \left(1 - \frac{\beta}{i\alpha}\right)e^{\beta d} & \left(1 + \frac{\beta}{i\alpha}\right)e^{-\beta d} \end{vmatrix}. \quad (11.17)$$

Combining Eqs. (11.13) through (11.17), the probability of finding a particle in an interval dx located at x, that is, $\psi\psi^* dx$, can be evaluated in terms of U_0, E, and d. Specifically, the probability of finding a particle in an interval dx located in region 3 is given by

$$\psi_3 \psi_3^* \, dx = \frac{4\left(\frac{U_0}{E} - 1\right) A^2 dx}{\left(\frac{U_0}{E} - 2\right)^2 \sinh^2 \frac{\sqrt{2m(U_0 - E)}}{\hbar} d + 4\left(\frac{U_0}{E} - 1\right) \cosh^2 \frac{\sqrt{2m(U_0 - E)}}{\hbar} d}. \quad (11.18)$$

A number of interesting interpretations of this equation can be made. First, $\psi\psi^*$ decreases with increasing barrier thickness d but is finite for all finite values of d. In other words, there is a finite probability of finding a particle in region 3 even though its energy E is less than U_0. We thus interpret this situation to mean that a particle in region 1 with $E < U_0$ can penetrate or *tunnel through* the potential energy barrier of region 2 into region 3. Furthermore, as shown in Fig. 11.3, $\psi\psi^*/A^2$ decreases rapidly with increasing d/d_0, where we have defined d_0 as

$$d_0 = \frac{\hbar}{2}\sqrt{\frac{1}{2m(U_0 - E)}} = \frac{1}{2\beta}. \qquad (11.19)$$

The factor 2 in the $d_0 = 1/2\beta$ relation is introduced so that $\psi\psi^*$ in region 2 decreases exponentially as e^{-x/d_0} for the case where the barrier is of infinite thickness. A second important observation in connection with Eq. (11.18) and Fig. 11.3 is that $\psi\psi^*$ decreases relatively slowly with decreasing E/U_0, indicating that particles at all energy levels $E\,(< U_0)$ can partake in the tunneling process. It follows that, because of this large supply of electrons, the tunneling current can be quite respectable. This fact clearly has important practical implications in devices employing the tunneling phenomenon.

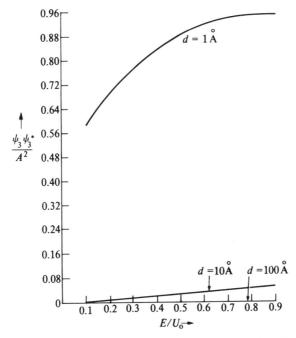

Fig. 11.3. Transmission coefficient vs. E/U_0 for fixed values of barrier thickness d. Barrier height U_0 is set equal to 1eV.

For a typical p–n junction diode with an impurity concentration of one part in 10^8, the width of the depletion layer, which constitutes a potential barrier at the junction, is on the order of $5\,\mu$ or 5×10^{-4} cm, as we can estimate by means of Eq. (11.10). If the concentration of impurities is greatly increased to, say, one part in 10^3, corresponding to a density in excess of $10^{19}/\text{cm}^3$, as in the case of a tunnel diode, d is reduced to about 100 Å or 10^{-6} cm. As a consequence, the device characteristics change drastically.

According to Eq. (11.18), the probability of an electron being found in region 3 is determined, for a given E/U_0 ratio, by the characteristic decay distance d_0 given by Eq. (11.19). Assuming that

$$(U_0 - E) = 1.60 \times 10^{-20} J \quad \text{or} \quad 0.1 \ eV,$$

d_0 turns out to be about 3 Å, a very small distance indeed. For this reason, the barrier width d must be extremely small if a significant number of electrons are to tunnel through the barrier. In practice, we can obtain a depletion layer width only as small as 100 Å by large-scale impurity diffusion, as alluded to above.[2] Since d_0 is on the order of 3 Å, d/d_0 is still quite large. Under these circumstances, Eq. (11.18) reduces simply to

$$\frac{\psi_3 \psi_3^*}{A^2} = 16 \frac{E}{U_0} \left(1 - \frac{E}{U_0}\right) e^{-d/d_0}. \tag{11.20}$$

Typically, the coefficient of e^{-d/d_0} is fairly close to unity, so that $\psi_3 \psi_3^*/A^2$ is, to a first approximation, simply equal to e^{-d/d_0}. We should not be surprised at the simple relationship between $\psi_3 \psi_3^*/A^2$ and d/d_0 once we recall that the reflected wave in region 2 was neglected in the derivation of Eq. (11.20). Accordingly, $\psi_2 \psi_2^*$ should vary as e^{-x/d_0}; because of the continuity of ψ at $x = d$, $\psi_3 \psi_3^*$ associated with the transmitted wave should likewise be proportional to e^{-d/d_0}. A little reflection will show that the only reason the coefficient of e^{-d/d_0} in Eq. (11.20) is not unity is that $\psi_3 \psi_3^*$ is normalized to the square of the incident amplitude AA^* rather than to the square of the total amplitude in region 1.

Equation (11.20) was derived for the case in which the potential barrier of region 2 is of constant height. It can be generalized to include cases which the potential energy U is a function of x even in region 2. If $U(x)$ is a slowly varying function of x in region 2, it is reasonable, to a first approximation, for us to replace the exponent $-d/d_0$ in Eq. (11.20) by the integral

$$-2 \int_0^d \frac{\sqrt{2m[U(x) - E]}}{\hbar} dx.$$

Of course, the factor multiplying e^{-d/d_0} in Eq. (11.20) should also be different, dependent upon E and $U(x)$. Therefore, let us write this general factor as $F(E, U)$. Equation (11.20) thus generalizes to

$$T = F(E, U) e^{-2 \int_0^d \frac{\sqrt{2m[U(x) - E]}}{\hbar} dx}, \tag{11.21}$$

where again $F(E, U)$ can be set equal to unity to a first approximation. Here T, known as the transmission coefficient, replaces $\psi_3 \psi_3^*/A^2$. It is interesting to note that expression (11.21) is the same as that derived from the Schrödinger equation

---

[2] Ultimate impurity concentration is determined by the solid state solubility of impurities in the host material. Because of their high solubility, Ga and Al are used as the p-impurity while As is used as the n-impurity in a tunnel diode using a Ge host.

using the WKB (Wentzel-Kramers-Brillouin) approximation. This expression will be used in the next section to calculate the tunneling current.

2. TUNNEL DIODES

A. Energy Band of Heavily Doped Semiconductor

The requirement that the barrier thickness d must not be overwhelmingly large compared to the characteristic decay distance d_0 defined by Eq. (11.19) is a necessary but not sufficient condition for tunneling to occur. In addition, there must also be filled energy states on the side from which particles tunnel and allowed empty states on the other side into which particles penetrate at the same level of energy. Therefore, we must now study the energy-band picture of a heavily doped p–n junction. If a semiconductor is heavily doped, its Fermi level E_F lies outside rather than inside the forbidden energy gap, in contrast to the lightly doped case.[3] Based on this assumption, the energy band picture of a heavily doped semiconductor under open-circuit condition will be as shown in Fig. 11.4(a). If we assume that the p- and n-type materials of the junction are in intimate contact on an atomic scale, then the Fermi level must be constant throughout the specimen at equilibrium. If this were not so, electrons on one side of the junction would have an average energy higher than those on the other side, and there would be a transfer of electrons and energy until the Fermi levels on the two sides of the junction were equalized. We further observe that because of the equalization of the Fermi levels in the p- and n-type regions, there are no filled states on one side of the junction that are at the same energy as empty allowed states on the other side. Therefore, there can be no net flow of charge across the junction and the current is zero, as expected of an open-circuited diode.

Let us now consider the case of reverse bias. In this case, the n-side of the semiconductor is biased positively with respect to the p-side. Under an open-circuited condition, the potential barrier across the junction is given by the contact potential V_0 indicated in Fig. 11.4(a). When a reverse-bias voltage V is applied to the diode, the height of the potential barrier is increased by an amount eV, where e is the electronic charge. Hence the n-side levels must shift downward

---

[3] For an intrinsic semiconductor, the Fermi level E_F is at the center of the energy gap. This is reasonable since at 0°K the conduction and valence bands are, respectively, completely empty and completely occupied and E_F is defined as the energy of the state whose probability of occupation is one-half. If donor impurities are introduced, there will be additional impurity states just below the bottom of the conduction band. Since these states are filled at 0°K, and in view of the definition of E_F, we would expect E_F to move from the center of the energy gap toward the conduction band. Indeed, if the impurity concentration is sufficiently large, as in the case of tunnel diodes, E_F can move inside the conduction band itself. Similar reasoning will show that E_F can lie inside the valence band for p-type semiconductors of high impurity concentration. For a mathematical justification, see e.g., J. Millman and C. C. Halkias, reference 23 at the end of this chapter.

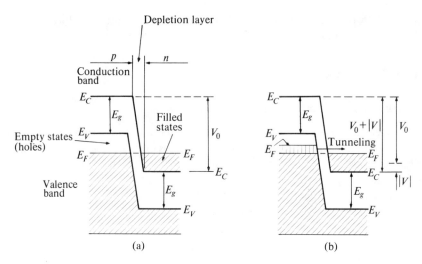

Fig. 11.4. Energy-band diagrams for a tunnel diode: (a) under open-circuited condition, and (b) with an applied reverse bias of magnitude $|V|$.

with respect to the p-side levels, as shown in Fig. 11.4(b). We see that there are now some energy states (in the heavily shaded region) in the valence band of the p-type region which lie at the same levels as some allowed empty states in the conduction band of the n-type region. Electrons in the heavily shaded band can therefore tunnel from the p-side to the n-side, giving rise to a reverse diode current flowing from n to p. As the reverse bias V increases, the heavily shaded area of Fig. 11.4(b) expands in size with an accompanying increase in reverse current, as sketched in Fig. 11.5(a).

Now consider forward bias, in which case the p-side is made positive with respect to the n-side; accordingly, the potential barrier is decreased to below V_0.

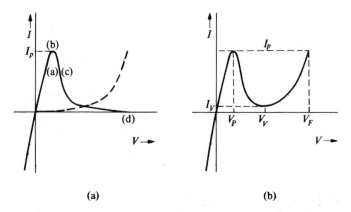

Fig. 11.5. (a) Tuneling current (solid curve) and injection current (dashed curve) as function of bias voltage. (b) I–V characteristic of a tunnel diode.

Thus the *n*-side levels must shift upward with respect to those on the *p*-side, as depicted in Fig. 11.6(a). Since there are now occupied states in the conduction band of the *n*-material (in the heavily shaded region) which are at the same energy as allowed empty states or holes in the valence band of the *p*-side, electrons will tunnel from *n* to *p*, giving rise to a forward current flowing from *p* to *n* as shown in sector (a) of Fig. 11.5(a).

As the forward bias voltage V is increased, the situation depicted in Fig. 11.6(b) will be reached. Here the maximum number of electrons can leave occupied states on the *n*-side and tunnel through the barrier to empty states on the *p*-side, giving rise to peak current I_p at the point marked by (b) in Fig. 11.5(a).[4] If V is further increased, we see from Fig. 11.6(c) that although the number of allowed empty states (holes) on the *p*-side and filled states on the *n*-side are independent of V, the extent of the overlapping region (heavily shaded area) decreases with increasing V. Thus, the tunneling current decreases with increasing V (negative dynamic conductance) in the sector marked (c) in Fig. 11.5(a). Finally, at an even higher value of V, the band structure corresponds to that of Fig. 11.6(d). Since there are now no empty *allowed* states on the *p*-side of the junction at the same energy as occupied states on the *n*-side, the tunneling current drops to zero at the point marked (d) in Fig. 11.5(a). The current remains zero with further increase in V as the occupied region on the *n*-side is further elevated with respect to the empty allowed region on the *p*-side.

In addition to the quantum-mechanical tunneling current previously discussed, the ordinary *p–n* junction injection current I_{in} is also being collected. It is well known that this current increases with V in a manner indicated by the dashed curve of Fig. 11.5(a).[5] The total current, given by the sum of the tunneling current and the injection current, is shown in Fig. 11.5(b). We can see from this figure that the total current reaches a minimum value I_v, or valley current, somewhere in the region where the tunneling characteristic meets the forward-diode characteristic. The ratio of this theoretical peak to the valley current, I_p/I_v, can be very high, on the order of 50 to 100. However, for practical diodes this ratio is much lower, in the neighborhood of 15 or less. The excess valley current has a number of origins [3].[6]

B. Tunneling Current

We shall now proceed to utilize expression (11.21) for the transmission cofficient T to calculate the tunneling current. To begin with, were it not for the electric field

---

[4] For simplicity, we have assumed here that the doping of the *p*- and *n*-regions is such that $(E_{vp} - E_F) = (E_F - E_{cn})$.

[5] Note that $I_{in} = I_0(e^{eV/\eta kT} - 1)$ where I_0 is the reverse saturation current, k is Boltzmann's constant and T is the absolute temperature. For Ge diodes, η equals 1. For Si diodes, $\eta \simeq 2$ for small (rated) currents and $\eta \simeq 1$ for large currents. See reference 23 at the end of this chapter.

[6] Bracketed numbers refer to references listed at the end of this chapter.

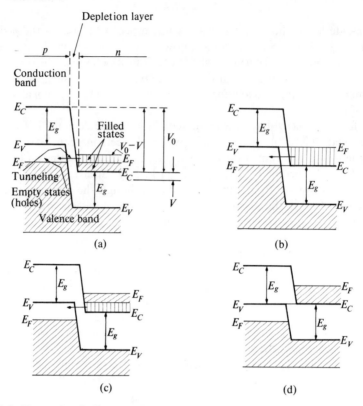

Fig. 11.6. Energy-band diagrams for a tunnel diode under forward bias. As the bias is increased, the band structure changes progressively from (a) to (d).

set up across the depletion layer, an electron in a heavily doped semiconductor would have to acquire an energy at least equal to $E_c - E$ to go from the valence band into the conduction band. Because of the limiting solubility of impurities in the host material it is possible only to have the Fermi level a few kT inside the valence and conduction bands of a tunnel diode. This implies that an electron at or near the Fermi level must surmount an energy barrier roughly equal to the gap energy E_g to go from the valence band into the conduction band. In a heavily doped p–n junction, however, the energy gap barrier is drastically modified by the electric field established across the depletion layer as a result of the locally unneutralized acceptor and donor ions. Although this field is not uniform according to Eqs. (11.4) and (11.7), we can nevertheless use an average field $\mathscr{E}$ for the layer in the calculation to follow, with little loss in physical insight. Accordingly, $\mathscr{E}$ is equal to V_B/d, where V_B is the barrier voltage and d is the depletion layer thickness as before. It then follows that the potential energy $U(x)$ in the depletion region is given by

$$U(x) = E_g - e \frac{V_B}{d} x. \tag{11.22}$$

Since E is assumed to coincide with the zero of $U(x)$ in the spirit of the preceding discussion, Eq. (11.21) becomes

$$T \simeq e^{-2\int_{x_1}^{x_2} \frac{\sqrt{2m[E_g(eV_B/d)x]}}{\hbar} dx}, \qquad (11.23)$$

where we have assumed that $F(E, U)$ is nearly unity.

Before we evaluate the integral (11.23), let us simplify our task by making one more observation. From Fig. 11.6, we see that only particles residing in the energy band between the Fermi level and the band edge participate in the forward tunneling process. Since the width of this band is small compared to the energy gap E_g or the barrier energy eV_B, Figs. 11.4(a) and 11.6 show that $E_g \simeq eV_B$ within the forward tunneling region. It then follows from Eq. (10.22) that $E_g - (eV_B x/d) = 0$ at $x = d$. In other words, the limits of integration x_1 and x_2 in Eq. (10.23) should be replaced respectively by zero and depletion layer width d. Carrying out this integration, we find that [7]

$$T = \exp\left[-\frac{4}{3}\frac{\sqrt{2m^*}}{\hbar}\frac{E_g^{3/2}}{e\mathscr{E}}\right], \qquad (11.24)$$

where we have replaced m by the reduced effective mass $m^* = m_c m_v/(m_c + m_v)$ and m_c, m_v are respectively the effective masses of condution and valence-band electrons.

We can obtain the number of particle transitions N_T per second across the depletion layer by multiplying T by the number of electrons striking the band edge per second. Thus we have

$$N = \frac{e\mathscr{E}a}{\hbar}\exp\left[-\frac{4}{3}\frac{\sqrt{2m^*}}{\hbar}\frac{E_g^{3/2}}{e\mathscr{E}}\right], \qquad (11.25)$$

since particles strike the band edge at the rate of $e\mathscr{E}a/h$ per second, where a is the lattice constant [31].

We can now complete our calculation of the tunneling current. To begin with, at zero bias, the tunneling current density $i_{v\to c}$ from the valence band to the empty states of the conduction band and the current density $i_{c\to v}$ from the conduction band to the empty states of the valence band should be equal. It is reasonable to surmise that $i_{c\to v}$ and $i_{v\to c}$ are of the form [5]

$$i_{c\to v} = A_1 \int_{E_c}^{E_v} f_c(E)\rho_c(E)N_{c\to v}[1 - f_v(E)] \, dE, \qquad (11.26)$$

$$i_{v\to c} = A_1 \int_{E_c}^{E_v} f_v(E)\rho_v(E)N_{v\to c}[1 - f_c(E)] \, dE, \qquad (11.27)$$

where $N_{c\to v}$ and $N_{v\to c}$ are the probabilities of penetrating the gap per second

---

[7] Actually, the assumed potential barrier (11–22) does not conserve momentum. However, the expression for T by using a correct potential differs little from that given by Eq. (11–24). In particular, it has the same $\bar{\varepsilon}$ dependence. See Wang, [30].

(these could be assumed to be nearly equal); $f_c(E)$ and $f_v(E)$ are the Fermi-Dirac distribution functions, namely, the probabilities that a quantum state is occupied in the conduction and valence bands respectively; and $\rho_c(E)$ and $\rho_v(E)$ are the density of states in the conduction and valence bands respectively.

When the junction is slightly biased either in the forward or reverse direction, the net current i will be given by

$$i = i_{c \to v} - i_{v \to c} = A_1 \int_{E_c}^{E_v} [f_c(E) - f_v(E)] N \rho_c(E) \rho_v(E) dE. \qquad (11.28)$$

Since the range of applied voltage involved in the tunneling process is small (< 0.1 V) compared to the contact potential $V_0 (\simeq 0.5$ V), the electric field $\mathscr{E} = (V_0 - V)/d$ will be nearly constant. It follows from Eq. (11.25) that N can be assumed to be nearly constant in Eq. (11.28). If we further assume that the density of states are parabolic functions of energy, as for free electrons in a metal [12], Eq. (11.28) becomes

$$I = A_2 \int_{E_c}^{E_v} [f_c(E) - f_v(E)] N \sqrt{(E - E_c)(E_v - E)} \, dE. \qquad (11.29)$$

The tunneling current calculated from this equation as a function of applied voltage V agrees fairly well with that shown in Fig. 11.5(a), which was physically deduced.

3. TUNNEL-DIODE CHARACTERISTICS

A. General Characteristics

Referring to the tunnel-diode characteristic of Fig. 11.5(b), we see that the diode is an excellent conductor in the reverse direction and also for small forward bias (up to 50 mV for Ge); the resistance is also small, on the order of 5 Ω. In the region between peak current I_p, corresponding to voltage V_p, and minimum or valley current I_v, corresponding to valley voltage V_v, the dynamic conductance $g = dI/dV$ is negative. It is this negative resistance characteristic which enables the tunnel diode to function as a microwave oscillator or amplifier in the operating region between I_p and I_v. At the so-called peak forward voltage V_F the current is again equal to I_p. For current values between I_v and I_p, the tunnel-diode characteristic is triple valued in that the same current can be obtained at three different applied voltages. For this reason it is classified as a voltage-controlled (in contrast to current-controlled) device. It is this multivalued feature which makes the tunnel diode useful in pulse and digital circuitry.

The small-signal equivalent circuit for a tunnel diode operated in the negative-resistance region is shown in Fig. 11.7(a), where $-R_n$ and R_s are respectively the diode negative resistance and series ohmic resistance. The value of the series inductance L_s depends upon the lead length and the geometry of the diode package. The junction capacitance C is a function of the bias and is usually measured at the valley point. Typical values for these parameters for a tunnel diode of peak current $I_p = 10\ mA$ are $-R_n = -30\ \Omega$, $R_s = 1\ \Omega$, $L_s = 5\ \text{nH}$, and $C = 20\ \text{pF}$.

Most commonly, tunnel diodes are made from germanium or gallium arsenide. It is difficult to manufacture a silicon tunnel diode with a high I_p/I_v ratio. For a germanium tunnel diode, typical values are $I_p/I_v = 8$, $V_v = 0.06$ V, $V_p = 0.35$ V, $V_F = 0.50$ V. The voltage values are determined principally by the particular semiconductor used and are almost independent of the current rating. On the other hand, I_p is determined by the impurity concentration and the junction area. For computer applications, devices with I_p in the range of 1 to 100 mA are most common. However, it is possible to obtain diodes whose I_p is as small as 100 μA or as large as 100 A. Significantly, tunnel diodes are found to be several orders of magnitude less sensitive to nuclear radiation than are transistors. This insensitivity to radiation appears to result from the plentiful supply of donors and acceptors and the fact that quantum-mechanical tunneling is independent of carrier lifetime.

The advantages of the tunnel diode are low cost, low noise, simplicity, high speed, environmental immunity, and low power.[8] Its disadvantages are its low output-voltage swing and the fact that, like the parametric amplifier, it is a two-terminal device. Because of the latter, there is no isolation between input and output, which leads to serious circuit design difficulties at frequencies below a few hundred megahertz where ferrite devices are not generally available. Therefore, at frequencies below about 1 GHz or for switching times longer than several nanoseconds, transistors, essentially unilateral devices, are usually employed instead of or in conjunction with tunnel diodes.

At microwave frequencies, 1 GHz or higher, a ferrite circulator can be used to isolate the input and output of the tunnel diode. Briefly, a ferrite circulator is a three-terminal device operated on the circulation principle. Energy entering port 1 emerges from port 2 while energy entering port 2 emerges from port 3. Thus if

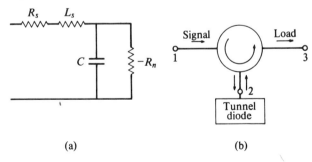

Fig. 11.7. (a) Equivalent circuit of a tunnel diode. (b) Tunnel diode input-output isolation using a ferrite circulator.

---

[8] The chief source of noise in tunnel diodes is shot noise, which is proportional to the dc current. Since tunnel diodes are widely used at low signal levels, they contribute negligible noise to the circuit.

the signal source is connected to port 1, the tunnel diode to port 2, and the load to port 3, as shown in Fig. 11.7(b), we can construct an amplifier with input-output isolation. To be more explicit, signals entering port 1 of the circulator travel toward and emerge from port 2 and go into the tunnel diode. The amplified signal from the diode enters port 2 and proceeds to port 3, finally emerging from port 3 into the load. If there is any reflection from the load, the reflected energy will proceed to and emerge from port 1. Under these circumstances, the source load isolation will not be complete. The ferrite circulator accomplishes this $1 \rightarrow 2 \rightarrow 3 \rightarrow 1$ circulation by having the ferrite sample, a ferrimagnetic material, magnetized by a static magnetic field of proper magnitude so that the phase shifts between different ports are such that wave cancellation occurs for all ports but one. Energy entering a particular port will thus emerge from only one other port, as hypothesized.

B. Stability

Let us now examine the operation of the tunnel diode as an amplifier and as a bistable switch. As with other negative resistance devices, the circuit behavior is intimately related to the nature of the external load. Assume that a tunnel diode is connected between a source voltage V_0 with source resistance R_g and load resistance R_L. The equivalent circuit of such an arrangement is shown in Fig. 11.8(a), where we have utilized the tunnel diode equivalent circuit of Fig. 11.7(a). For simplicity, we shall neglect the inductance L_s and the capacitance C in our analysis. Corresponding to this condition, we can sketch the I–V plot of Fig. 11.8(b), where a load line has been superimposed upon the tunnel-diode characteristic. From the circuit of Fig. 11.8(a), we see that $V_0 = I(R_g + R_L + R_s) + V_d$, where I is the circuit current and V_d is the voltage across the diode, or

$$I = \frac{V_0 - V_d}{R}. \tag{11.30}$$

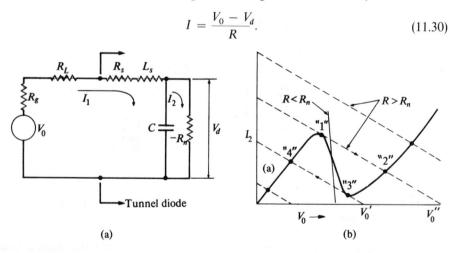

Fig. 11.8. (a) Tunnel diode circuit. (b) Direct-current I–V characteristic of a tunnel diode and associated load lines.

To secure an operating point in the negative dynamic resistance region, the bias resistance $R = R_g + R_L + R_s$ must be such that it intersects the I–V curve at only one point. Since the voltage across a tunnel diode is a multivalued function of current, the bias resistance has to be smaller than the negative resistance to intersect the curve at only one point. Otherwise there will be three intersections, as shown by the dashed load lines of Fig. 11.8(b), and the diode will be in a bistable state. From this bias requirement we have the stability criterion

$$R < R_n. \tag{11.31}$$

On the other hand, if $R > R_n$, there is no unique operating point. The particular point chosen depends on whether the applied voltage has been reached by increasing the voltage from a value below or decreasing the voltage from a value above, the ambiguous or negative resistance range. For example, raising the voltage from zero will cause the operating point to stay on the low voltage part of the curve [sector (a) of Fig. 11.8(b)], and only when the peak current I_p is reached (corresponding to an applied voltage of V_0'') will the operating point be forced into the high-field part of the curve (jumping from point 1 to point 2). Similarly, reducing the applied voltage from a high value, say V_0'', will cause the operating point to proceed along the high voltage part of the curve, reach the point of minimum current (point 3) corresponding to applied voltage V_0', and only then jump to the low-voltage sector (from point 3 to point 4). This hysteresis effect prevents the operating point from being on the negative resistance part of the curve and thus functioning as an amplifier. A sudden application of a voltage of appropriate magnitude may allow the operating point to be in the negative-resistance region. However, since the stability condition embodied by Eq. (11.31) is not satisfied, the system would be extremely unstable. Nevertheless, it is this very instability property that enables the diode to function as a binary computer logic element or as an extremely high-speed bistable switch. Since tunneling takes place at the velocity of light, the transient response is limited only by the total shunt capacitance (junction plus stray wiring capacitance) and peak current.

If we do not neglect L_s and C in the equivalent circuit of Fig. 11.8(a), the analysis is still straightforward but more tedious. Solving the two loop equations of Fig. 11.8(a), we find for the load current

$$i_1 = Ae^{\lambda_1 t} + Be^{\lambda_2 t} + \frac{I_0 - \dfrac{V_0 - V_d}{R_n}}{1 - \dfrac{R}{R_n}}, \tag{11.32}$$

where $R = R_g + R_L + R_s$ and

$$\lambda_{1,2} = \tfrac{1}{2}\left(\frac{1}{R_n C} - \frac{R'}{L}\right) \pm \sqrt{\tfrac{1}{4}\left(\frac{1}{R_n C} - \frac{R'}{L}\right)^2 - \frac{1 - (R'/R_n)}{LC}}. \tag{11.33}$$

In the derivation of Eq. (11.33), we have assumed that the diode is biased at the quiescent point (I_d, V_d), and the negative dynamic resistance at this point is $-R_n$, that is, for a small region around this point, the I–V curve is assumed to be practically linear. Thus the instantaneous current I and voltage V of a small-signal swing within the linear region are related by the approximate diode expression

$$I = I_d + \frac{(V - V_d)}{R_n}. \qquad (11.34)$$

Depending on the values of the circuit components and the parameters of the diode, λ_1 and λ_2 may be real, complex, or imaginary. If the λ's are real and positive, an initial disturbance will grow exponentially with time. If they are real and negative the disturbance will exponentially decay. If they are complex, the transients will be growing or decaying sinusoids. In any event, if either λ_1 or λ_2 or both are real and positive, the circuit is unstable.

If we set $R_g = R_L = 0$, we can obtain from Eq. (11.33) the criteria for the intrinsic stability of a tunnel diode. Setting $R_g = R_L = 0$ in Eq. (11.33), we find

$$\frac{\lambda_{1,2}}{\omega_0} = \frac{1}{2}\left(\frac{1}{Q_d} - \frac{R_s Q_d}{R_n}\right) \pm \left[\frac{1}{4}\left(\frac{1}{Q_d} - \frac{R_s Q_d}{R_n}\right)^2 - \left(1 - \frac{R_s}{R_n}\right)\right]^{1/2} \qquad (11.35)$$

where we have made the substitutions

$$\omega_0 = 1/\sqrt{L_s C}$$

and

$$Q_d = \omega_0 R_n C. \qquad (11.36)$$

A stability diagram based on the nature of the roots of Eq. (11.35) is plotted in Fig. 11.9. The regions of stability and instability are clearly shown. In region 1, the diode is unstable. We note that for $R_s/R_n > 1$ the diode is always unstable; this condition is consistent with the dc stability condition (11.31). In regions 2 and 3, the diode is inherently stable. In region 4, the diode will oscillate sinusoidally.

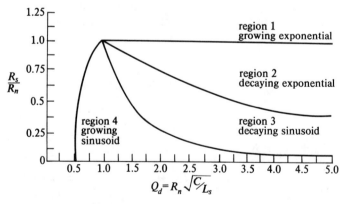

Fig. 11.9. The inherent stability diagram of a tunnel diode.

C. Amplifier Gain

For simplicity, let us temporarily neglect L_s and C in Fig. 11.8(a). In this case, the generator is connected in series with the positive resistance R and negative diode resistance $-R_n$. The I–V curve for the tunnel diode and the resistance R are shown dotted and dashed respectively in Fig. 11.10. Since this is a series circuit, the current must be the same for the diode and positive resistance. The voltages across these elements are in general different, as we can see from the figure. Thus we can find the composite characteristic representing the I–V relationship of the source V_g by adding the resistance and diode voltages at the same current. The resulting curve is shown solid in Fig. 11.10. From this composite curve we can see that a small change in the source voltage ΔV_{in} about an operating point located in the negative resistance region can lead to a large change in output voltage, or ΔV_{out}. Since this is a series circuit, no current gain is possible and the voltage gain $\Delta_{out}/\Delta V_{in}$ is equal to the power gain.

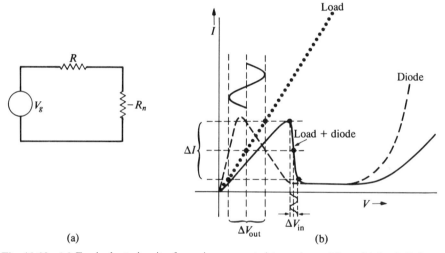

Fig. 11.10. (a) Equivalent circuit of a series-connected tunnel amplifier. (b) Its I–V characteristics.

Tunnel diodes can also be connected in parallel with both a load and a current source. From an analysis entirely analogous to the series-connected case, we can show that current gain is possible and that the current gain is equal to the power gain.

If we do not neglect L_s and C, solving the network problem of Fig. 11.8(a) gives us the complex gain of the amplifier as a function of signal frequency ω:

$$A(\omega) = \frac{R_L\left(-\dfrac{1}{R_n} + j\omega C\right)}{\left(1 - \dfrac{R'}{R_n} - \omega^2 L_s C\right) + j\omega\left(R'C - \dfrac{L_s}{R_n}\right)}. \tag{11.37}$$

If L_s and C are set equal to zero, Eq. (11.37) reduces to

$$A = \frac{R_L}{R' - R_n}. \tag{11.38}$$

Again, we see that in this limiting case, the circuit will become unstable as $R' \to R_n$. Conversely, for high gain, R' and R_n should be nearly equal.

4. ZENER DIODES AND AVALANCHE BREAKDOWN

Figure 11.11 shows a typical I–V characteristic of an ordinary highly-doped p–n junction. According to our observation in footnote 5, the injection current increases with magnitude of the reverse bias $|V|$ but saturates at the reverse saturation current I_0 as $e|V|$ becomes much larger than kT. For comparison this ideal characteristic is also shown (dashed) in Fig. 11.11. It is seen that the typical reverse characteristic departs greatly from the ideal one for voltages larger than V_a. The rapid rise in current with slight increase in voltage above V_a indicates the possible occurrence of some avalanche or particle multiplication phenomenon.

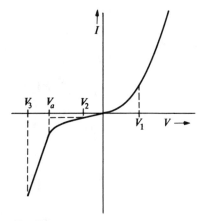

Fig. 11.11. I–V characteristic of an avalanche breakdown diode.

A. Zener Diodes

Two distinct processes may be responsible for the diode breakdown at V_a. The first is the so-called Zener breakdown resulting from the tunneling of electrons between the valence and conduction bands at high reverse bias. We may recall from our discussion of tunnel diodes that a high electric field across the depletion layer, obtained by heavy doping of the semiconductor, is necessary for tunneling to occur. In a diode exhibiting the avalanche phenomenon, high doping is also used to achieve a high electric field across the depletion layer. However, the doping level is considerably smaller than that in a tunnel diode, so that its depletion width is several hundred rather than about 100 Å thick. At this larger width, the Fermi level lies outside the valence and conduction bands so that tunneling in the forward direction does not occur. In the reverse direction, however, tunneling is possible at high reverse-bias voltages since the energy levels on the n-side of the semiconductor move lower relative to those on the p-side as the reverse bias increases. Thus, with high electric fields across the junction, energy levels on the p- and n-sides can become degenerate. It has been found that for Ge and Si diodes the Zener breakdown voltage is usually less than 4 V.

To be more explicit, the reverse current in a tunnel diode increases rapidly as the bias voltage increases from zero, while that in the Zener diode does not show a rapid increase until the breakdown voltage is reached. In the tunnel diode, the doping is so heavy that the Fermi level is inside the valence and conduction bands so that some of the energy levels in the two bands are degenerate even at zero bias. On the other hand, for the Zener diode, the doping level is smaller and the Fermi level lies inside the energy gap. Therefore, no degeneracy between energy levels in the two bands exist until the reverse bias has reached a value sufficient to depress the conduction-band levels on the n-side relative to those on the p-side.

B. Avalanche Breakdown

The second process by which the reverse diode current can increase rapidly beyond a certain applied voltage is attributed to the phenomenon known as *avalanche* or *dielectric breakdown*. The process is essentially the solid state analogue of the Townsend discharge in gases. In a gaseous discharge, electron and positive-ion pairs are produced in an elastic collision process between gas molecules. Similarly, electron-hole pairs can be created in an inelastic collision process involving energetic electrons or holes in semiconductors. In the more sophisticated language of quantum mechanics, we can explain the dielectric breakdown process in semiconductors as follows: In the absence of an external electric field, the few electrons in the conduction band resulting from thermal ionization are in the lowest energy states of this band; under the action of an electric field, they are raised to higher levels. When one of these electrons reaches a sufficiently high level, it will give up energy to an electron in the valence band. This process will then be repeated and the number of electrons in the conduction band will therefore increase exponentially with time as long as the electric field is maintained.

Let us briefly examine the avalanche breakdown process in quantitative terms. For simplicity, consider the case of two parallel plate electrodes a distance d apart, one emitting electrons and the other emitting holes at a rate of n_0 electrons and p_0 holes per unit area per unit time respectively. In a differential distance dx, the number of electron-hole pairs created per square meters per second by electrons and holes respectively is given by

$$dn = dp = \beta_1 n \, dx \qquad (11.39)$$

and
$$dn = dp = \beta_2 p \, dx, \tag{11.40}$$

where β_1 and β_2 are the ionization coefficients for the two processes.

For simplicity and with no loss of generality, we shall assume that only electron emission takes place from electrode A and no hole emission takes place from electrode B. Refer now to Fig. 11.12. Because of ionization, the number of electrons entering the region dx from its left boundary per unit area per unit time is equal to n_1, the number of electrons generated per unit area per unit time between planes A and C plus n_0. Similarly, because of the electron-hole pair created between B and D, holes are crossing the right boundary dx from the right at the rate of p_2 holes per unit area per unit time. Accordingly, Eq. (11.39) becomes

$$dn_1 = \beta_1(n_0 + n_1) \, dx + \beta_2 p_2 \, dx. \tag{11.41}$$

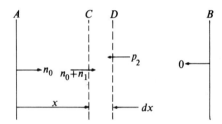

Fig. 11.12. Avalanche multiplication of carriers between electrodes.

Under steady-state conditions, electrons must arrive at electrode B at a rate equal to the total rate at which electrons leave electrode A and are created within the space between the electrodes. Thus, if n_f is the number of electrons arriving at electrode B per unit area per unit time, then

$$n_f = n_0 + n_1 + n_2, \tag{11.42}$$

with $n_2(p_2)$ being the electrons created in the region BD. Eliminating p_2 from Eq. (11.41) by means of Eq. (11.42) gives

$$\frac{dn_1}{dx} = (\beta_1 - \beta_2)(n_0 + n_1) + n_f \beta_2. \tag{11.43}$$

Applying the boundary conditions—$n_1 = 0$ at $x = 0$, and $n_1 + n_0 = n_f$ at $x = d$—to the solution of Eq. (11.43), we find

$$M_e = \frac{n_f}{n_0} = \frac{1}{1 - \int_0^d \beta_1 \exp - \int_0^x (\beta_1 - \beta_2) \, dx' \, dx}, \tag{11.44}$$

where M is generally known as the multiplication factor. In the derivation of Eq. (11.44), we have implicitly assumed that only the n-region is heavily doped, since we set the rate of hole emission from the right electrode, p_0, equal to zero. In this case, the quantity $en_0 S$, where S is the cross-sectional area of the semiconductor, is just the reverse saturation current of the diode with a heavily doped n-region. Because of impact ionization, however, the electron current will be proportional not to n_0 but to n_f. In other words, M_e defined by Eq. (11.44) is the ratio of the diode current to reverse saturation current.

We can determine the breakdown condition by setting the denominator of Eq. (11.44) equal to zero, yielding

$$\int_0^d \beta_1 e^{-\int_0^x (\beta_1 - \beta_2) dx'} dx = 1. \qquad (11.45)$$

We can best gain physical insight into the implications of this equation by considering the case $\beta_1 = \beta_2$. In this case, Eq. (11.45) reduces to

$$\int_0^d \beta_1 dx = 1.$$

This expression indicates that an electron in going from A to B creates one electron-hole pair and the hole thus generated in turn creates another electron-hole pair on its way from B to A. As this process continues, the carriers inside the region A to B multiply themselves and the current increases indefinitely. In other words, avalanche breakdown occurs. We would expect the ionization constants to be a function of the electric field; the electric field in the depletion layer is in turn a function of x. It follows that the β's are also functions of x. Thus, to determine the average electric field at which particle multiplication occurs, we must first determine the β's as a function of $\mathscr{E}$. Since $\mathscr{E}(x)$ can be ascertained from Eqs. (11.4) and (11.7) for fixed concentrations of donor and acceptor ions, Eq. (11.45) can be applied to determine the breakdown electric field. Unfortunately, β_1 and β_2 cannot be measured directly. Instead the multiplication factors M_e and M_h must be measured as a function of $\mathscr{E}$.[9] Then, using Eq. (11.44), we can deduce $\beta_1(\mathscr{E})$ and $\beta_2(\mathscr{E})$. Recalling that $\mathscr{E}$ is a function of x, we can use Eq. (11.45) to determine the breakdown condition, i.e., the electric field distribution at which the integral of Eq. (11.45) is unity. There are a number of ways of measuring M_e and M_h. For example, we can measure the current of a diode in the reverse direction as a function of applied voltage in the breakdown region, or we can measure the photo response of the diode as a function of the applied reverse voltage.

C. Applications

We now consider briefly the applications of a diode known interchangeably as an *avalanche breakdown* or a *Zener diode*. To begin with, the breakdown phenomenon resulting from Zener's tunneling effect generally occurs at applied reverse voltages less than 4 V. If the breakdown voltage is higher than 8 V, the breakdown process

---

[9] For a diode with a heavily doped p-region, the expression for the hole amplification factor M_h has the same form as that for M_e given by Eq. (11.44), except that β_1 and β_2 are interchanged.

is avalanche in nature. If the breakdown voltage is between 4 and 8 V, the two processes are presumably competitive. Due to historical precedence, however, breakdown diodes are sometime called Zener diodes though the breakdown is avalanche rather than tunneling in character. The fact that the avalanche process occurs at a higher voltage than the tunneling process can be appreciated physically as follows: In the tunneling process, the electric field in the depletion region need only be large enough to detach the electron from its parent atom and set it free. In the breakdown process, however, the electric field must be high enough not only to detach an electron from its parent atom but also to accelerate it to a sufficiently high velocity between collisions to enable it to dislodge other electrons from their parent atoms. Whereas the Zener breakdown voltage is rather limited in range (< 4 V), the avalanche breakdown voltage can range from 8 to more than 100 V. We can easily adjust the breakdown voltage by varying the doping of the junction over a wide range. Since the slope of the I–V characteristic beyond breakdown is extremely large, the diodes can be used as voltage regulators to maintain the voltage across the diode terminals at essentially the breakdown value regardless of diode current. The diode can also act as a switch between the points of forward conduction (voltage V_1 of Fig. 11.11) and a point in the reverse saturation current region (V_2 of Fig. 11.11). Alternatively, it can switch between the states of forward conduction (V_1) and avalanche breakdown (V_3). Using the latter method, an ultrahigh-speed diode switch in the millimeter wave region utilizing avalanche breakdown in varactor diodes has been developed [28].

5. AVALANCHE–TRANSIT-TIME DIODES

In the last section, we discussed in some detail the process of avalanche breakdown in a heavily doped semiconductor under high reverse bias. We found that beyond some critical voltage (> 8 V) determined by the doping level of the semiconductor, the current increases rapidly with only a slight additional increase in bias. Since the slope of the associated static I–V characteristic is thus positive in the avalanche region, the question of microwave generation and amplification using avalanche diodes did not arise in the last section. We hasten to add, however, that these well-known observations do not exclude the possibility that the dynamic I–V characteristic may well have a negative slope for some appropriate junction doping profile and external circuit parameters. If this is the case, operating the avalanche diode with the quiescent point in the breakdown region can in principle generate energy at microwave frequencies. In this case, energy is transferred from the dc bias to the oscillations.

A. Comparison Between Read Diodes and p–n Junctions

A possible avalanche-transit-time negative-resistance diode structure as proposed by Read [27] is shown in Fig. 11.13(a). Basically, a reverse bias is applied to establish a space charge or depletion layer of fixed width in a relatively weakly-doped region bounded by highly doped end regions. The structure shown in the

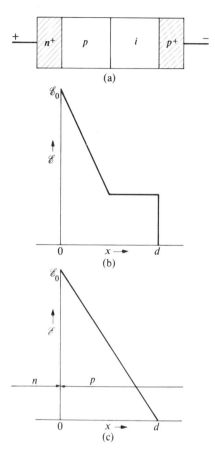

Fig. 11.13. (a) n^+–p–i–p^+ structure proposed by Read. (b) Field distribution in the depletion layer of (a). (c) Corresponding field distribution of n–p diode with a heavily doped n-region.

figure is known as the n^+–p–i–p^+ structure, where + and i denote high doping and intrinsic regions respectively. The electric field distribution in the diode under reverse bias is sketched in Fig. 11.13(b). The bias voltage is always well above the punch-through voltage, so that the space-charge region always extends from the n^+–p junction through the p- and i-regions to the i–p^+ junction.

We can justify the field distribution of Fig. 11.13(b) by our findings in Section 1. First, according to Eq. (11.3), other factors being equal, the width of a heavily doped region is small. Thus in the limit of very high doping in the n^+- and p^+-regions, the field distribution at the n^+–p and i–p^+ boundaries will be very sharp, as depicted in Fig. 11.13(b). Second, according to Eqs. (11.3) and (11.6), the electric field $\mathscr{E}$ ($= -\partial V/\partial x$) is proportional to x as measured from the metallurgical junction, except for an additive constant. Hence $\mathscr{E}$ should vary linearly with x in the p-region, as sketched in Fig. 11.13(b). Recalling that $\nabla \cdot \mathbf{D} = \rho$ from Eq.

(2.3), we see that $\mathscr{E}$ must be a constant in the intrinsic region since there is no net charge there.

Before we examine the physical reasons for the occurrence of negative resistance in a Read avalanche diode, it would be instructive to contrast its field distribution to that of an asymmetrically doped ordinary $n-p$ junction. The electric field distribution for the latter case can be easily sketched in Fig. 11.13(c) with the aid of Eqs. (11.1), (11.4), and (11.7). However, a slight modification of these equations is necessary to fit the $n-p$ (n-region on the left and p-region on the right) instead of the $p-n$ geometry of Fig. 11.1. In the limit of relatively high doping density in the n-region, $\mathscr{E}$ merely rises abruptly at the n^+-p interface and decreases linearly with x in the p-region. Note that this behavior is very much akin to that of the left half of the $\mathscr{E}-x$ distribution in a Read diode, as depicted in Fig. 11.13(b)! In the right half of the $\mathscr{E}-x$ distribution, however, $\mathscr{E}$ in a Read diode is constant while $\mathscr{E}$ in an ordinary $n-p$ junction varies linearly, with x reaching zero at the plane $x = d \simeq d_p$.

Aside from the $\mathscr{E}$ saturation in the intrinsic region, there is another fundamental difference between the field distribution in a Read diode and in an $n-p$ junction. Whereas the width of the depletion layer d in an $n-p$ junction increases with increasing reverse bias according to Eq. (11.10), the width of the depletion layer in a Read diode is constant, independent of bias. This is so because boundaries of the depletion layer cannot easily extend into the n^+- and p^+-regions because of the extremely high doping of these regions. Thus if the voltage is above the punch-through voltage, an increase in bias voltage simply raises the whole field distribution throughout the depletion layer but leaves the shape of the $\mathscr{E}-x$ plot of Fig. 11.13(b) unchanged.

In spite of these obvious differences between the $n-p$ junction and the Read diode, however, microwave radiation can in principle be generated in either type of diode when the diodes are reverse-biased into the avalanche condition. Indeed, it turns out that many different doping profiles are permissible, although the detailed structure of a particular profile does influence such parameters as the oscillation frequency, efficiency, and electronic tuning characteristics. That this is indeed the case will become evident in the discussion to follow.

B. Read Diodes

We can now return to a more detailed discussion of the physical processes leading to the negative resistance in a Read diode. In Fig. 11.13(a), carriers moving in the high field region near the n^+-p junction acquire enough energy to knock valence electrons into the conduction band, thus producing hole-electron pairs. Because of the reverse bias, the electrons generated move immediately into the n^+-region. Holes, on the other hand, move to the right across the space-charge region to the p^+-collector.

The rate of pair production or multiplication is a sensitive nonlinear function of the field, in accordance with Eq. (11.45). With a proper doping profile, the field can have a relatively sharp peak near the n^+–p junction; in this case, avalanche multiplication will be confined to a very narrow region near this junction. Furthermore, at the high fields under consideration the carriers, once created, emerge from the region of creation with a scattering-limited velocity characteristic of a given semiconductor [9]. For holes in Ge, this saturation velocity is 6×10^6 cm/sec at an $\mathscr{E}$ larger than 9×10^3 V/cm. Similarly, it is $\simeq 5 \times 10^6$ cm/sec at an $\mathscr{E} > 2 \times 10^4$ V/cm for holes in Si. Thus the holes, once created, will drift from left to right in the depletion layer with the saturation velocity. In other words, the layer can be divided into two regions: a narrow region on the p-side of the n^+–p junction that supplies the carriers (holes), and a wide region in which the carriers drift at a constant velocity.

The hole current $I_0(t)$ generated at the n^+–p junction will induce a current $I_e(t)$ in the external circuit. Since the velocity v is constant, I_e is simply $v/d = 1/\tau$ times the total charge of the moving holes, according to Eq. (6.25). Here d is the width of the depletion layer and τ is the associated transit time; the width of the generation region is assumed to be much smaller than d. If a pulse of holes of charge ΔQ is suddenly generated at the n^+–p junction, a constant current $I_e = \Delta Q/\tau$ will flow immediately in the external circuit and continue to do so until the holes are collected by the p^+-region τ sec later. On the average, therefore, the circuit current $I_e(t)$ is delayed by $\tau/2$ sec relative to the hole current $I_0(t)$ generated near the n^+–p junction.

Since multiplication is confined to a very narrow high-field region at the n^+–p junction, we can regard the multiplication rate as a function of the peak field $\mathscr{E}_0$. At a critical field $\mathscr{E}_c$, breakdown will occur. Thus any current will be self-sustaining, since every pair of carriers produced will on the average produce one other pair.[10] When the field is above $\mathscr{E}_c$, $I_0(t)$ will be more than self-sustaining and will build up. When the field is below $\mathscr{E}_c$, the current is less than self-sustaining and will die down.

During operation, the diode is biased so that the peak field is above $\mathscr{E}_c$ during the positive half of the voltage cycle and below $\mathscr{E}_c$ during the negative half. Therefore, $I_0(t)$ builds up during the positive half and dies down during the negative half. In other words, $I_0(t)$ reaches its maximum in the middle of the voltage cycle, or one quarter of a cycle later than the voltage.

When the dc bias is sufficiently small that we can neglect carrier space charge, the peak field varies in phase with the voltage since, if the voltage is above the punch-through value, an increase in voltage simply raises the whole field distribution throughout the depletion layer. The peak field in phase with the voltage generates a current $I_0(t)$ at the n^+–p edge of the depletion layer that is delayed from the voltage by $90°$. This current in turn gives rise to an $I_e(t)$ in the external circuit delayed another $\tau/2$ sec from the voltage. To obtain a pure negative resistance, we want a phase angle of $180°$ between the circuit current $I_e(t)$ and V.

---

[10] For simplicity, we shall neglect the much smaller thermally generated reverse saturation current.

To satisfy this condition, therefore, we must have an additional delay of 90° between $I_e(t)$ and $I_0(t)$. In other words, the cavity into which the diode is inserted should be tuned to give a resonance frequency ω_0 in radians,

$$\omega_0 \frac{\tau}{2} = \frac{\pi}{2}, \tag{11.46}$$

or a frequency of $1/2\tau$ cps.

When the current is too large, we can no longer neglect the effect of the carrier space charge. Since the holes are positively charged and the fixed charges in the p region are negatively charged, the effect of the holes will be to oppose the fixed negative charges in the depletion layer and hence flatten the field distribution shown in Fig. 11.13(b).

In particular, the space charge tends to reduce the peak field at the n^+–p junction and hence the rate of carrier multiplication or current generation. Therefore, instead of building up throughout the positive half of $V(t)$, $I_0(t)$ builds up until the carrier space charge has reduced the peak field below the sustaining field $\mathscr{E}_c$ and then decreases. In other words, $I_0(t)$ reaches its peak before the middle of the voltage cycle. Reducing the delay between $I_0(t)$ and $V(t)$ to less than 90° and hence also the power output of the diode. Increasing the current, therefore, increases the ac power only up to the point at which carrier space charge begins to spoil the phase relations. In practice, this effect limits the dc bias current. Presumably, the cavity should be retuned to maintain resonance at high current levels.

C. Diode Characteristics

Since Read proposed his avalanche transit-time structure shown in Fig. 11.13(a), various attempts have been made to realize it in the laboratory. It was not until 1965, however, that a low-frequency version of the semiconductor diode was successfully constructed [20]. Subsequently, a microwave version of the Read diode oscillator was developed and operated at the 5-GHz region [4]. This instrument was an epitaxial version of the Read structure; it had an efficiency of 1 or 2% with output power in the 5-to-10-mW range.

In 1965, however, it was realized that a conventional p–n junction can also have a negative resistance associated with its avalanche-transit-time properties. This discovery led to the construction of avalanche p–n microwave amplifiers and oscillators [4,16]. Differences between the avalanche p–n junction and the Read diode have been described by Misawa [25] and by Hoefflinger [14]. Indeed, we should not be at all surprised to learn that the Read diode and the p–n junction behave in a similar fashion in the avalanche region, at least as far as their negative resistance properties are concerned. We can best understand this behavior by comparing the field distribution plots of Fig. 11.13(b) and (c) for Read and p–n junction diodes. In both cases, there is a region within which the field varies linearly. Since the carriers, once created, drift with the saturation velocity characteristic of the material, the existence of the constant field region in the Read diode but not in the p–n junction is of no real consequence, as long as the region of creation is much smaller compared to the width of the depletion layer.

The behavior of avalanche junction devices with a variety of doping profiles has been studied theoretically by many researches [6,10,11,15,24]. Except for Read's original treatise, which emphasizes large-scale nonlinear behavior, most of the subsequent theoretical work deals mainly with the small signal impedance of the diodes. Electronic tuning effects [10] and amplification [4,17,26] in avalanche microwave diodes have also been theoretically studied; the noise properties of diodes have been studied theoretically [13,22] and experimentally [4,17]. These results show that the noise figure is very high, about 11,000 or 40 db. Therefore, it is unlikely that these diodes will be useful for low-level signal amplification. However, as a high-level amplifier or oscillator, such a noise figure might be useful in some applications. Other studies of microwave generation in avalanche-transit-time diodes include microwave generation using varactor diodes [1]., millimeter wave generation [2,18], frequency conversion with gain [7], frequency locking and modulation [8], high power output [21,29], and efficiency [19].

The devices described can also operate on the Zener tunneling principle, although their power and efficiency will be much lower. Since a variety of depletion-layer doping profiles can give rise to negative dynamic resistance, the abbreviation IMPATT (IMPact ionization Avalanche Transit Time) is sometimes used to describe this class of active devices.

REFERENCES

1. F. A. Brand, V. J. Higgins, J. J. Baranowski, and M. A. Druesue, *PIEEE* **53,** 1276 (1965).

2. C. A. Burrus, *PIEEE* **53,** 1256 (1965).

3. W. F. Chow, *Principles of Tunnel Diode Circuits,* John Wiley and Sons, New York, 1964; p. 17.

4. B. C. DeLoach, Jr. and R. L. Johnston, *IEEE Trans. Electron Devices* **ED-13,** 181 (1966).

5. L. Esaki, *Phys. Rev.* **109,** 603 (1958).

6. S. T. Fisher, *IEEE Trans. Electron Devices* **ED-14,** 313 (1967).

7. Y. Fukatsu and H. Kato, *PIEEE* **57,** 342 (1969).

8. H. Fukui, *PIEEE* **54,** 1475 (1966).

9. W. W. Gartner, *Transistors: Principle, Design and Applications,* D. Van Nostrand Co., New York, 1960, p. 48.

10. M. Gilden and M. E. Hines, *IEEE Trans. Electron Devices* **ED-13,** 169 (1966).

11. H. K. Gummell and D. L. Scharfetter, *Bell Sys. Tech Jour.* **45,** 1797 (1966).

12. C. L. Hemenway, R. W. Henry, and M. Caulton, *Physical Electronics,* John Wiley and Sons, New York, 1963; pp. 245–51.

13. M. E. Hines, *IEEE Trans. Electron Devices* **ED-13,** 158 (1966).

14. B. Hoeffinger, *IEEE Trans. Electron Devices* **ED-13,** 151 (1966).

15. K. M. Johnson, *IEEE Trans. Electron Devices* **ED-15,** 141 (1968).

16. R. L. Johnston, B. C. DeLoach, and B. G. Cohen, *Bell Sys. Tech. Jour.* **44,** 369 (1965).

17. J. G. Josenhans and T. Misawa, *IEEE Trans. Electron Devices* **13,** 206 (1966).

18. S. Kito, *PIEEE* **54,** 1992 (1966).

19. S. R. Kovel and G. Gibbens, *PIEEE* **55,** 2066 (1967).

20. C. A. Lee, R. L. Batdorf, W. Wiegmann, and G. Kaminsky, *Appl. Phys. Lett.* **6,** 89 (1965).

21. S. G. Liu, *IEEE J. Solid State Circuits* **SC-3,** 213 (1968); *PIEEE* **57,** 707 (1969).

22. R. J. McIntyre, *IEEE Trans. Electron Devices* **ED-13,** 164 (1966).

23. J. Millman and C. C. Halkias, *Electronic Devices and Circuits,* McGraw-Hill Book Co., New York, 1967; pp. 149–50.

24. T. Misawa, *IEEE Trans. Electron Devices* **ED-13,** 137 (1966).

25. T. Misawa, *Proc. IEEE* **53,** 1236 (1965).

26. L. S. Napoli and R. J. Ikola, *PIEEE* **53,** 1231 (1965).

27. W. T. Read, *Bell Sys. Tech. Jour.* **37,** 401 (1958).

28. S. Sugimoto, *IEEE Trans. Micro. Theo. Tech.* **MTT-16,** 1017 (1968).

29. C. B. Swan, T. Misawa, and L. Marinaccio, *IEEE Trans. Electron Devices* **ED-14,** 584 (1967).

30. S. Wang, *Solid State Electronics,* McGraw-Hill Book Co., New York, 1966; p. 377.

31. C. Zener, *Proc. Royal Soc. (London)* **145-A,** 523 (1934).

PROBLEMS

11.1 Evaluate the constant A of Eq. (11.16) by normalization of the wave function ψ. If constant F in this problem were not zero, how would you evaluate the arbitrary constants $A, B, C, D, E,$ and F?

11.2 What difficulties would you expect if you were to diffuse, into an intrinsic semiconductor, impurities to the extent of one impurity atom per 10 host atoms?

11.3 a) Referring to Fig. 11.6, sketch the I–V characteristics of a diode for which $(E_F - E_{Cn}) > (E_{Vp} - E_F)$.
b) Repeat (a) for $(E_F - E_{Cn}) < (E_{Vp} - E_F)$.
c) Is it likely that $(E_F - E_{Cn})$ be very different from $(E_{Vp} - E_F)$ for a tunnel diode?

11.4 What is a hole in an acceptor semiconductor and how does it move?

11.5 Referring to Section 5B, justify in more detail the phase relationship between the various quantities in a Read diode.

CHAPTER 12

MICROWAVE GENERATION AND AMPLIFICATION IN BULK SEMICONDUCTORS

Up to this point, we have discussed the process of microwave generation and amplification in junction semiconductors. The question naturally arises as to whether microwave oscillation and amplification can also occur in bulk semiconductors. The answer is yes; in fact, bulk semiconductor devices such as the Gunn and LSA (Limited Space-Charge Accumulation) diodes have been successfully built in recent years. These are bulk devices in the sense that microwave power or gain is derived from the bulk negative-resistance property of uniform semiconductors such as gallium arsenide (GaAs) and indium phosphide (InP), rather than from the negative resistance associated with junctions between semiconductors of different doping density. Compared to junction devices, bulk semiconductors are capable of higher power output, operate at higher frequencies, and switching at higher speeds.

1. GUNN EFFECT

The simplest possible bulk semiconductor device is composed of a bar of uniform n-type GaAs with ohmic contacts at the end faces. It is shown in Fig. 12.1. Microwave emission can occur if a sufficiently high voltage is applied between faces of the sample. To understand the processes involved, we must first examine briefly the Gunn effect [5,6].[1]

In 1963 J. B. Gunn, while studying electrical noise emitted by semiconductors, observed a rather puzzling phenomenon [6]. He found that when he applied a pulse voltage of amplitude V between the ends of a rectangular bar of GaAs or InP, of length L, the pulse current at first rose linearly as V increased from zero; however, if he increased the voltage to a threshold value V_{th}, the current I began to fluctuate with time. Initially, I decreased to some value I_{min} less than the steady state value I_{th} corresponding to V_{th}. This decrease was followed by a rise to a value I_{max} which was usually equal to I_{th}, and then by further fluctuations between I_{min} and I_{max}. For $V > V_{th}$, I continued to fluctuate, with I_{max} usually remaining almost unchanged. For long samples ($L > 0.2$ mm), the fluctuation was almost completely random, resembling white noise with a bandwidth on the order of 1 GHz. Short specimens ($L < 0.2$ mm) behaved similarly with high external

---

[1] Bracketed numbers refer to references listed at the end of this chapter.

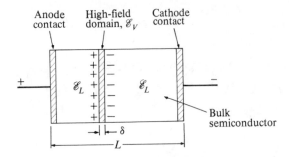

Fig. 12.1. A simple Gunn diode with a high field domain.

circuit impedance, but generated coherent current oscillations with low circuit impedance. Gunn found that the period of these oscillations was equal to the transit time of the electrons calculated from the threshold current I_{th}. These oscillation, with frequencies ranging from 0.5 to 6.5 GHz and I_{min}/I_{max} equal to 0.7 to 0.8, normally built up to full amplitude within one cycle of the oscillation.

Gunn also discovered that the threshold electric field $\mathscr{E}_{th}$ varied with the length and type of material, ranging from 1250 V/cm to 6000 V/cm. The magnitude of $\mathscr{E}_{th}$ in both GaAs and InP and the nature of the instability in GaAs were unaffected by the nature of the contacts, specimen surface condition, light irradiation, or application of a moderate static magnetic field. Gunn found that in InP, the nature of the instability was partly dependent on the contact material; for heavily-doped n^+-contacts, the instability closely resembled that in GaAs. For lightly-doped contacts, however, the initial decrease of current was rapidly followed by an increase to a value larger than I_{th}, and hole injection was found to have occurred.

Before we consider the theoretical explanation of the Gunn effect, we shall further examine the related experimental results. Since the period of the observed oscillations is equal to the transit time of the electrons through the sample, it seems reasonable to associate the Gunn effect with the propagation of an electric field between the sample contacts. Indeed, we can explore the changing potential distribution $V(x, y, t)$ over the plane surface of the sample by a novel capacitive-probe technique [5]. The separation of the probe from the sample is kept small and constant, but its position can be changed by a micrometer stage. The signal from the probe is fed into a sampling oscilloscope. However, since it is difficult to interpret the normal sampling oscilloscope presentation of the sample voltage as a function of time (t), we must utilize alternative forms of display. In one such method the probe position is held fixed while the signal is electronically integrated and then displayed on an oscilloscope. The display thus represents $V(x, y, t)$ with x and y held constant. In another method, the instant of sampling t it held fixed while x is varied over the length of the specimen. The signal is applied to the vertical plates of an oscilloscope whose horizontal deflection is proportional to x. This display gives $\partial V(x, y, t)/\partial t$ as a function of x, the distance along the specimen, with y and t held fixed. Experimental results of the first and second kinds of display are given in Figs. 12.2, 12.3 and 12.4 respectively.

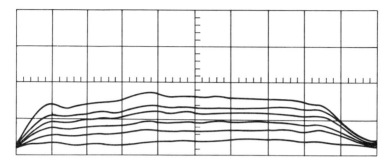

Fig. 12.2. Potential V as a function of time t at points (reading upward) $x = 0, 40, 80, 120, 180, 240\,\mu$ when $I < I_{th}$. Time scale: 1 major division $= 5 \times 10^{-10}$ sec. (After J. B. Gunn, [5].)

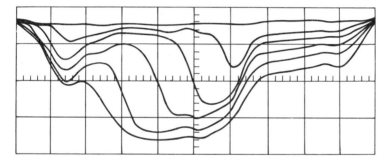

Fig. 12.3. Same data as Fig. 12.2 except that $I > I_{th}$. (After J. B. Gunn, [5].)

Gunn made measurements on a specimen of n-type GaAs of length $L = 210\,\mu$ and cross-sectional area $3.5 \times 10^{-3}\,\text{cm}^2$ with a low-field resistance of 16 Ω. He used a circuit with an impedance of 50 Ω to apply rectangular positive pulses of a few nanoseconds' duration, achieving nearly constant-current conditions, though current instabilities occurred above a specimen threshold voltage V_{th} of 59 V. Figure 12.2 shows the waveform $V(t)$ at equal intervals of x for V just less than V_{th}. We can see that under these circumstances $V(t)$ at any point x along the length of the specimen merely reproduces that at $x = L$ multiplied by x/L, as expected of a homogeneous conductor. For $V > V_{th}$, Fig. 12.3 shows that we obtain very different results which exhibit current instability. When the instability begins at about the middle of the pulse, $V(t)$ at L rises sharply, remains high until almost the end of the pulse, and then drops rapidly again. This variation does not occur at other points in the specimen. A roughly equal rise in potential of about 55 V occurs simultaneously at all points, but the drop takes place earlier at smaller values of x. At a given value of x, the values of V before the rise and after the fall are approximately equal.

216 Gunn effect [1

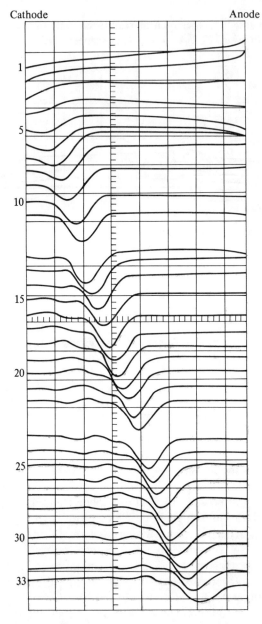

Fig. 12.4. $\partial V/\partial t$ as a function of x. Top trace represents a measurement at a fixed time near the beginning of the applied pulse. Subsequent traces represent the same quantity at successively later instants, separated by intervals of 6.6×10^{-11} sec. Horizontal scale: 1 major division = $26.5\ \mu$. (After J. B. Gunn,[5].)

A more startling display of the information contained in Fig. 12.3 is shown in Fig. 12.4. Here $\partial V/\partial t$ is a function of x, with t a constant parameter. Since V

is slightly larger than it is in Fig. 12.3, the instabilities start at the beginning rather than at about the middle of the current pulse. In trace 1, $\partial V/\partial t$ with t fixed increases approximately linearly with x, indicating that the electric field ($= -\partial V/\partial x$) is building up uniformly in the specimen. In trace 2, delayed from trace 1 by 6.6×10^{-11} sec, this linear distribution is beginning to undergo a distortion. By trace 11, the distortion has taken the form of a well-defined negative spatial pulse extending over a distance of about 30 μ. Elsewhere in the crystal, $\partial V/\partial t = 0$ (the zeros of $\partial V/\partial t$ for various values of time are separated from each other for clarity of presentation). From this instant on, the negative maximum of the negative pulse propagates in the $+x$-direction or in the direction of electron flow, with no change in slope, until it reaches the anode shortly after trace 33.

The constant vertical separation between successive traces corresponds to a fixed time interval of 6.6×10^{-11} sec, and the horizontal coordinate is proportional to x. Thus the inverse slope of a line joining the negative pulse maxima is equal to the velocity of the pulse, 8×10^6 cm/sec. The value is about equal to that estimated for the drift velocity of electrons at the threshold. Also, we can roughly estimate the average electric field within the "shock" by taking the ratio of the specimen voltage and the pulse width, yielding a value of about 2×10^4 V/cm. Since the $\partial V/\partial t$ pulse travels at about the same velocity as the drifting electrons in the GaAs semiconductor and $\partial V/\partial t$ does not depend on the transverse coordinate y, we might surmise that the Gunn effect is related to a high-field domain drifting in the axial direction of the specimen with about the same velocity as that of the electron stream. However, it is intriguing to note that the shape of the spatial pulse remains unchanged as it travels toward the anode. We shall now, therefore, consider possible explanations of the phenomenon observed by Gunn.

2. TWO-VALLEY MODEL

Many explanations have been offered for the Gunn effect. In 1964, Kroemer [12] pointed out that Gunn's observations were in complete agreement with the earlier prediction of Ridley, Watkins, and Hilsum based on the so-called two-valley energy band model of a semiconductor. Ridley and Watkins had considered the possibility of finding negative resistivity in certain bulk semiconductors in 1961 [15]. Ridley later correctly predicted that the instability of space charge in a bulk semiconductor would cause it to break up into domains of high and low electric fields, the former drifting along with the carrier stream and producing oscillations [14]. Almost at the same time, Hilsum hypothesized that bulk negative resistivity existed in n-type GaAs and calculated a threshold field of 2800 V/cm, very close to what is now observed [10].

At this point, let us pause to summarize Gunn's observations in order to understand what a valid theory must encompass. His measurements showed that when the sample voltage was above the threshold value, a high-field domain formed near the cathode that reduced the electric field in the rest of the specimen and caused the current to drop to about two-thirds of its maximum value. The high-field domain then drifted with the carrier stream across the sample and disappeared at the anode contact. As the old domain disappeared at the anode, the electric field behind it increased (to keep the voltage $= -\int \mathscr{E}\, dx$ constant) until the threshold field was reached and the current regained its initial value. A new domain would then form at the cathode, accompanied by a current drop, and the cycle would begin anew. A current waveform produced by this type of operation is shown in Fig. 12.5 [2].

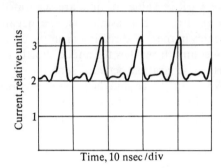

Fig. 12.5. Current waveform due to Gunn oscillations. (After J. A. Copeland, [2].)

Let us now return to a discussion of the two-valley model of semiconductor. First, "two-valley" refers to a minima in a total energy versus wave number (E–k) diagram.[2] These minima appear because of the periodic nature of the crystal lattice. If it were not for the discrete nature of the lattice, the potential energy V of an electron would be a constant and E and k would be simply related by

$$E = \frac{\hbar^2 k^2}{2m} + U, \qquad (12.1)$$

where the momentum p or mv is equivalent to $\hbar k$ in quantum mechanics. Except for the arbitrary constant potential reference U, E is a continuous parabolic function of k. We can easily verify this relationship by noting that Eq. (12.1) satisfies Schrödinger's time-independent equation:

$$\nabla^2 \psi + \frac{2m}{\hbar^2}(E - U)\psi = 0, \qquad (12.2)$$

if ψ is of the form $e^{j\mathbf{k}\cdot\mathbf{r}}$. If U is a function of the spatial coordinate $\mathbf{r}$, however, the E–k relation will in general be transcendental. Furthermore, in contrast to the E–k relation given by Eq. (12.1) where any value of E is permissible, there are ranges of E in this case for which there is no permissible solution of Eq. (12.2) that also satisfies the proper boundary conditions of the lattice. In other words,

---

[2] Since $E = \hbar\omega$ in quantum mechanics, where $\hbar$ is Planck's constant h divided by 2π, and k is synonymous with the phase constant β, an E–k diagram is equivalent to an ω–β diagram of microwave periodic structures discussed in Chapter 8.

the E–$\mathbf{k}$ diagram now contains regions of permissible energy separated by forbidden zones. This situation is analogous to the case of periodic microwave structures discussed in Chapter 8, for which we found that only certain ranges of ω are permissible solutions of Maxwell's equations satisfying appropriate boundary conditions. Recall that, in that case, the forbidden zones occur when the spacing between obstacles in a guide corresponds to an integral multiple of a half-wavelength. Under such circumstances, the reflections from the various obstacles are all in phase, resulting in constructive interference in the reverse direction and no propagation in the forward direction.

Since there are many possible crystal lattice structures and spacings, we can expect the E–$\mathbf{k}$ diagram to be different for different substances and different directions of $\mathbf{k}$ with respect to the crystalline axes. An important phenomenon in real crystals which a one-dimensional model does not exhibit is the "overlapping" of bands, i.e., the lowest energy of one band may be lower than the highest energy of the next lower band. When overlap occurs, the crystal behaves essentially as if there were no energy gap, and the allowed energy values are practically continuous. Overlap always occurs among the higher lying bands of real crystals. For example, the conduction and valence bands in semiconductors consist of a number of subbands; for normal conduction, only the subbands with the lowest carrier energy are effective, while the next highest subband may be separated from the lowest subband by a large fraction of an electron volt. Carriers in the subbands have different resulting effective masses dependent upon the curvature of the bands and the direction of the applied electric field. The average effective mass and hence the conductivity of the crystal depends upon the relative population of each subband, which is in turn a function of carrier temperature. The carrier temperature can then be varied with an applied electric field. Under certain conditions, we can use this effect to obtain a negative resistance in semiconductors.

The conduction band in the E–k diagram of Fig. 12.6 consists of only two subbands, a and b. We have also assumed that the $\mathbf{k}$ direction is along a principal axis of the crystal and parallel to the electric field $\mathscr{E}$. The effective masses and mobilities are scalar quantities denoted by m_a, m_b and μ_a, μ_b for the subbands a and b respectively.[3] If the electron density in the subbands is n_a and n_b, the conductivity of the crystal is

$$\sigma = e(\mu_a n_a + \mu_b n_b), \qquad (12.3)$$

where e is the electronic charge. When we apply a sufficiently high electric field $\mathscr{E}$ to the crystal, the electrons are accelerated and their effective temperature rises above the lattice temperature, and in general, we must also expect the lattice temperature itself to increase. As a result, not only the mobilities but also the

---

[3] The scalar effective mass m^* is defined as equal to $\hbar^2/(\partial^2 E/\partial k^2)$. (See Hemenway, et al., [9].) In general, it is a tensor implying that its value is dependent upon the relative directions of $\mathbf{k}$, ε, and the crystalline axes.

electron densities in the bands will be altered. In other words, μ_a, μ_b and n_a, n_b are all functions of the applied field $\mathscr{E}$. By a straightforward differentiation of Eq. (12.3) with respect to $\mathscr{E}$, we have

$$\frac{d\sigma}{d\mathscr{E}} = e\left(n_a \frac{d\mu_a}{d\mathscr{E}} + n_b \frac{d\mu_b}{d\mathscr{E}}\right) + e\left(\mu_a \frac{dn_a}{d\mathscr{E}} + \mu_b \frac{dn_b}{d\mathscr{E}}\right). \tag{12.4}$$

If we let $n_a + n_b = n$, where n is the constant total electron density of the two subbands, and assume that μ_a, μ_b are proportional to $\mathscr{E}^p$, where p is a constant, Eq. (12.4) becomes

$$\frac{d\sigma}{d\mathscr{E}} = e\left(\mu_a n_a + \mu_b n_b\right)\frac{p}{\mathscr{E}} + e\left(\mu_a - \mu_b\right)\frac{dn_a}{d\mathscr{E}}. \tag{12.5}$$

Differentiating Ohm's law $i = \sigma\mathscr{E}$ given by Eq. (4.24), where i is the current density, we find the relation

$$\frac{1}{\sigma}\frac{di}{d\mathscr{E}} = 1 + \frac{d\sigma/d\mathscr{E}}{\sigma/\mathscr{E}}. \tag{12.6}$$

Negative resistance occurs if i decreases with increasing $\mathscr{E}$, or with $di/d\mathscr{E}$ negative. This would be the case if

$$-\frac{d\sigma/d\mathscr{E}}{\sigma/\mathscr{E}} > 1, \tag{12.7}$$

according to Eq. (12.6). Combining Eqs. (12.3) and (12.5) with Eq. (12.7), we arrive at the condition for negative resistance in terms of the n's, μ's, and $\mathscr{E}$

$$\left[\left(\frac{\mu_a - \mu_b}{\mu_a + f\mu_b}\right)\left(\frac{\mathscr{E}}{n_a}\right)\left(\frac{dn_a}{d\mathscr{E}}\right) - p\right] > 1, \tag{12.8}$$

where $f = n_b/n_a$.

Let us examine the various terms in Eq. (12.8). First, we expect $dn_a/d\mathscr{E}$ to be negative, since we have seen that more and more carriers will be transferred from the lower energy subband a to the higher energy subband b as $\mathscr{E}$ is increased. It follows that implementation of the inequality (12.8) will be facilitated if $\mu_a > \mu_b$. In other words, *electrons must begin in a low mass (high μ) band and transfer to a high mass (low μ) band when heated by the applied field.* On the other hand, the field exponent p is a function of the scattering mechanism and should be negative and large. This makes impurity scattering quite undesirable, since when it is dominant the mobility rises with increasing field, rendering p positive. When lattice scattering is dominant, however, p is negative and will depend on the lattice and carrier temperatures.

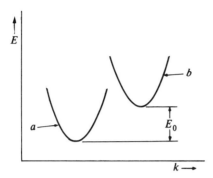

Fig. 12.6. Energy-band diagram for a two-valley semiconductor.

On the basis of the above discussion, a little reflection will show that the band structure of a semiconductor must satisfy three criteria in order to exhibit negative resistance [2]. They are:

1. The energy difference between the bottoms of the upper and lower subbands (E_0 of Fig. 12.6) must be several times larger than the thermal energy, which is about 0.025 eV at room temperature. Otherwise the upper subband will be highly populated as a result of thermal excitation, even in the absence of an electric field. In other words, other factors being equal, the electric field will not be as effective in transferring electrons between the subbands.
2. The energy difference between the subbands must be smaller than the energy difference between the conduction and valence bands. Otherwise the semiconductor will break down (become highly conductive because of hole-electron pair generation) before the electrons begin to transfer to the upper subband.
3. The electron velocities dE/dk must be much smaller in the upper than in the lower subband. When the electric field is above the threshold value defined in connection with the Gunn effect, electrons are mostly in the upper subband and the average electron velocity decreases to a more or less constant value which is 8×10^6 cm/sec for GaAs.

Unfortunately, the two most common semiconductors, silicon (Si) and germanium (Ge) do not meet all of these criteria. They do appear to be fulfilled by some compound semiconductors, including gallium arsenide (GaAs), indium phosphide (InP), and cadmium telluride (CdTe), but not by others, such as indium arsenide (InAs), gallium phosphide (GaP), and indium antimonide (InSb). Since this field of microwave generation in bulk semiconductors is still in its infancy, the search for appropriate semiconductors will undoubtedly continue.

We can now show that negative differential resistance does indeed occur within a certain range of field values. When $\mathscr{E}$ is below the threshold value $\mathscr{E}_{th}$, most of the electrons are in subband a. Thus, according to Eq. (12.3), the conductivity σ is proportional to the a subband mobility μ_a, which is in turn equal to the drift velocity per unit $\mathscr{E}$. On the other hand, we can see from Fig. 12.2 that σ

is a constant. It therefore follows that μ_a is constant or that the drift velocity v_d is porportional to $\mathscr{E}$ for $\mathscr{E} < \mathscr{E}_{th}$, as depicted by sector 1 of Fig. 12.7. For $\mathscr{E} > \mathscr{E}_{th}$, electrons in subband a will be transferred to subband b, which has a higher effective mass and hence a lower mobility. It follows that v_d must now decrease with increasing $\mathscr{E}$ as indicated by sector 2 of Fig. 12.7. As $\mathscr{E}$ increases, however, a point will be reached at which all the electrons have been transferred to subband b. Beyond this point, v_d again increases more or less linearly with increasing $\mathscr{E}$ since the electrons, now all in subband b, should behave much the same as when they were all in subband a, except that they will have a lower mobility. Thus the curve in sector 3 of Fig. 12.7 rises with a smaller slope than that of sector 1. We hasten to point out that Fig. 12.7 definitely indicates that negative conductivity or resistance exists within a certain range of $\mathscr{E}$, since the conductivity σ is merely proportional to the differential slope of the v_d-$\mathscr{E}$ curve.

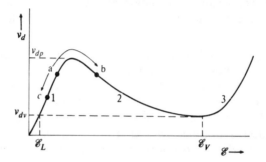

Fig. 12.7. Drift velocity of conduction band electrons in a two-valley semiconductor vs. electric field.

3. HIGH-FIELD DOMAINS

In the last section, we showed how differential negative resistance can occur when an electric field within a certain range of values is applied to an appropriate multivalley semiconduction such as that provided by GaAs. Within this range of electric field values, hot electrons are transferred from a lower subconduction band to a higher subconduction band, giving rise to a negative differential mobility, as illustrated in Fig. 12.7. We shall now demonstrate how a decrease in drift velocity with increasing electric field can lead to the formation of domains of high electric field, the propagation of which in turn gives rise to microwave generation.

Consider the effect of a dipole layer which may be nucleated due to some local fluctuation of carrier density in the semiconductor, as sketched in Fig. 12.1. Assume that the sheets of charge lie along equipotentials, so that we need consider field changes only along the length of the sample. Since an electric field extends from a positive to a negative charge layer, the field inside the dipole layer is increased above the value in the absence of the layer. If the applied voltage V

is such that the electric field $\mathscr{E}$ before domain formation is above the threshold value $\mathscr{E}_{th}$, both the differential resistance and differential drift velocity should be negative. In this case, the initial increase in electric field within the layer decreases the drift velocity v_d of the electrons within the layer.

On the other hand, since the applied voltage $V = \int_0^L \mathscr{E}\, dx$ is constant, an increase in electric field within the layer must be accompanied by a decrease in electric field in the rest of the sample. Because of the negative differential drift velocity, the electrons outside the dipole layer will initially speed up. Thus there will temporarily be more current flowing into the left boundary of the dipole layer than leaving it, and the amount of positive charge at this boundary will build up with time. Similarly, the amount of negative charge at the right boundary will also increase with time, since temporarily less current is flowing into the left boundary than leaving it. As a consequence of this buildup of positive and negative charges, the electric field in the dipole layer increases further with time; correspondingly, the electric field outside the layer decreases further with time, since the applied voltage $V = -\int_0^L \mathscr{E}\, dx$ is assumed fixed. This process presumably continues until the current inside and outside of the dipole region is again equalized.

More specifically, let us refer again to Fig. 12.7 and assume that before the appearance of the dipole layer, the uniform electric field inside the sample is everywhere equal to that corresponding to point a on the v_d-$\mathscr{E}$ curve. As $\mathscr{E}$ increases in the dipole layer, the operating point b of the dipole layer moves along the curve to the right, while that of the rest of sample c moves, also along the curve, to the left. This process will continue until point b is at the valley corresponding to the drift velocity v_{dv} of the v_d-$\mathscr{E}$ curve, as shown in Fig. 12.7. Correspondingly, the location of point c is such that v_d and therefore the currents inside and outside of the domain are equalized. Since this equalization can only occur if points b and c are on opposite sides of the peak point v_{dp}, the region outside the dipole layer must have positive resistance. The result is a stable situation, although markedly inhomogeneous fields are set up in the crystal. Inside the dipole layer of negative resistivity is the high-field domain, where the electric field may be greater than 60,000 V/cm. Outside the domain, however, the resistivity is positive and the field is below threshold (< 3000 V/cm).

If the initial operating point is in the positive rather than the negative differential resistance region, a smaller current initially flows outside the dipole layer than inside it; consequently, the charge layer decays until a uniform field is established again. From the above discussion, we can see that a high-field domain is merely a manifestation of a local fluctuation in the carrier charge density. Since carriers travel down the sample with a drift velocity v_d, we would expect a domain to drift along the sample at the same velocity, as has been experimentally verified [5]. Thus a domain, once created at some nucleation center, will move continuously along the crystal with a drift velocity characteristic of the material and disappear at the anode of the device if the majority carriers are electrons, as in the case of an n-type material. For a p-type material, the domain movement is in the opposite direction because the majority carriers are now holes. As soon as the domain disappears at the anode, the electric field is again uniform throughout the sample. A domain may then be nucleated again and the propagation process will repeat.

4. MICROWAVE GENERATION

Figure 12.5 shows a current waveform produced by the type of operation described above. The waveform's flat valley forms as the domain drifts across the sample, the upward spikes form as the domain reaches the anode, and a new domain forms at the cathode. The small current fluctuation in the valley occurs as a domain passes through regions in the diode of varied doping or cross-sectional area. The nucleation centers are probably located where there is a locally high field resulting from either a geometric irregularity (near the contact regions, say) or an inhomogeneity in the resistivity. In sufficiently short samples (< 0.2 mm) only one nucleation center will be active, causing the coherent oscillations shown in Fig. 12.5 with a frequency approximately equal to v_d/L, where L is the length of the sample. With a typical drift velocity of 10^7 cm/sec, the corresponding frequency is larger than 0.5 GHz. Diodes can be made with $L = 10$ μm or 10 μ with a corresponding oscillation frequency of 10 GHz. However, we cannot stabilize voltage-controlled bulk-type negative conductance by loading it with a sufficiently low impedance, as we can the interface-type negative conductance of a single-tunnel diode. As may be expected, the former behaves essentially like a large number of series-connected tunnel diodes.

To increase the power content at the fundamental frequency, we must make the wave shape shown in Fig. 12.5 more symmetrical. For example, if the diode is made shorter, flat current valleys will become narrower, since the domains will spend less time traveling through the sample. We can also show that the width of the upward spike remains unchanged. From Figs. 12.1 and 12.7, we have

$$\mathscr{E}_v \delta + \mathscr{E}_L(L - \delta) = V, \tag{12.9}$$

where δ is the width of the domain. Solving for δ/L, we have

$$\frac{\delta}{L} = \frac{\mathscr{E} - \mathscr{E}_L}{\mathscr{E}_v - \mathscr{E}_L}, \tag{12.10}$$

where $\mathscr{E} = V/L$ is the average field within the sample. For a sample of given doping, $\mathscr{E}_v$ and $\mathscr{E}_L$ are both fixed. It follows that δ/L and hence the width of the upward spike will remain unchanged with changing L if V is adjusted for each L to give the same $\mathscr{E}$. We can also improve the waveform symmetry without changing the frequency by reducing sample doping. This widens the upward spike at the expense of the valley because the domain becomes wider and thus takes longer to build up and disappear. Note also that high efficiency can be obtained from the diode if we prevent the formation of a new domain by decreasing the voltage slightly below threshold. Thus, if new domains are inhibited from starting for a time equal to the domain transit time, the waveform will approximate a symmetrical square wave. The highest efficiencies reported, which approach 20%, have been obtained in this manner by operating diodes in a resonant circuit tuned to half the transit-time frequency. The output power obtained on a pulse basis was approximately 100 W at 1 GHz [4].

Large Gunn-effect oscillators are produced by cutting wafers (thin disks) and then strips from a piece of bulk-grown material; the contacts are made by alloying balls of indium or tin to the ends. The best devices for practical applications are produced by growing thin films of n-type GaAs on a substrate of either high-conductivity or high-resistivity GaAs. Contacts to the epitaxial film are best made after growing thin high-conductivity layers on the surface. Devices with an active layer thickness of from 20 to 80 μ can produce 0.1 W c.w. power with 3% efficiency at frequencies from 6 to 15 GHz and with a tuning range as high as 2 to 1 [1]. The power and efficiency of thinner diodes decrease rapidly at higher frequencies.

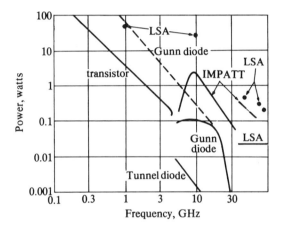

Fig. 12.8. Maximum power vs. frequency for a number of solid state oscillators as of 1966. Solid lines are for continuous room-temperature operation; dashed lines and points are for pulsed operation. Gunn and LSA results are from bulk n-GaAs diodes. (After J. A. Copeland [2].)

For frequencies below 6 GHz, a GaAs diode is too thick for heat removal through contact surfaces without excessive temperature rise in the center of the sample. This effect can be alleviated by making diodes from thin films on high-resistivity substrates so that heat flow can be perpendicular to the current. Figure 12.8 shows maximum power curves obtained from Gunn diodes under both c.w. and pulsed conditions as a function of frequency. For comparison, similar characteristics for other solid-state oscillators such as transistors, tunnel diodes, IMPATT (avalanche) diodes, and LSA devices (to be discussed in the next section) are also given.

5. LIMITED SPACE-CHARGE ACCUMULATION MODE (LSA)

By properly designing a bulk n-GaAs diode and the corresponding resonant circuit, we can prevent domains and other types of space charge from building up within the diode. This mode of operation, called Limited Space-charge Accumulation (LSA), makes it possible to build oscillators with higher frequencies, and with higher power at a given frequency, than can be obtained with a transit-time device [3].

Devices such as transistors and IMPATT (avalanche) diodes, as well as Gunn-effect oscillators, must be thin enough so that carriers can move through the active region in a time corresponding to one cycle or less. This means that we must decrease the thickness and voltage to raise the frequency of operation. To maintain a reasonable impedance, the current must also be decreased. The result of these considerations is that the maximum power for a given transit-time or subtransit-time device falls off faster than $1/(\text{frequency})^2$, as shown in Fig. 12.8. The LSA oscillator is the first practical solid-state oscillator that is free of this limitation, since in principle it can be made thick compared with the distance a carrier drifts in one period of the oscillation.

For LSA operation of a bulk n-GaAs diode, the voltage must swing below the threshold voltage long enough during each cycle to dissipate space charge. Also, the part of each cycle during which the voltage is above threshold must be too short for space charge to build up and form a domain. Since the speed of space-charge dissipation and growth is proportional to the doping, the ratio of doping to frequency must be within 2×10^4 to 2×10^5 cm^3/sec. The circuit must be lightly loaded to achieve the necessary r.f. voltage swing. For high efficiency, the doping should be uniform within 10%.

Preliminary experiments with epitaxial LSA oscillators have produced 0.7 W with 9% efficiency at 50 GHz on a pulse basis and 0.02 W with 2% efficiency at 88 GHz on a continuous basis. Experiments with bulk-grown diodes have yielded 33-W pulses at 10 GHz, leading to predictions that 250-kW pulses from a single block of n-GaAs are theoretically possible up to 100 GHz [11]. The upper frequency limit for LSA operation of a GaAs diode has not been determined, but it seems certain that an appreciable amount of power can be produced at frequencies of several hundred GHz, well into the millimeter-wave range.

6. MICROWAVE AMPLIFICATION

Not nearly as much work has been done on Gunn-effect amplifiers as on oscillators. As is the case with IMPATT (avalanche) diodes, the minimum noise figure obtained to date is quite high, on the order of 20 db above the thermal limit. This noise is largely the result of internal amplification of thermal noise because of negative resistivity [16].

In 1965, it was discovered that an n-type GaAs diode can amplify signals in the vicinity of the transit-time frequency without oscillation if the product (doping) × (length) is of the order of 10^{12}/cm^2 or less [18], consistent with the result of theoretical calculations [13]. In this mode of operation, the diode is filled with an excess of carriers and the electric field continuously increases from the cathode to the anode. The dc resistance is always positive, but an applied r.f. signal will set up space-charge waves that grow at subthreshold voltages because of negative resistivity [8]. Beyond a certain maximum bias, the space-charge waves may become traveling domains which produce oscillations, as described in Section 4. As larger bias is applied, but before the traveling domain is fully grown, the semiconductor medium outside the domain may once again become active, giving rise to the possibility of postthreshold amplification "external" to the traveling domain. Indeed, researchers have operated a GaAs diode simultaneously as a microwave amplifier, mixer, and oscillator under c.w. conditions [7] by biasing the sample at postthreshold conditions, i.e., beyond the bias at which the sample oscillated. They found that even when the device was oscillating at a certain frequency under these conditions, it still retained the properties of a linear active element capable of mixing and amplification at other frequencies. The frequencies at which the device exhibited simultaneous amplification, mixing, and oscillation were not harmonically related but rather determined jointly by the semiconductor and microwave circuit. In this particular experiment, a maximum amplification of some 40 db was obtained at frequencies of 3.622 and 2.492 GHz for the subthreshold and postthreshold cases respectively. For bias above the subthreshold region, the oscillation frequency was 3.622 GHz, the same as the subthreshold amplification frequency.

If the product (doping) × (length) is too large ($> 10^{12}/cm^2$ for GaAs) for stable amplification, oscillations occur and the diode exhibits a negative resistance at direct current and at frequencies above and below the oscillation frequency. This is due to the fact that if additional voltage is applied to a device containing a domain, the domain will increase in size, according to Eq. (12.10), and absorb more voltage than was added with a consequent decrease in current. This effect can be utilized to construct amplifiers that are linear until the output power approaches the value that can be obtained from the oscillation [17]. Five-db amplification at 6 GHz was achieved with 1-db compression when the output power reached 60 mW, which is an approximately 15-db increase in linear range over the stable amplifier diode. [For GaAs, an increase of this type requires that (doping) × (length) $< 10^{12}/cm^2$ as previously discussed.]

REFERENCES

1. D. P. Brady, S. Knight, K. L. Lawley, and M. Uenohara, *Proc. IEEE* **54**, 1479 (1966).

2. J. A. Copeland, *IEEE Spectrum* **4**, 71 (1967).

3. J. A. Copeland, *Proc. IEEE* **54**, 1497 (1966); *Bell Sys. Tech Jour.* **46**, 284 (1967); *Jour. Appl. Phys.* **38**, (1967).

4. D. G. Dow, C. H. Mosher, and A. B. Vane, *IEEE Trans. Electron Devices* **ED-13,** 105 (1966).
5. J. B. Gunn, *Plasma Effects in Solids* (ed. J. Bok), Academic Press, New York, 1965; p. 199.
6. J. B. Gunn, *Solid State Commun.* **1,** 88 (1963); *IBM Jour. Res. and Develop.* **8,** 141 (1964).
7. B. W. Hakki, *Proc. IEEE* **54,** 299 (1966).
8. B. W. Hakki and S. Knight, *IEEE Trans. Electron Devices* **ED-13,** 94 (1966).
9. C. L. Hemenway, R. W. Henry, and M. Caulton, *Physical Electronics,* 1st ed., John Wiley and Sons, New York, 1962; p. 231.
10. C. Hilsum, *Proc. Inst. Radio Engrs.* **50,** 185 (1962).
11. W. K. Kennedy, Jr. and L. F. Eastman, *Proc. IEEE* **55,** 434 (1967).
12. H. Kroemer, *Proc. Inst. Elec. Engrs.* **52,** 1736 (1964).
13. D. E. McCumber and A. G. Chynoweth, *IEEE Trans. Electron Devices* **ED-13,** 4 (1966).
14. B. K. Ridley, *Proc. Phys. Soc. (London)* **82,** 954 (1963).
15. B. K. Ridley and T. B. Watkins, *Proc. Phys. Soc. (London)* **78,** 293 (1961).
16. W. Shockley, J. A. Copeland, and R. P. James, "The Impedance Field Method of Noise Calculation in Active Semiconductor Devices," in *Quantum Mechanics of Atoms, Molecules, and the Solid State,* ed. P. O. Londin, Academic Press, New York, 1966; pp. 537–563.
17. H. W. Thim, *Proc. IEEE* **55,** 446 (1967).
18. H. W. Thim, M. R. Barber, B. W. Hakki, S. Knight, and M. Uenohara, *Appl. Phys. Letters* **7,** 167 (1965).

PROBLEMS

12.1 It has been established that for electrons in a given energy band to carry current, the band must be only partially filled. From your knowledge of quantum mechanics, explain why this is so.

12.2 a) Draw the unit cell of a GaAs crystal.
b) Construct the Fermi surfaces, i.e., surfaces of constant energy in k-space, for a GaAs crystal.

12.3 a) Gunn effect devices are classified as voltage-controlled devices. Sketch their I–V characteristic. Sketch also the I–V characteristic of current-controlled devices.
b) Compare the inherent stability of the two classes of devices mentioned in (a).

12.4 Devise a means of tuning the frequency of a Gunn oscillator by connecting it in some way to an external circuit.

CHAPTER 13

QUANTUM ELECTRONICS

The study of masers and lasers at present constitutes, to a large measure, the field of *quantum electronics*, which utilizes the quantum character of matter in the generation or amplification of microwave, infrared, or optical radiation. This rather exciting field derives its vitality from a number of related, yet seemingly distinct, disciplines. Thus, some knowledge of quantum and statistical mechanics, optics, and material chemistry, as well as some microwave know-how, is required for the study of the material in this chapter. To study masers and lasers we need not, of course, have a comprehensive knowledge of all these fields; indeed, as we shall demonstrate, the amount of material based on the aforementioned disciplines that is pertinent to masers and lasers is not unduly large. Starting from rather fundamental considerations, we shall develop the pertinent concepts of quantum and statistical mechanics before applying them to a detailed discussion of maser action and its associated devices.

1. INTRODUCTION

The word *maser* is an abbreviation for *microwave amplification by stimulated emission of radiation*; laser is an abbreviation for *light amplification by stimulated emission of radiation*. Because the maser was invented first, a laser is sometimes called an optical maser. Similarly, an infrared amplifier operating on the stimulated-emission principle is called an infrared maser. Unlike the operation of ordinary microwave tubes such as Klystrons and traveling-wave tubes, which can be understood in terms of classical mechanics, the behavior of masers and lasers can be completely analyzed only in terms of quantum and statistical mechanics, that is, we usually deal with the interaction of electromagnetic fields with bound aggregates of charges rather than with free charges as in the case of electron beams in microwave tubes.

It is appropriate at this point to inquire into the advantages of masers and lasers over conventional amplifiers. The chief advantage of the maser is its extremely low noise figure. In contrast to conventional microwave amplifiers, which always contain shot noise from the electron beam, the only significant noise in the maser is that of the spontaneous emission from the paramagnetic specimen which provides the amplification. This type of noise is extremely small at microwave frequencies and liquid helium temperatures. Since the noise figure of an amplifier chain is determined mainly by that of the leading amplifier, providing its gain is sufficiently high, radio astronomers find that a maser placed just behind the receiving antenna is an invaluable device for amplifying weak signals. Because of its narrow bandwidth and extreme frequency stability, the maser operating as an oscillator is a very accurate frequency or time standard (atomic clock).

The laser is capable of producing optical radiation which is powerful, monochromatic, coherent, and unidirectional. In contrast, emission from other light sources is usually neither monochromatic nor coherent. The high directivity of the laser output enables us to focus it into a very small spot whose diameter is on the order of a wavelength of light or a few thousand angstroms. We can use this intense localized beam to study nonlinear effects of matter at optical frequencies or for surgical purposes. When modulated, a laser beam can become a communication medium.

2. THERMODYNAMIC EQUILIBRIUM AND STATISTICAL STEADY STATE

For a material at thermal equilibrium, we can deduce from statistical mechanics that the average number of particles N_i^e in a quantum state in which E_i is the atomic or molecular energy is

$$N_i^e = C\, e^{-E_i/kT}. \tag{13.1}$$

The derivation of the Boltzmann distribution, Eq. (13.1), is given in Appendix B. Note that the total number of atoms or molecules N must be equal to the total of N_i^e summed over all quantum states of the system. Accordingly, we have

$$N = \sum_i N_i^e = C\sum_i e^{-E_i/kT}, \tag{13.2}$$

using Eq. (13.1). Dividing Eq. (13.1) by Eq. (13.2), we find

$$N_i^e = \frac{N e^{-E_i/kT}}{\sum_i e^{-E_i/kT}} = \frac{N}{Z} e^{-E_i/kT}, \tag{13.3}$$

where

$$Z = \sum_i e^{-E_i/kT}$$

is known as the partition function of the system.

To recapitulate, if a system is at *thermal equilibrium*, the population of the various quantum states is given by Eq. (13.3). Of course, Eq. (13.3) does not hold if the system is not at thermal equilibrium. A particular state, called the *statical steady state*, is of special significance to our discussion of masers and lasers. In this state, the mean properties of the macroscopic system are time independent, but the microscopic state populations are not necessarily given by Eq. (13.3). An example of such a state is macroscopic system in contact with a thermal reservoir and subjected to a steady, monochromatic radiation field of frequency f near some resonance frequency of the constituent microscopic system.

According to Eq. (13.3), the population of the various quantum states decreases with increasing energy, that is, $N_j^e < N_i^e$ if $E_j > E_i$. It turns out that even in the presence of an external electromagnetic field, the sample containing these energy levels can only absorb energy from the field if the inequalities are not disturbed. However, if population inversion, that is, $N_j^e > N_i^e$ for $E_j > E_i$, occurs, the sample can be made to emit rather than absorb radiation. Indeed, because the operation of masers and lasers is dependent on just such population inversion, we shall now study ways to achieve it, called *pumping methods*.

3. POPULATION INVERSION

In the last section, we indicated that to obtain maser action we need a population inversion. We shall now show that in the absence of such an inversion, energy will always be absorbed by the microscopic system from the external electromagnetic field. We shall then describe several means for achieving population inversion.

To illustrate the first point, let us consider the simple problem of an isolated electron spin in a static magnetic field. Associated with the electron spin **S** is a magnetic moment $\boldsymbol{\mu}$ proportional to **S** and directed opposite to it. Since the energy of interaction between the magnetic moment $\boldsymbol{\mu}$ and the static field $\mathbf{H}_0$ is $-\boldsymbol{\mu} \cdot \mathbf{H}_0$ and the electron spin can have one of two possible projections along the static field axis, that is, $\pm \frac{1}{2}$, the original energy level of the electron E_0 is split into two levels separated by $2|\mu_z|H_0$, as shown in Fig. 13.1, where μ_z is the component of $\boldsymbol{\mu}$ in the direction of $\mathbf{H}_0$.

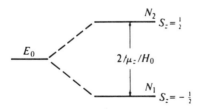

Fig. 13.1. Energy splitting of an electron spin in a static magnetic field.

If there are a number of uncoupled electron spins such as those existing in a dilute paramagnetic sample, let us represent the populations in the minus-spin and plus-spin states by N_1 and N_2 respectively. Also, let P_{12} be equal to the probability per unit time that an electron in state 1 makes a transition to state 2 due to the interaction of the electron with its surroundings; conversely, let P_{21} be equal to the probability per unit time that an electron in state 2 makes a transition to state 1 due to the interaction of the electron with its surroundings.

At thermal equilibrium, we would expect the rate of upward transition to be exactly equal to that of downward transition, that is,

$$N_1^e P_{12} = N_2^e P_{21}. \tag{13.4}$$

Equation (13.4) is a statement of the *principle of detail balance*, which in reality is merely a mathematical statement of thermal equilibrium in which the number of upward transitions is exactly canceled by the number of downward transitions so that the populations of both states are independent of time.[1] It now follows from Eqs. (13.3) and (13.4) that

$$P_{12}/P_{21} = e^{-(E_2 - E_1)/kT}. \tag{13.5}$$

For the population of the system to reach the equilibrium distribution given by Eq. (13.3) from some nonequilibrium initial distribution, or vice versa, there must exist a mechanism for inducing transitions between N_1 and N_2 which arise from the coupling of the spins to some other system. It would be instructive to calculate the final population and the characteristic time required to reach the final state in terms of the quantities P_{12}, P_{21} and N_1, N_2 already introduced.

To begin with, let us replace N_1 and N_2 by two new variables n and N via the relations

$$N = N_1 + N_2,$$
$$n = N_1 - N_2. \tag{13.6}$$

Solving for N_1 and N_2, we find

$$N_1 = \tfrac{1}{2}(N + n),$$
$$N_2 = \tfrac{1}{2}(N - n). \tag{13.7}$$

Note that N_1 and N_2 are not in general equal to N_1^e and N_2^e, the values at thermal equilibrium. Indeed, the rate equations[2] are

$$\frac{dN_1}{dt} = N_2 P_{21} - N_1 P_{12},$$
$$\frac{dN_2}{dt} = N_1 P_{12} - N_2 P_{21}. \tag{13.8}$$

If we use Eq. (13.7) and note that $N = N_1 + N_2 = N_1^e + N_2^e$ is independent of time, Eq. (13.8) becomes

$$\frac{dn}{dt} = N(P_{21} - P_{12}) - n(P_{21} + P_{12}), \tag{13.9}$$

---

[1] A more detailed discussion of the phenomena of thermal transitions involving both electron spin states and energy states of the reservoir can be found in Slichter, reference 21 at the end of the chapter.

[2] We have implicitly assumed that the values of P_{12} and P_{21} are the same as those corresponding to thermal equilibrium [see Eq. (13.4)]. This is permissible providing the system never deviates much from thermal equilibrium.

which we can rewrite as

$$\frac{dn}{dt} = \frac{n_0 - n}{T_1}, \qquad (13.10)$$

where

$$n_0 = N\left(\frac{P_{21} - P_{12}}{P_{21} + P_{12}}\right), \qquad (13.11)$$

$$\frac{1}{T_1} = P_{21} + P_{12}.$$

The solution of Eq. (13.10) is

$$n = n_0 + Ae^{-t/T_1}, \qquad (13.12)$$

where A is an arbitrary constant. Thus n_0 represents the thermal equilibrium population difference and T_1 is a characteristic time associated with the approach to thermal equilibrium called the *spin-lattice relaxation time*. For example, if a sample is initially unmagnetized, the magnetization process is described by the expression

$$n = n_0(1 - e^{-t/T_1}), \qquad (13.13)$$

since $n = 0$ at $t = 0$. From Eq. (13.13), we see that T_1 is a time constant characterizing the time required to magnetize an initially demagnetized sample.

If we apply an alternating magnetic field $\mathbf{H}_1$ to the sample in addition to the static one, further transitions between the $+\frac{1}{2}$ and $-\frac{1}{2}$ spin states will occur. Let us denote the probability per unit time of inducing the transition of an electron from state 1 to state 2 by C_{12} and that of the reverse process by C_{21}. Then the complete rate equations become

$$\frac{dN_1}{dt} = N_2 P_{21} - N_1 P_{12} + N_2 C_{21} - N_1 C_{12}, \qquad (13.14)$$

$$\frac{dN_2}{dt} = N_1 P_{12} - N_2 P_{21} + N_1 C_{12} - N_2 C_{21}.$$

Using the transformations of Eq. (13.7), Eq. (13.14) becomes

$$\frac{dn}{dt} = -2Cn + \frac{n_0 - n}{T_1}, \qquad (13.15)$$

where we have assumed that $C_{12} = C_{21} = C$. The equality of C_{12} and C_{21} is shown on very general grounds in Appendix C.

At steady state, the populations of states 1 and 2 remain unchanged with time. It follows that $dn/dt = d(N_1 - N_2)/dt = 0$. Thus we find the expression for n at statistical steady state from Eq. (13.15) as

$$n = \frac{n_0}{1 + 2CT_1}. \qquad (13.16)$$

The induced transition probability C, as might be expected, is linearly proportional to the r.f. power. Thus, we see from Eq. (13.16) that as long as H_1 is sufficiently small so that $2CT_1 \ll 1$, $n \simeq n_0$, or the population of the states is essentially equal to that at thermal equilibrium.

The rate of absorption of energy, dE/dt, is given by the expression

$$\frac{dE}{dt} = (N_1 - N_2)C\hbar\omega = nC\hbar\omega, \tag{13.17}$$

where $N_1 C$ is the number of electron spins per unit time that goes from the lower energy state to the higher one, $N_2 C$ is the number of electron spins per unit time that goes in the reverse direction, and $\hbar\omega$ is the energy of a photon corresponding to a frequency ω. It follows that $N_1 C\hbar\omega$ corresponds to energy absorption while $N_2 \hbar\omega$ corresponds to energy emission. Thus a positive dE/dt represents net energy absorption while a negative dE/dt represents net energy emission. Maser action corresponds to energy emission which, according to Eq. (13.17), requires that $N_2 > N_1$ or, equivalently, $n < 0$. However, according to Eq. (13.16), $n > 0$ since n_0, C, and T are all positive. Thus, not only is $N_2^e < N_1^e$ at thermal equilibrium according to Eq. (13.3), but $N_2 < N_1$ even at statistical steady state in the presence of an alternating magnetic field. Therefore, if we wish to obtain maser action, we must find some means of inverting the population so that $N_2 > N_1$. We will discuss possible methods in detail in the next section.

Before we proceed further, we shall combine Eqs. (13.16) and (13.17) to obtain a final expression for the energy absorption dE/dt:

$$\frac{dE}{dt} = n_0 C\hbar\omega \frac{1}{1 + 2CT_1} \tag{13.18}$$

So long as $2CT_1 \ll 1$, dE/dt will be proportional to C or to the r.f. power. However, if C is comparable to $1/2T_1$, the absorption levels off despite an increase in the r.f. power. This effect is known as saturation. If we can compute C, as is often the case, then we can obtain T_1 from Eq. (13.18) by an absorption measurement. This is one of the common ways of measuring the spin-lattice relaxation T_1 in a paramagnet.

4. INVERSION METHODS

For a two-level system, *inversion* and *amplification* are usually spatially separated. In the ammonia maser, for example, the two processes are spatially separated in that the NH_3 molecules are passed through a velocity selector to accomplish inversion before they enter the cavity to perform amplification. The velocity selector is so constructed that the ammonia molecules belonging to the lower energy state are deflected out of the vapor beam by an inhomogeneous electric field. The remaining molecules of the upper inversion energy state are in turn directed through a cavity. The electromagnetic field in this cavity induces transitions of the molecules from the upper to the lower state, emitting energy in the transition processes. The electromagnetic field associated with this energy is in phase with the cavity field that induced the transition, thereby amplifying it.

For a three-level maser, we can obtain population inversion without spatial separation of the inversion and amplification processes as in the two-level maser. How this process is accomplished can best be understood by reference to Fig. 13.2.

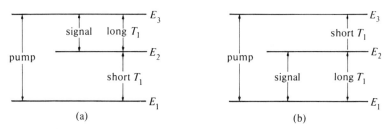

Fig. 13.2. Inversion scheme for a three-level maser.

Consider first Fig. 13.2(a). Particles are induced to go from energy level E_1 to E_3 by a pumping force which is relatively strong compared to the amplified signal. If the relaxation time between E_3 and E_2 is relatively long compared to that between E_2 and E_1, a population inversion may develop between levels 3 and 2 because a particle on level E_2 will quickly relax to level E_1, thus helping to deplete E_2; a particle in E_3, however, will take a relatively longer time to relax toward E_2, thus adding to the population excess of E_3. We shall now prove these statements mathematically.

First, we note from the state equation (13.14) that if the pumping field is sufficiently strong for the C_{12} terms to predominate (recalling that $C_{21} = C_{12}$), $N_1 \simeq N_2$ in the steady state in which the populations of N_1 and N_2 are independent of time. Therefore, under the so-called saturation-pumping condition, the transition probability resulting from the pumping field may be much larger than those resulting from natural relaxation processes. As a consequence, the populations of the two states, across which the pump is connected, are equalized after the pump has been turned on for a length of time sufficient for the condition of statistical steady state to be attained. (Note, however, that in Fig. 13.2 the pump is connected between levels 1 and 3; thus, instead of $N_1 \simeq N_2$ referred to in connection with Eq. (13.14), we have $N_1 \simeq N_3$.) If the pumping action does not disturb level 2, the total population of levels 1 and 3 must be independent of time, and indeed should equal that at thermal equilibrium. Thus, we have

$$N_1 + N_3 = N_1^e + N_3^e. \tag{13.19}$$

Or, recalling that $N_1 \simeq N_3$, we find from Eq. (13.19) that

$$N_1 \simeq N_3 \simeq \tfrac{1}{2}(N_1^e + N_3^e). \tag{13.20}$$

To demonstrate population inversion, we must now show that

$$N_3 > N_2 \simeq N_2^e, \tag{13.21}$$

where we again assumed that level 2 is not sufficiently disturbed. Combining Eqs. (13.20) and (13.21), we must show equivalently that

$$N_1^e/N_2^e + N_3^e/N_2^e > 2. \tag{13.22}$$

Substituting the expressions for N_1^e, N_2^e, etc. by means of Eq. (13.3), we find that Eq. (13.22) becomes

$$e^{(E_2-E_1)/kT} + e^{(E_2-E_3)/kT} > 2. \tag{13.23}$$

Now, let $(E_2 - E_1) = (E_3 - E_2) + \delta$. Then Eq. (13.23) reads

$$e^{[(E_3-E_2)+\delta]/kT} + e^{-(E_3-E_2)/kT} > 2. \tag{13.24}$$

An examination of Eq. (13.24) or Fig. 13.3 shows that the inequality of Eq. (13.24) is satisfied if δ is positive for all values of $E_3 - E_2$. Equivalently, we have

$$(E_3 - E_1) > 2(E_3 - E_2),$$

or the pump frequency must be more than twice the signal frequency if we are to obtain population inversion. We shall substantiate this inequality by a detailed calculation of the emissive power of the three-level maser.

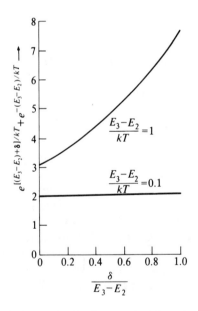

Fig. 13.3. Relationship between energy level separations in a three-level maser.

Another three-level maser scheme is shown in Fig. 13.2(b). Here, the signal is applied across levels 2 and 1 instead of between levels 3 and 2 as in Fig. 13.2(a). Correspondingly, the relaxation time between E_3 and E_2 is shorter than that between E_2 and E_1, while the converse is true for Fig. 13.2(a). Thus, referring to

Fig. 13.2(b), a particle in E_3 will relax quickly to E_2 because of the short relaxation time between them. On the other hand, since the relaxation time between E_2 and E_1 is long, it is possible to maintain a population inversion between E_2 and E_1.

5. THE THREE-LEVEL MASER

Refer now to the energy diagram of the three-level maser shown in Fig. 13.2(a). We can immediately write the rate equations for this system by analogy with Eq. (13.14) for the two-level maser:

$$\dot{N}_3 = (P_{13}N_1 - P_{31}N_3) + (P_{23}N_2 - P_{32}N_3) + C_p(N_1 - N_3) + C_s(N_2 - N_3),$$
$$\dot{N}_2 = (P_{32}N_3 - P_{23}N_2) + (P_{12}N_1 - P_{21}N_2) + C_s(N_3 - N_2), \quad (13.25)$$
$$\dot{N}_1 = -\dot{N}_2 - \dot{N}_3,$$

where $C_p = C_{13} = C_{31}$ is the induced transition probability between pump levels 1 and 3 and $C_s = C_{23} = C_{32}$ that between signal levels 2 and 3.

Under statistical steady-state conditions, $\dot{N}_1 = \dot{N}_2 = \dot{N}_3 = 0$. Solving Eqs. (13.25), we find that the normalized population difference between levels 3 and 2 is

$$\frac{N_3 - N_2}{N_1} = -\frac{C_1 + (P_{12} - P_{21} + P_{32} - P_{23})C_p}{(P_{31} + C_p)(P_{23} + P_{21} + C_s) + P_{21}(P_{32} + C_s)}, \quad (13.26)$$

where

$$C_1 = P_{12}(P_{31} - P_{23}) - P_{13}(P_{21} + P_{23}) + P_{32}(P_{12} + P_{13}). \quad (13.27)$$

In a similar manner, we also find

$$\frac{N_3 + N_2}{N_1} = \frac{C_2 + (P_{12} + P_{21} + P_{32} + P_{23} + 2C_s)C_p + 2(P_{12} + P_{13})C_s}{(P_{31} + C_p)(P_{23} + P_{21} + C_s) + P_{21}(P_{32} + C_s)} \quad (13.28)$$

where

$$C_2 = P_{12}(P_{31} + P_{23}) + P_{13}(P_{21} + P_{23}) + P_{32}(P_{12} + P_{13}). \quad (13.29)$$

Furthermore, since the total number of particles conserved is $N_1 + N_2 + N_3 = N$, it follows that

$$N_1 = \frac{N}{1 + (N_3 + N_2)/N_1}. \quad (13.30)$$

These equations can be simplified considerably if we assume that the pump is sufficiently strong, the $C_p \gg C_s$ or $C_p \gg P$. In this case, Eqs. (13.26) and (13.28) become

$$\frac{N_3 - N_2}{N_1} \simeq -\frac{P_{12} - P_{21} + P_{32} - P_{23}}{P_{23} + P_{21} + C_s}, \quad (13.31)$$

$$\frac{N_3 + N_2}{N_1} \simeq \frac{P_{12} + P_{21} + P_{32} + P_{23} + 2C_s}{P_{23} + P_{21} + C_s}. \quad (13.32)$$

If we further stipulate that $h\nu_{21}$, $h\nu_{32} \ll kT$, as may be the case with masers, we may further simplify Eqs. (13.31) and (13.32) by noting that in this case

$$P_{21} = P_{12} e^{h\nu_{21}/kT} \simeq P_{12}\left(1 + \frac{h\nu_{21}}{kT}\right),$$

$$P_{32} = P_{23} e^{h\nu_{32}/kT} \simeq P_{23}\left(1 + \frac{h\nu_{32}}{kT}\right),$$

(13.33)

where we have used Eq. (13.5). We have also noted that $h\nu_{21} = E_2 - E_1$ and $h\nu_{32} = E_3 - E_2$. Using approximation (13.33) and the statement of particle conservation, $N_1 + N_2 + N_3 = N$, we find from Eqs. (13.31) and (13.32) that

$$N_3 - N_2 \simeq \frac{hN}{3kT} \frac{P_{12}\nu_{21} - P_{23}\nu_{32}}{P_{12} + P_{23} + C_s}.$$

(13.34)

In order to achieve stimulated emission, we have shown previously that first there must be a population inversion, that is, $N_3 > N_2$. Since all quantities appearing in Eq. (13.34) are positive, it follows that population inversion is possible if and only if

$$P_{12}\nu_{21} > P_{23}\nu_{32}.$$

(13.35)

By definition, we have

$$\nu_{21} = \nu_{31} - \nu_{32}.$$

(13.36)

Combining Eqs. (13.35) and (13.36), we find

$$\nu_{31} > \left(1 + \frac{P_{23}}{P_{12}}\right)\nu_{32}.$$

(13.37)

According to Eq. (13.11), P_{23}/P_{12} is, to a first approximation, nearly equal to $(T_1)_{12}/(T_1)_{23}$, where $(T_1)_{12}$ and $(T_1)_{23}$ are respectively relaxation times between levels 1 and 2, and 2 and 3. Since $(T_1)_{12}$ must be smaller than $(T_1)_{23}$ according to the discussion in the last section, Eq. (13.37) yields the inequality

$$\nu_{31} > K\nu_{32},$$

where the constant K for a given medium has a value between 1 and 2. However, if the relaxation times are equal, Eq. (13.37) shows that $\nu_{31} > 2\nu_{32}$, as stated in connection with Eq. (13.24), which was derived by assuming that level 2 is not disturbed. Thus, the three-level maser requires a source of saturating power at a frequency approximately double that of the signal frequency.

The emitted power is equal to the total number of emitted photons per second, $C_s(N_3 - N_2)$, times the energy of each photon, $h\nu_{32}$. Thus we have

$$P_e = C_s(N_3 - N_2)h\nu_{32} = \frac{Nh^2\nu_{32}C_s}{3kT} \frac{P_{12}\nu_{21} - P_{23}\nu_{32}}{P_{12} + P_{23} + C_s}.$$

(13.38)

Let us now examine the various factors involved in the above expression for the emitted power. First, we have already discussed the relationship between the pump and signal frequencies. Second, we note that P_e is directly proportional to N. It turns out, however, that as N increases, P_e will reach some maximum value and then decrease as N is further increased. The nonoccurrence of a finite optimum value of N in the above equation is due to the fact that the various particles are not isolated from one another as assumed in its derivation. As the number of paramagnetic ions in the diamagnetic salt N increases, the mutual interaction between them increases, which eventually leads to a decrease in P_e with further increase in paramagnetic ion concentration.

We also note from Eq. (13.38) that P_e is inversely proportional to T, indicating the desirability of using low temperatures. Of course, the (2,3) signal transitions must be allowed.

Finally, we note that if $C_s \ll (P_{12} + P_{23})$, then the emitted power is proportional to the induced signal transition probability C_s, signifying linear amplification. If the signal transition is saturated, that is, C_s cannot be neglected compared to $P_{12} + P_{23}$, then P_e is no longer proportional to C_s, signifying nonlinear amplification.

The three-level maser was first proposed by Bloembergen [1].[3] Scovil, Feher, and Seidel [20] constructed the first experimental device, using the spin resonance of the Gd^{4+} paramagnetic ion[4] of diluted gadolinium ethyl sulfate at 1.2°K. The ground state of Gd^{4+} ion is ^8S, indicating that there are eight electrons with a total orbital angular momentum L equal to zero and a total spin angular momentum S equal to $\frac{7}{2}$. In the absence of the crystalline field, the ground state is split into eight ($= 2S + 1$) equally spaced energy levels characterized by the quantum number S_z having half-integral values $-\frac{7}{2}, -\frac{5}{2}, ..., +\frac{5}{2}, +\frac{7}{2}$ when a static magnetic field is applied in the direction z. The interaction between the ionic spin and the crystalline field of the solid may cause the energy levels to be nonequidistant, as exemplified by the actual Gd^{4+} energy-level diagram of Fig. 13.4. Here a signal of 9 GHz is applied between the $S = -\frac{5}{2}$ and $S = -\frac{3}{2}$ states, while a pump of 17.5 GHz is applied across the $S = -\frac{5}{2}$ and $S = -\frac{1}{2}$ states.

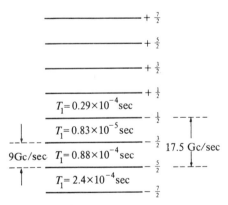

Fig. 13.4. Energy levels for the Gd^{+4} ion (after Scovil, Feher, and Seidel [20]).

---

[3] Bracketed numbers refer to references listed at the end of the chapter.
[4] An ion is paramagnetic if its effective spin is nonzero.

In maser action, we are concerned with the resonance of the paramagnetic ion embedded in a diamagnetic crystalline lattice. Two complications, already mentioned, deserve further attention. First, in paramagnetic ions, several electronic spins may be strongly coupled together in such a way that the total spin is larger than $\frac{1}{2}$. In this case, the number of possible energy levels is equal to $2S + 1$, where S is equal to the total spin in a given direction when the individual spins are all parallel. Thus, for the Gd^{4+} ion mentioned above, $S = \frac{7}{2}$ since there are seven electrons giving rise to the effective spin; since $2S + 1 = 8$, it follows that there are eight energy levels. The second complication arises as a result of the crystalline field. When a paramagnetic ion is embedded in a crystalline lattice, the neighboring ions produce an electric field which acts upon it and distorts the motion of its paramagnetic electrons. Since the electron orbits in general differ among the states with different components of S along the magnetic field, the energy levels of the corresponding electron states will be unequally perturbed.

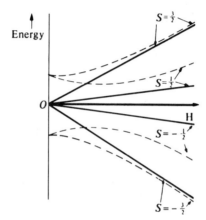

Fig. 13.5. Cr^{+3} ions in ruby in a magnetic field.

Thus the energy of interaction between the electron spins and the magnetic field is no longer linear in H, as is the case with the free electron of Fig. 13.1. A typical example of crystal field distortion is given in Fig. 13.5 for a ruby crystal containing a small amount of chromium in aluminium oxide (Al_2O_3). The solid lines indicate the energy levels in the absence of a crystalline field while the dashed lines represent the same levels in the presence of a crystalline field. Note that here the crystal field has split the fourfold degenerate levels in zero field into a pair of twofold degenerate levels; the application of a magnetic field further changes the levels in a nonlinear manner. Furthermore, the energies, except for the case of cubic crystals, are dependent upon the direction of the magnetic field relative to the crystalline axes. Note that we can change the frequency to be amplified by changing the magnetic field and adjusting the pump frequency accordingly.

It is important to note that since the splittings in a magnetic field are purely Zeeman in character, the energy levels are equally spaced. It follows that it is impossible to pump and amplify only among the selected three levels. However, due to the presence of the superimposed crystalline field, the levels corresponding to different spin states are, in general, unevenly spaced, thus rendering maser level selection possible.

6. TWO-LEVEL MASER

The first successful demonstration of maser action, by Gordon, Zeiger, and Townes [8], utilized the so-called inversion levels of ammonia gas molecules. In this type of maser, a molecular beam of ammonia is passed through an inhomogeneous electrostatic field which effectively focuses the molecules in the upper energy state and defocuses those in the lower energy state. The focused beam composed of upper state molecules then passes through a microwave cavity which has a mode with a resonance frequency equal to the inversion frequency of the ammonia molecule. If the cavity is excited to oscillate at this particular mode by a microwave signal in the presence of the molecular beam, the ammonia molecules can be induced to go from the upper state to the lower state. Since the radiation from these molecules is in phase with the microwave signal inducing it, the signal is amplified. For a sufficiently intense beam, oscillation may also occur. There are no molecules in the lower energy state as the beam enters the microwave cavity; thus population inversion is accomplished by preselection rather than by pumping as in the case of the three-level maser.

Population inversion can also be obtained by a process known as adiabatic rapid passage. Consider the case of a rotating magnetic field H_1 of frequency ω, perpendicular to a static field H_0. Far below resonance, the magnetization M is nearly parallel to the effective field H_{eff}, with $|H_{eff}|$ given by[5]

$$|H_{eff}| = \sqrt{H_1^2 + [|\omega/\gamma| - H_0]^2}, \qquad (13.39)$$

where γ is the gyromagnetic ratio. As resonance is approached, both magnitude and direction of H_{eff} change, but if resonance is approached sufficiently slowly, M will remain parallel to H_{eff}. Exactly at resonance, that is, at $|\omega/\gamma| = H_0$, $H_{eff} = H_1$ according to the expression for H_{eff} above so that the magnetization will lie along H_1, making a 90° angle with H_0. If we vary H_0 through resonance, M will end up pointing in a direction opposite to H_0. This technique of inverting M so that the higher energy spin state is relatively more populated is called *adiabatic inversion*.

7. MASER STRUCTURES

For a three-level maser, both the pump and the signal must be coupled to the paramagnetic material. To maximize the magnetic interaction between the pump and signal and the material, it is desirable to place the material at the position of a

---

[5] H_{eff} is an effective field as seen by an observer in a coordinate system in which H_1 is stationary. For a detailed derivation, see Slichter [21], pp. 18–22.

high-Q cavity at which the r.f. magnetic field is maximum. Of course, if the material dielectric losses are sufficiently low at microwave frequencies, as is the case with ruby, we can fill the entire cavity with the maser material since the cavity field is uniform in time phase. Furthermore, a maser material such as ruby can be used as a dielectric cavity and metal walls need not be used. Normally, the pump and signal are coupled to the cavity via separate coupling holes, as schematically shown in Fig. 13.6.

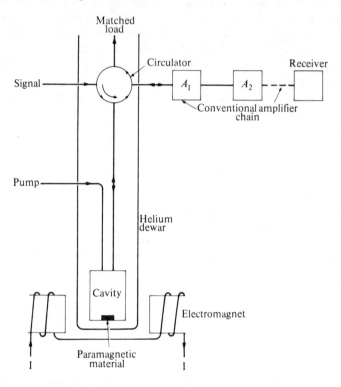

Fig. 13.6. Schematic diagram of a maser.

To prevent the maser from amplifying the noise from the next stage, amplifier A_1, we must use a ferrite circulator in the manner indicated in Fig. 13.6. The construction of the circulator is such that power entering the signal terminal of the circulator will emerge from the maser terminal, while the amplified signal will return from the maser via the same maser-to-circulator waveguide to emerge from the amplifier terminal to enter the next stage, A_1. Any reflection from A_1 will go into the matched load and be dissipated. In order to maximize the maser gain-bandwidth product and minimize the noise level, we must frequently enclose not only the cavity but also the circulator in the helium dewar since ferrite losses are lower at lower temperatures. We can adjust the magnetic field appropriate for a given combination of pump and signal frequencies by changing the current I

in the electromagnet; alternatively, we can use a superconducting solenoid immersed in the dewar to provide the required magnetic field.

Since the field in a resonsant cavity is enhanced by a factor of Q at resonance as compared to the field off resonance, the cavity maser has the advantage of a large field-material interaction in a reasonably small volume. On the other hand, the high-cavity Q also limits the usable bandwidth of the maser. We can alleviate this effect by putting the material in a slow-wave rather than a cavity structure. In this device, known as the traveling-wave maser [5], the electromagnetic field is weakly coupled to the structure since the r.f. field is no longer enhanced by a high Q, as in the case of the cavity structure. Nevertheless, we obtain high gain since the interaction takes place for an extended time and distance while the signal travels through the material placed in a slow-wave structure. Since the slow-wave structure is a nonresonant device, the bandwidth is large compared to that of the cavity maser. To avoid possible oscillations, we must provide for unilateral gain in the direction of wave propagation, either by building ferrite isolation into the structure or by using maser materials of high paramagnetic concentration.

Slow-wave structures are used for a traveling-wave maser solely to reduce the physical length of the waveguide required to obtain useful gain. We can understand this concept based on the following relation between the guide wavelength λ_g and the phase velocity v_p:

$$\lambda_g = v_p/f, \tag{13.40}$$

where f is the frequency. Thus, if $v_p = 0.1c$ where c is the velocity of light, the slow-wave structure need only be $\frac{1}{10}$ as long as a corresponding normal waveguide operated far above cutoff for the same gain.

From the energy conservation standpoint, the signal in a traveling-wave maser will be amplified if the energy gain by the signal in its interaction with the paramagnetic material exceeds the losses resulting from the slow-wave structure, material dielectric losses, and possibly ferrite losses. Similarly, for the cavity maser, the signal is amplified if the gain exceeds all losses in the structure. As with all other types of amplifiers, oscillation can be achieved if the gain is sufficiently large. An example is the ammonia maser, which acts as an oscillator at sufficiently high gas-beam density, also serves as a very accurate frequency standard or atomic clock.

8. LASER STRUCTURES

The principle of laser operation is similar to that of masers; the major difference, conceptually, lies in the cavity structure.[6]

---

[6] It should be pointed out, however, that for a laser, optical frequencies are sufficiently high such that the approximation $h\nu_{21}, h\nu_{32} \ll kT$ for a maser is no longer valid. In this case, Eqs. (13.31) and (13.32) cannot be simplified by means of Eq. (13.33).

At microwave frequencies, we can construct cavities of laboratory dimensions such that they support only the dominant resonant mode since the wavelengths at these frequencies are on the order of a few centimeters. On the other hand, the wavelength at optical frequencies is so short, on the order of 5000 Å or 5×10^{-5} cm, that any physically realizable cavity configuration is likely to be sufficiently large compared to a wavelength that it is capable of supporting multiple modes over a rather small frequency band. This operation may give rise to some mode interference problems seldom found in a microwave cavity. Furthermore, because of the extremely high frequencies in the visible spectrum, the propagation effects are such that, unlike a microwave cavity, a laser cavity need not be completely enclosed, as we shall demontrate.

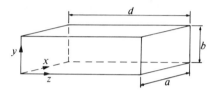

Fig. 13.7. A rectangular cavity.

Consider first a rectangular cavity with cross-sectional dimensions a, b and length d as shown in Fig. 13.7. If the electromagnetic field across the cross-sectional area is uniform, then to satisfy the boundary conditions of zero electric field at the end plates, d must be an integral multiple of a wavelength, that is,

$$d = l\frac{\lambda}{2}, \tag{13.41}$$

where l is an integer. If the end fields are not uniform but instead have an r and s number of half-wavelength variations in the x- and y-directions respectively, then, by induction we can generalize Eq. (13.41) to read

$$\lambda = \frac{2}{[(l/d)^2 + (m/a)^2 + (n/b)^2]^{1/2}}, \tag{13.42}$$

where m and n are integers. Factoring out l/d in the denominator of Eq. (13.42) we have

$$\lambda = \frac{2d}{l}\left[1 + \frac{(m/a)^2 + (n/b)^2}{(l/d)^2}\right]^{-1/2}. \tag{13.43}$$

If $d \gg a, b$ but $l \gg m, n$ ($d/a, b > 10$ and $l/m, n > 100$, say), a condition which is often met for laser cavities, then Eq. (13.43) becomes

$$\lambda \simeq \frac{2d}{l}\left[1 - \frac{1}{2}\left(\frac{dm}{al}\right)^2 - \frac{1}{2}\left(\frac{dn}{bl}\right)^2\right], \tag{13.44}$$

where we have made use of the binomial expansion theorem and retained only the leading terms. Since factors $\frac{1}{2}(dm/al)^2$ and $\frac{1}{2}(dn/bl)^2$ are small compared to unity, we can see from Eq. (13.44) that the spacing between adjacent modes of the same l but different m or n on the frequency or wavelength scale will be considerably less than that between adjacent modes whose l-values differ by unity but have the same m and n. This is illustrated by the line spectra of a few possible modes in Fig. 13.8. Since l is usually a very large number for laser cavities (10^6, say), all these modes are extremely close in wavelength; thus interference effects between modes may be very serious since the Q's of these modes are necessarily finite because of cavity and other losses.

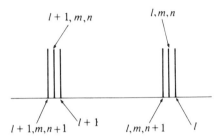

Fig. 13.8. Line spectra of representative laser cavity modes.

We shall now examine in some detail the field configuration in a cavity at extremely high optical frequencies. For our purposes, it is best to examine the case of TE_{nlp} modes. For definiteness, let us examine the field of the TE_{01p} cavity mode, that is, a mode for which $n = 0$ and $l = 1$. The field expressions for this mode can be obtained from the TE_{01} waveguide mode field expression given by Eqs. (3.20), (3.30), and (3.31) as

$$\begin{aligned} H_z &= AJ_0(k_c r), \\ E_r &= H_\phi = 0, \\ E_\phi &= jZ_0 \frac{f}{f_c} AJ_0'(k_c r), \\ H_r &= -E_\phi/Z_{TE}, \end{aligned} \qquad (13.45)$$

where $Z_{TE} = Z_0/\sqrt{1 - (f_c/f)^2}$ is the wave impedance corresponding to the medium filling the cavity and $Z_0 = \sqrt{\mu/\varepsilon}$ is the wave impedance of infinite medium, assumed lossless here for simplicity. The quantities μ and ε are the permeability and dielectric constant of the medium respectively, f_c is the cutoff frequency, and k_c is the cutoff wave number.

Let us now form a cylindrical cavity by inserting two shorting plates perpendicular to the guide axis at $Z = 0$ and d. The field in the guide is now composed of two waves traveling in the $+z$- and $-z$-directions, or

$$H_z = (Ae^{-j\beta z} + Be^{j\beta z})J_0(k_c r). \qquad (13.46)$$

Since the normal component of the magnetic field, H_z, must be zero at the perfect conductors located at $z = 0$, we readily find from Eq. (13.46) that

$$H_z = -2jAJ_0(k_c r) \sin \frac{p\pi z}{d}, \qquad (13.47)$$

where p is an integer. Similarly, E_ϕ and H_z are given by

$$E_\phi = 2Z_0 \frac{f}{f_c} AJ'_0(k_c r) \sin \frac{p\pi z}{d},$$

$$H_r = -2jA\sqrt{(f/f_c)^2 - 1}\, J'_0(k_c r)\cos \frac{p\pi z}{d}, \qquad (13.48)$$

where we have made use of Eqs. (3.80) and (3.81).

We must now determine the frequency dependence of the arbitrary constant A for a given value of z-directed transmitted power W_T in a waveguide propagating the TE_{01} mode. From Eq. (3.136), we have

$$\begin{aligned} W_T &= \int_0^a \mathbf{z} \cdot (\mathbf{E} \times \mathbf{H}^*)\, 2\pi r dr \\ &= -\int_0^a E_\phi H_r^*\, 2\pi r dr, \end{aligned} \qquad (13.49)$$

where a is the radius of the guide. Substituting Eq. (13.48) into Eq. (13.49) and integrating, we find the expression for the square of the arbitrary constant A in terms of W_T and f:

$$A^2 = \frac{W_T}{\pi a^2 Z_0(f/f_c)^2 \sqrt{1-(f_c/f)^2}\,[J_0^2(k_c a) + J_1^2(k_c a) - (2/k_c a)J_0(k_c a)J_1(k_c a)]}. \qquad (13.50)$$

For a finite value of W_T, it is clear that A and therefore H_z approach zero as $f/f_c \to \infty$. Correspondingly, E_ϕ and H_r become frequency-independent as $f/f_c \to \infty$. Since the tangential component of an electric field must be zero at the surface of a perfect conductor [see the derivation leading to Eq. (3.44)], E_ϕ is zero at the cylindrical surface $r = a$, that is, $J'_0(k_c a) = 0$ according to Eq. (13.48). Inasmuch as H_z, E_ϕ, and H_r are all zero at $r = a$ as $f/f_c \to \infty$, indicating that neither charge nor current exists there, we can completely remove the cylindrical surface of the cavity without affecting the field configuration in the cavity. What is left is a Fabry-Perot *open-structure resonator* which is composed simply of two cylindrical parallel reflectors coaxially placed at a spacing that is large compared to a wavelength. Since the wavelength in the visible spectrum is extremely small compared to typical laser cavity dimensions, letting $f/f_c \to \infty$ in the above equations is quite justified. Of course, since f/f_c is in practice large but not infinite, there will be some energy loss due to the fact that radiation out of the open portion of the resonator accompanies the slight field perturbation when the cylindrical wall is removed.

Other features of the above calculation are worth noting. For example, aside from the fact that E_ϕ and H_r are in space and time quadrature, E_ϕ/H_r is equal to

$Z_0 = \sqrt{\mu/\varepsilon}$, the impedance of the unbounded medium as $f/f_c \to \infty$, as an examination of Eqs. (13.48) and (13.50) will readily show. This fact, coupled with the vanishing of H_z as $f/f_c \to \infty$, indicates that here energy is propagated only in the axial direction and in the form of a nonuniform plane wave. Furthermore, we observe from these results that even for $n = 0$, $l = 1$, p could be extremely large at optical frequencies. Indeed, for an end plate spacing d of 25 cm, the relation $d = p(\lambda/2)$ shows that p is approximately equal to 10^6 at a wavelength of 5000Å, even though $n = 0$ and $l = 1$. This finding is consistent with the assumption leading to the derivation of Eq. (13.44).

An examination of field expressions derived from Eqs. (3.20) and (3.28) will readily show that for all TE_{0l} modes, H_z approaches zero at the cylindrical boundary as $f/f_c \to \infty$. We conclude, therefore, that the removal of the cylindrical boundary of the cavity will not affect the field distribution within the cavity, as was the case for the TE_{01} mode. However, for the TE_{nl} modes where $n \neq 0$, the fields do not vanish at the cylindrical boundary as $f/f_c \to \infty$; removal of the boundary will cause radiation losses which may greatly damp oscillations corresponding to the modes.

Similarly, an examination of field expressions derived from Eqs. (3.20) and (3.29) will show that the fields at the cylindrical boundary of the cavity are not zero even as $f/f_c \to \infty$. For this reason, we would expect substantial radiation losses to be associated with these TM_{nl} modes.

Actual laser resonators may employ prisms and/or reflectors; thus their cavity structure may be somewhat more complicated than the simple Fabry-Perot configuration. It is common for gas lasers, for example, to use two windows placed at the Brewster angle within the cavity formed by the reflectors to allow only the polarization with an electric field in the plane of incidence to pass without reflection. Thus only the wave with this polarization will reach the reflectors in force to sustain oscillations in the composite cavity.

9. LASER CAVITY FIELD

Schawlow and Townes [18], Prokhorov [16], and Dicke [6] have suggested the use of the Fabry-Perot interferometer [14], composed of two reflecting plates separated by an arbitrary distance, as a resonator for infrared and optical masers. These open structures have two sources of loss: diffraction loss due to energy escape out of the region between the two plates, and the losses resulting from absorption in and transmission through the reflectors. Although the losses of a parallel plate resonator are reasonably small, its configuration does not yield the highest possible Q. It has been found that a resonator formed by two identical spherical reflectors separated by their radii of curvature gives diffraction losses of orders of magnitude lower than that of the parallel plane resonator [2,7]. Since the focal length of a spherical reflector is one-half its radius of curvature, the focal points of the reflectors are coincident; therefore the resonators are termed *confocal*. Intuitively, we might think of the shape of the spherical surface as being more effective in confining the energy between them than are two parallel plates, thus yielding lower diffraction losses. The use of confocal spherical reflectors as an interferometer has been described by Connes [4]. In contrast with the case of the Fabry-Perot interferometer, parallelism between spherical reflectors is not a strict requirement; the only fine adjustment required is the spacing between the reflectors. When $d \neq b$, the resonator is termed *nonconfocal* [23].

In the last section, we demonstrated that as $f/f_c \to \infty$, the TE_{0l} modes have low radiation or diffraction losses out of the open portion of the Fabry-Perot resonator. This analysis, which is intellectually satisfying, demonstrates how an optical Fabry-Perot open-structure resonator can be evolved from a completely enclosed cavity of the type used at microwave frequencies. But for finite frequencies, for other modes, and for different reflector configurations, we must develop a more general method of attack to obtain the field distribution in an optical resonator.

Our calculation of the electromagnetic field in an open-structure cavity such as the Fabry-Perot resonator is based on the method of selfconsistent fields. We shall start by computing the fields in a given space in terms of specified fields on the boundaries of that space.

By a series of vector manipulations, we can show directly from Maxwell's equation that [19, 24]

$$\mathbf{E}' = -\frac{1}{4\pi} \int_S \{-j\omega\mu(\mathbf{n} \times \mathbf{H})\psi + (\mathbf{n} \times \mathbf{E}) \times \nabla\psi + (\mathbf{n} \cdot \mathbf{E})\nabla\psi\}\, ds,$$

$$\mathbf{H}' = \frac{1}{4\pi} \int_S \{-j\omega\mu(\mathbf{n} \times \mathbf{E})\psi - (\mathbf{n} \times \mathbf{H}) \times \nabla\psi - (\mathbf{n} \cdot \mathbf{H})\nabla\psi\}\, ds,$$

(13.51)

where, as shown in Fig. 13.9, S is the boundary surface; $\mathbf{E}'$ and $\mathbf{H}$ are fields at any point inside the surface S; $\mathbf{E}$ and $\mathbf{H}$ are the fields on the surface, $\psi = e^{-jkr}/r$, where r is the distance from the differential elements ds to the point at which $\mathbf{E}'$ and $\mathbf{H}'$ are being evaluated; and $\mathbf{n}$ is the surface normal.

In the derivation of Eq. (13.51), it is assumed that $\mathbf{E}$ and $\mathbf{H}$ are continuous functions of the space coordinates. This is unlikely to be the case for an open-structure resonator used for lasers, since line charges and currents at the conduct-

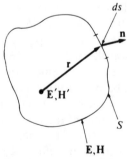

Fig. 13.9. Boundary surface for field calculation.

ing boundaries of the resonator may exist, giving rise to discontinuities in the electromagnetic fields. Therefore, Eq. (13.51) must usually be augmented by contour integrals representing effects due to the fact that line charges collect at the discontinuities:

$$\mathbf{E}'' = \frac{1}{4\pi j\omega\varepsilon} \oint \nabla\psi \, \mathbf{H} \cdot \mathbf{dl}, \tag{13.52}$$

$$\mathbf{H}'' = -\frac{1}{4\pi j\omega\mu} \int \nabla\psi \, \mathbf{E} \cdot \mathbf{dl};$$

that is, the total fields inside the surface S are given by $\mathbf{E}' + \mathbf{E}''$ and $\mathbf{H}' + \mathbf{H}''$. However, we may note that as $\omega \to \infty$, $\mathbf{E}'' \ll \mathbf{E}'$ and $\mathbf{H}'' \ll \mathbf{H}'$, so that eqs. (13.51) alone represent the fields inside S in this limiting case.

If we consider a plane wave propagating in the $\mathbf{n}$-direction, we find that $\mathbf{n} \times \mathbf{H} = -\sqrt{\varepsilon/\mu}\, \mathbf{E}$ and $\mathbf{n} \times \mathbf{E} = \sqrt{\mu/\varepsilon}\, \mathbf{H}$, since $E/H = \sqrt{\mu/\varepsilon}$ for a plane wave, where μ and ε are respectively the permeability and dielectric constant of the medium. Since $\psi = e^{-jkr}/r$, $\nabla\psi \simeq (\mathbf{r}/r)jke^{-jkr}/r$ as $\omega \to \infty$. Thus, Eq. (13.51) becomes

$$\mathbf{E}' = -\int_S \frac{jke^{-jkr}}{2\pi r} \mathbf{E} \, ds. \tag{13.53}$$

Let us now apply Eq. (13.53) to the problem of a typical laser resonator composed of two perfectly reflecting mirrors depicted in Fig. 13.10.[7] Let the path of integra-

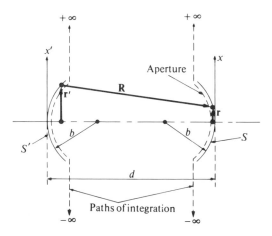

Fig. 13.10. A typical laser resonator.

---

[7] In practice, the reflectors are made of a number of dielectric layers of different dieletric constants and thicknesses. Alternatively, the reflection could be provided for by thin metallic coatings. In any event, the transmission through one of the reflectors is extremely small (0·1%, say), while that for the other is in the neighborhood of 10%, say. In this way, a high Q for the laser cavity can be obtained while it is still possible to couple out of the cavity via small transmission through one of the reflectors.

tion be as shown by the dashed line and for definiteness, let the reflectors be portions of a sphere with square sides of dimension $2a$ and radius of curvature b. The use of Eq. (13.53) implicitly assumes that $2a$ and the mirror separation d are both large compared to the wavelength. For the configuration of Fig. 13.10, Eq. (13.53) simplifies to

$$E_y = \int_{S'} \frac{jk}{2\pi R} e^{-jkR} E_0 F_m(x') G_n(y') \, ds'. \tag{13.54}$$

We can obtain the eigenfunctions of the resonator by solving Eq. (13.54) with $E_y = E_1 F_m(x) G_n(y)$, thus requiring that the field distribution over the x', y' aperture reproduce itself over the x, y aperture within a constant. It follows from Eq. (13.54) that

$$\sigma_m \sigma_n F_m(x) G_n(y) = \int_{-a}^{+a} \int_{-a}^{+a} \frac{jk}{2\pi R} e^{-jkR} F_m(x') G_n(y') \, dx' \, dy', \tag{13.55}$$

where $\sigma_m \sigma_n = E_1/E_0$ is the eigenvalue belonging to the eigenfunction $F_m(x) G_n(y)$ and is in general complex. Thus $F_m(x) G_n(y)$ represents the electromagnetic field distribution over the x,y aperture, while $\sigma_m \sigma_n$ is related to the phase shift and diffraction loss of the normal mode m,n. Note that if it were not for diffraction losses, the amplitude of the reflected wave E_1 from the right surface S would be exactly equal to $+E_0$ or $-E_0$, making $\sigma_m \sigma_n = \pm 1$. Accordingly, the diffraction loss α_d is equal to $1 - |\sigma_m \sigma_n|^2$.

For the square spherical mirror configuration shown in Fig. 13.10 and for $a \ll d$, analytical solutions for Eq. (13.55) can be obtained [2,3,23]. The corresponding form for polar coordinates has also been given for this case [9]. For the Fabry-Perot interferometer configuration, Fox and Li [7] solved Eq. (13.53) by the process of interaction using a high-speed computer. In the solution of this problem, they assumed an initial distribution over one of the apertures, and then calculated the distribution at the second aperture by summing over the elemental Huygens sources of the first aperture field. With the resulting distribution over aperture 2, the calculation was reversed, yielding a new distribution for aperture 1. This process was repeated again and again for hundreds of transients by high-speed computation, with final convergence to a stable field distribution.

Not all combinations of radii of curvature and reflector spacings give stable solutions for Eq. (13.53). For some radii of curvature, the bouncing rays may leave the reflector systems for certain reflector spacings after relatively few passes, yielding a high loss or unstable solution. For example, for the spherical reflector system depicted in Fig. 13.10, it is easy to show by geometrical considerations that light rays reflected from one reflector will not be intercepted by the other reflector if $d > 2b$. For the general case in which the radii of curvature of the reflectors b_1, b_2 are not equal, it is convenient to represent this stability condition in a diagram shown in Fig. 13.11, as in Boyd and Kogelnik [3].

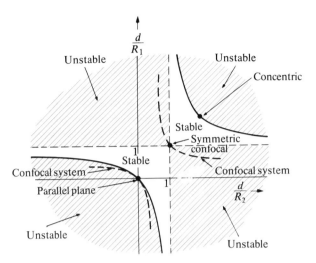

Fig. 13.11. Stability diagram for laser resonators.

10. TYPES OF LASERS

Maiman [13] constructed a successful experimental laser a year and a half after Schawlow and Townes [18] published their theoretical analysis of the device. The theory emphasized gases and continuous operation; the experiment used a solid ruby with pulse excitation. As in masers, there must be population inversion in lasers, which in the case of solid-state lasers requires a pumping power of several hundred watts. The heating effect of this large amount of pumping power makes continuuous laser operation difficult to achieve. In the case of a ruby (Cr^{3+} in Al_2O_3) laser, there is a broad absorption band in the green and others in the ultraviolet. When excited through these bands, the crystal emits a moderate number of sharp lines in the deep red (~ 7000 Å). There are, however, a number of competing lines in ruby and most atoms lose their excitation entirely without radiation. In Fig. 13.2(b), level 3 of ruby is a broad band of states while levels 1 and 2 are sharp.

Shortly after Maiman announced his invention, Javan, Bennett, and Herriott [11] succeeded in constructing a continuously operating gas laser of very high monochromaticity and small angular divergence. In this connection, note that laser action has also been achieved in liquids [12].

Figure 13.12(a) is a schematic diagram of a ruby or neodymium-glass (Nd^{3+} in glass) laser. If the ends of the ruby or glass rod are uncoated or coated with nonreflecting dielectric films, or if they are cut at Brewster angles, the device functions as an amplifier. If the ends are coated with reflecting dielectric layers or silver, or if one end is cut to make a 90° reflecting prism, the device functions as an oscillator. As shown in the figure, the ruby or glass rod is surrounded by a spiral xenon flash lamp used for pumping.

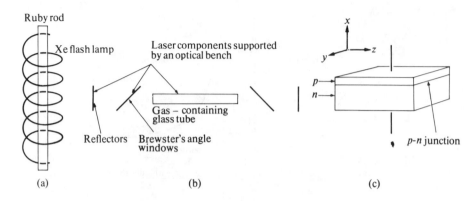

Fig. 13.12. (a) Solid state laser. (b) Gas laser. (c) Injection laser.

A schematic diagram of a gas laser is shown in Fig. 13.12(b). The mixed gases in this laser can be He-Ne, Ne-O, N_2-CO_2, or rare gases; helium neon is the most common. The amplifying gas medium is placed between windows oriented at the Brewster angle, and the whole assembly is placed between two dielectric-layer or silvered reflectors. Pumping is accomplished by passing an electric current of sufficient magnitude through the gas to cause a gaseous discharge; thus population inversion is achieved by collision excitation in which energy is transferred from one species to another, thereby raising the latter to a higher energy. If the active medium is a single vapor, say potassium vapor, a potassium discharge lamp shining on it will create inversion. In this way, ions are raised to a higher energy state and can emit radiation by returning to a level of lower energy.

For a semiconducting or injection laser, usually made of a GaAs crystal, inversion is obtained by electrical injection of charge carriers across the p–n junction. This type of laser has the advantages of small size and simple excitation requirements, but its monochromaticity (frequency purity) is of an order of magnitude lower than that of the solid state laser; the gas laser has the highest monochromaticity, an order of magnitude again better than that of the solid state laser. In general, monochromaticity is higher for an active medium whose atoms have smaller interactions. Thus the low-density gas in a gas laser should give rise to high monochromaticity because its atoms are far apart and their interactions are therefore small.

In all types of lasers, the light emitted by an atom in the general direction of the axis of the cavity will be amplified and grow as it bounces back and forth between the plates. Light traveling in any other direction will leave the system before making many passes and will not interact strongly with the medium. For this reason, laser outputs are unidirectional and thus can be focused to an extremely high intensity by optical lenses.

Successful injection laser experiments were first performed nearly simultaneously at three laboratories, General Electric, IBM, and the MIT Lincoln

Laboratory [10, 15, 17]. The operation of this kind of laser is sufficiently different from that of the solid state and gaseous types to warrant further discussion.

Figure 13.12(c) is a schematic diagram of an injection laser. The *p–n* junction is forward biased so that pumping current flows across the junction in the *x*-direction; the laser light propagates in the *z*-direction, or parallel to the plane of the junction. Laser action results from the radiative recombination of electrons and holes injected across the band gap of a semiconductor. However, for laser action to occur, the absorption of the laser radiation by the free carriers must be smaller than the induced emission. It turns out that for semiconductors such as GaAs, the intensity of induced emission from an inverted population (electron-hole combination) usually exceeds induced absorption of electrons into higher levels in the conduction band (free-electron absorption). Laser action has also been observed in InP, InAs, InSb, and alloys of GaAs-GaP, GaInAs, and InAsP.

Injection-diode lasers differ considerably from other lasers. They are very small; dimensions of $100\ \mu \times 100\ \mu \times 500\ \mu$ are typical. With proper preparation, the light comes out parallel to the long dimension and originates from an active region a few microns thick on the *p*-side of the *p–n* junction. The threshold current density for laser action varies from less than 100 A/cm^2 at temperatures below 10°K to 10^5 A/cm^2 at room temperature. At low temperatures, c.w. operation at input powers of 1 W has been achieved.

Finally, injection laser amplifiers as well as oscillators have been successfully constructed.[8]

11. LASER APPLICATIONS

The applications of the laser can be broadly divided into four categories: (1) those that combine the high directivity and monochromaticity of the laser beam; (2) those that utilize the intense power concentrated in the laser beam; (3) communications, and (4) coherent-light image and data processing. We shall now briefly describe some of these applications [22].

Since the power density and monochromaticity of lasers are of orders of magnitude higher than those of other optical sources, they can be used to examine large objects more precisely, or to reduce the sensitivity demands on the detectors. For example, the optical homogeneity and surface planarity of high-grade optical prisms, flats, etc., are now factory-tested by laser interferometry; surface figuring to compensate for a small amount of optical inhomogeneity is included in the manufacturing process if required. In other words, interferometric measurements using lasers can be coupled to the automatic control of machine tools fabricating the objects in question. The intense short light pulses from a laser can be utilized to record extremely short transient events on photographic plates. Lasers can also be used for long-distance and angle measurements.

At present, lasers are not highly tunable. To obtain new laser frequencies,

---

[8] For an extensive list of references on injection lasers, see Smith and Sorokin [22].

one can use the high-peak electric field available to generate harmonics by focusing the beam on, for example, a specimen of crystalline quartz. Alternatively, sum and difference frequencies can be obtained by mixing laser beams of different frequencies in a nonlinear medium.

In other applications, we are more interested in effects directly related to the power density than in the nonlinear effects resulting from the associated high electric field. Intense local heating by a laser beam can result in melting and partially vaporizing a small spot, plus punching out the spot by the tremendous radiation pressure in the beam. Thus, lasers can be used for the machining of small dimensions, as precision cutting tools, or as welding tools. Local heating is used in biological and medical applications, ranging from the destruction of accessible cancer to localized experiments on single cells or portions of cells.

We must be able to modulate lasers used as communication tools at a high, preferably microwave rate if they are to carry sufficient information to justify their application. The problem of laser modulation by a Kerr shutter or by a KDP crystal using the electro-optic deflection effect and other means is under intense investigation at present. The laser can be used for communication in outer space—between satellites, for example—although for terrestial or space-ground communication its usefulness is hampered by cloud and fog formations. For point to point private communication, however, the laser is clearly superior to radio communication; since it travels in a straight line and has a small cross section, the message contained in the beam by modulation cannot be easily intercepted, as is the case with relatively low-frequency radio waves.

A laser can be used to achieve three-dimensional image reconstruction by a process known as *holography*. In this connection, we note that the density of a photographic plate is a function of the light *intensity* rather than its *complex amplitude*. In other words, all phase information is lost in the recorded light of ordinary photography. However, if the reflected light from an object illuminated by a laser beam is superimposed on part of the illuminating laser beam serving as phase reference, the phase information of light arriving from the object is preserved. If the superimposed light is used to expose a photographic plate, a hologram is formed, composed mainly of groups of concentric circles. If the hologram is illuminated by a laser beam serving as phase reference, a three-dimensional image is formed between the laser output and the hologram; we can see this image by looking into the hologram toward the general direction of the laser. This three-dimensional image has several uses, among them data (image) processing and data (image) storage.

The most important sequential processes of logical operations in computers are essentially switching processes, where two input signals A and B proceed down two transmission lines to join at a switch and send a third output signal C down a third transmission line. Representative logic circuits have been constructed using lasers. A simple example is the invert circuit, in which laser A quenches laser B; when A is on, its radiation stimulates emission of radiation from B transverse to the lasing direction, thus depleting the population inversion of B and even causing it to fall below the level necessary to sustain laser oscillations. In other words, logic is performed in this case, since the output of laser B is either 0 or 1 (in computer technology notation) depending on whether or not laser A is on (in state 1 or 0). Thus the output of B represents the signal C discussed above. For size, economy, and fabrication reasons, injection lasers are used in these cases. It is not clear at this time whether or not 'laser logic' will eventually be compatible with that of electrically interconnected transistors.

REFERENCES

1. N. Bloembergen, *Phys. Rev.* **104**, 324 (1956).
2. G. D. Boyd and J. P. Gordon, *Bell Sys. Tech. Jour.* **40**, 489–508 (1961).
3. G. D. Boyd and H. Kogelnik, *Bell Sys. Tech. Jour.* **41**, 1347 (1962)
4. P. Connes, *Revue d'Optique* **35**, 37–42 (1956); *Jour. Phys. Radium* **19**, 262–269 (1958).
5. R. W. DeGrasse, E. O. Schulz-DuBois, and H. E. D. Scovil, *Bell Sys. Tech. Jour.* **38**, 305 (1959).
6. R. H. Dicke, *U.S. Patent No.* 2, 851,652, Sept. 9, 1958.
7. A. G. Fox and T. Li, *Bell Sys. Tech. Jour.* **40**, 469–472 (1961).
8. J. P. Gordon, H. J. Zeiger, and C. H. Townes, *Phys. Rev.* **99**, 1264–1274.
9. G. Gouban and F. Schwering, *Inst. Radio Engrs. Trans. Prof. Group Ant. Prop.* **AP-9**, 248 (1961).
10. R. N. Hall, G. E. Fenner, J. D. Kingsley, T. J. Soltys, and R. O. Carlson, *Phys. Rev. Letters* **9**, 366 (1962).
11. A. Javan, W. R. Bennett, Jr., and D. R. Herriot, *Phys. Rev. Letters* **6**, 106 (1961).
12. A. Lempicki and H. Samuelson, *Phys. Letters* **4**, 133 (1963).
13. T. H. Maiman, *Nature* **187**, 493 (1960); *Brit. Commun. Electron* **7**, 674 (1960).
14. K. W. Meissner, *Jour. Opt. Soc. Am* **31**, part I, 405–427; **32**, part II, 185–211 (1942).
15. M. I. Nathan, W. P. Dumke, G. Burns, F. H. Dill, Jr., and G. J. Lasher, *Appl. Phys. Letters* **1**, 62 (1962).
16. A. M. Prokhorov, *Soviet Physics JETP* **7**, (trans.) 1140–1141 (1958).
17. T. M. Quist, R. H. Rediker, R. J. Keyes, W. E. Krag, B. Lax, A. L. McWhorter, and H. J. Zeiger, *Appl. Phys. Letters* **1**, 91 (1962).
18. A. L. Schawlow and C. H. Townes, *Phys. Rev.* **112**, 1940–1949 (1958).
19. S. A. Schelkunoff, *Bell Sys. Tech. Jour.* **15**, 92 (1936); *Phys. Rev.* **56**, 308 (1939).
20. H. E. D. Scovil, G. Feher, and H. Seidel, *Phys. Rev.* **105**, 762 (1957)

21. C. P. Slichter, *Principles of Magnetic Resonance*, Harper and Row, New York, 1963; pp. 6–7.

22. W. V. Smith and P. P. Sorokin, *The Laser*, McGraw-Hill Book Co., New York, 1966; p. 408.

23. R. F. Soohoo, *Proc. Inst. Elec. Engrs.* **51,** 70 (1963).

24. J. Stratton and L. J. Chu, *Phys. Rev.* **56,** 99 (1939).

PROBLEMS

13.1 A paramagnetic salt contains 10^{21} ions/cm^3 with a magnetic moment of 1 Bohr magneton (μ_β) per ion. Calculate the magnetic field required to have 10% more magnetic moment aligned parallel than antiparallel to the field at 4°K (liquid helium temperature).

13.2 In a two-level maser, population inversion can be accomplished via velocity selection or adiabatic passage, as discussed in the text. Devise another method of population inversion in a two-level maser.

13.3 Equation (13.38) gives the expression for the emitted power P_e of a three-level maser in the derivation of which we assumed that $h\nu_{21}, h\nu_{32} \ll kT$. Derive the corresponding expression for a laser, for which the approximation $h\nu_{21}, h\nu_{32} \ll kT$ may be invalid.

13.4 a) Show how a push-pull maser can be constructed out of four nondegenerate energy levels; that is, in this scheme, the population of the upper signal level is being reinforced while the lower signal level is being depleted. State the requirements for the transition probabilities between different levels.

b) Write the equations of state for the push-pull maser of (a).

APPENDIX A

LORENTZ TRANSFORMATION

In Section 2 of Chapter 5, we showed that whereas Observer O in the unprimed system described an event by the equation

$$x^2 + y^2 + z^2 = (ct)^2, \tag{5.10}$$

Observer O' in the primed system described it by another equation;

$$x'^2 + y'^2 + z'^2 = (ct')^2. \tag{5.11}$$

Here x, y, z refer to the location of point P in the unprimed system and x', y', z' refer to that of point P' in the primed system, as depicted in Fig. 5.3. As noted, the primed and unprimed axes are respectively parallel, their origins being coincident at $t = 0$, and the x- and x'-axes slide along each other at a constant relative velocity u.

In accordance with the theory of relativity, we have not assumed that time intervals t and t' measured respectively in the unprimed and primed systems are identical. We will now show that t and t' are in general unequal and that the Galilean transformation (5.9) corresponds to the limiting case of the Lorentz transformation, where the relative velocity u is much less than the velocity of light c.

Inasmuch as the relativity motion of the two frames is assumed to be entirely in the x-direction, the Galilean transformation relations for y and y', and z and z' are still applicable even if u is comparable to c. Thus, from Eq. (5.9), we have immediately

$$\boxed{\begin{aligned} y' &= y, \\ z' &= z. \end{aligned}} \tag{A.1}$$

Furthermore, since every motion that is uniform and rectilinear in x, y, z must also appear uniform and rectilinear in x', y', z', the transformation from (x, t) to (x', t') takes straight lines into straight lines and is therefore *linear*. It follows that

$$x' = \alpha_1 x + \alpha_2 t, \tag{A.2}$$

$$t' = \alpha_3 x + \alpha_4 t, \tag{A.3}$$

where α_1, α_2, α_3, and α_4 are constants which must be evaluated.

Appendix A

First, consider a point P' (x', y', z') fixed with respect to observer O'. For this point $dx' = 0$. It follows from an implicit differentiation of Eq. (A.2) that

$$0 = \alpha_1 dx + \alpha_2 dt. \tag{A.4}$$

However, since point P' (x', y', z') appears to be moving in the positive x-direction at speed u when observed by O, we can set dx/dt in Eq. (A.4) equal to u. Thus Eq. (A.4) yields the relation between α_1 and α_2 as

$$\alpha_2 = -u\alpha_1. \tag{A.5}$$

It further follows from Eq. (A.2) that

$$x' = \alpha_1(x - ut). \tag{A.6}$$

We can determine the remaining three constants, α_1, α_3, and α_4, by requiring that Eqs. (5.10) and (5.11) transform into each other, as they must since they are describing the same event. Substituting Eqs. (A.1), (A.3), and (A.6) into Eq. (5.11), we have, after rearranging,

$$(\alpha_1^2 - c^2\alpha_3^2)x^2 + y^2 + z^2 = (c^2\alpha_4^2 - \alpha_1^2 u^2)t^2 + (2\alpha_1^2 u + 2c^2\alpha_3\alpha_4)xt. \tag{A.7}$$

If Eqs. (5.10) and (A.7) are to be identical, we must have

$$\begin{aligned}\alpha_1^2 - c^2\alpha_3^2 &= 1, \\ c^2\alpha_4^2 - \alpha_1^2 u^2 &= c^2, \\ 2\alpha_1^2 u + 2c^2\alpha_3\alpha_4 &= 0.\end{aligned} \tag{A.8}$$

Solution of these equations yields

$$\alpha_1^2 = \alpha_4^2 = \frac{1}{1 - (u^2/c^2)}, \tag{A.9}$$

$$\alpha_3 = -\frac{u}{c^2}\alpha_1. \tag{A.10}$$

Combining Eq. (A.9) with Eq. (A.6), we find

$$\boxed{x' = \kappa(x - ut),} \tag{A.11}$$

where $\kappa = 1/\sqrt{1-(u^2/c^2)}$. Similarly, combining Eqs. (A.9) and (A.10) with Eq. (A.3), we obtain

$$\boxed{t' = \kappa\left(t - \frac{u}{c^2}x\right).} \tag{A.12}$$

In summary, we have derived the four Lorentz transformations involving space and time, given by the bracketed equations (A.1), (A.11), and (A.12) [found in the text as Eq. (5.12).] Note that if u^2 and $u(x/t)$ are both much less than c^2, the Lorentz transformation (5.12) reduces to the Galilean transformation (5.9), as expected. These restrictions imply that, for the Galilean transformation to hold, both the relative velocity of the reference frames and the velocity of a point in a given frame must be small compared to c.

APPENDIX B

MAXWELL BOLTZMANN DISTRIBUTION

Let us first divide the range of energies that particles may have into a large number of n intervals, so chosen that each contains a large number of quantum states (degenerate or otherwise). For the ith interval, the number of quantum states and the average energy of states within the interval are denoted by S_i and E_i respectively. The number of particles occupying states in the ith interval is denoted by N_i. Since we assume that both the total number of particles and their total energy are conserved, we have

$$\sum_{i=1}^{n} N_i = \text{constant}, \tag{B.1}$$

$$\sum_{i=1}^{n} N_i E_i = \text{constant}. \tag{B.2}$$

Maxwell-Boltzmann statistics treats the particles as distinguishable and count each permutation of particles among levels as a different distribution. The statistics are applicable to weakly interacting systems such as particles in a low density gas or electron spin states in a dilute paramagnet dissolved in a diamagnetic host (as in the case of some maser materials). The number of ways W in which N_i distinguishable particles can be divided among S_i states is given by

$$W = N_n! \prod_i \frac{S_i^{N_i}}{N_i!} \tag{B.3}$$

and

$$\ln W = \sum_i (\ln N_n! + N_i \ln S_i - \ln N_i!). \tag{B.4}$$

For large N_i, we can simplify Eq. (B.4) by applying Stirling's formula:

$$\ln N_i! \simeq N_i \ln N_i - N_i.$$

Thus we have

$$\ln W = \sum_i (N_n \ln N_n - N_n + N_i \ln S_i - N_i \ln N_i + N_i). \tag{B.5}$$

We must now find the particular values of N_i that maximize W, subject to the constraints of (B.1) and (B.2). Using Lagrange's method, we take the differential of (B.1) with respect to N_i and multiply it by a constant $-\alpha$. Next we take the differential of (B.2) with respect to N_i and multiply it by another constant $-\beta$. We then add both resulting equations to $d(\ln W)/dN_i = 0$ to obtain

$$\ln \frac{N_i}{S_i} = -\lambda - \beta E_i. \tag{B.6}$$

Raising each side to the eth power, we find

$$\frac{N_i}{S_i} = C e^{-\beta E_i}, \tag{B.7}$$

where C is a constant. The ratio N_i/S_i is merely the probability P_i that a state in the ith energy interval is occupied (i.e., the time average of the number of particles in a state in the interval i). If all states within the interval i are degenerate, the ratio N_i/S_i of Eq. (B.7) is equivalent to N_i^e of Eq. (13.1). (For simplicity, however, we will not use the superscript e to denote thermal equilibrium distribution here.)

We must now show that $\beta = 1/kT$, where k is Boltzmann's constant and T is the absolute temperature. Consider one gram-molecule of a perfect gas, consisting of N monatomic molecules which interact only during collision; hence N is Avogadro's number. The energy E of this particle is solely kinetic:

$$E = \tfrac{1}{2} m v^2 = \tfrac{1}{2} m(v_x^2 + v_y^2 + v_z^2), \tag{B.8}$$

where m and v are the particle mass and velocity respectively. The normalized velocity distribution satisfying

$$\int_{-\infty}^{+\infty} \int_{-\infty}^{+\infty} \int_{-\infty}^{+\infty} p(v_x, v_y, v_z) \, dv_x \, dv_y \, dv_z = 1 \tag{B.9}$$

is

$$p(v_x, v_y, v_z) \Delta v_x \Delta v_y \Delta v_z = \left(\frac{m\beta}{2\pi}\right)^{3/2} e^{-1/2 m\beta(v_x^2 + v_y^2 + v_z^2)} \Delta v_x^2 \Delta v_y^2 \Delta v_z^2. \tag{B.10}$$

The total energy E_T of the system is equivalent to $N\bar{E}$, where $\bar{E}$ is given by (B.8) and $\overline{v_x^2}$, $\overline{v_y^2}$, and $\overline{v_z^2}$ can be calculated from the integrals

$$\overline{v_x^2} = \int_{-\infty}^{+\infty} \int_{-\infty}^{+\infty} \int_{-\infty}^{+\infty} v_x^2 \, p(v_x, v_y, v_z) \, dv_x \, dv_y \, dv_z, \tag{B.11}$$

etc., and

$$\bar{E} = \tfrac{1}{2} m(\overline{v_x^2} + \overline{v_y^2} + \overline{v_z^2}). \tag{B.12}$$

Carrying out this integration, we obtain

$$E_T = 3N/2\beta. \tag{B.13}$$

The specific heat of the system at constant volume is equal to dE_T/dT as well as to $\frac{3}{2}R$ from ideal-gas theory, where R is the gas constant.[1] Thus we find

$$\frac{3}{2}N\frac{d(1/\beta)}{dT} = \frac{3}{2}R \tag{B.14}$$

or

$$\frac{1}{\beta} = \frac{R}{N}T = kT. \tag{B.15}$$

---

[1] From kinetic theory, $E_T = N(\frac{3}{2}kT)$ so that $dE_T/dT = \frac{3}{2}R$.

APPENDIX C

EINSTEIN COEFFICIENTS

In this Appendix we shall give a classical derivation of the Einstein coefficients C_{ij} and C_{ji} used in Section 3 of Chapter 13. Consider the interaction of an atom with a blackbody radiation field at temperature T whose energy density per unit frequency range centered about frequency v_{ji} is given by[1]

$$\rho(v_{ji}, T) = \frac{8h^3_{ji}}{c^3} \frac{1}{e^{hv_{ij}/kT} - 1}, \qquad \text{(C.1)}$$

where h is Planck's constant, k is Boltzmann's constant, and c is the velocity of light. In general, we would expect the induced transition probability C_{ij}, defined as the induced probability per unit time that an electron in state i makes an upward transition to state j, to be proportional to ρ:

$$C_{ij} = C'_{ij}\rho(v_{ji}, T). \qquad \text{(C.2)}$$

Similarly, C_{ji}, the induced probability per unit time that an electron in state j makes a downward transition to state i, is given by

$$C_{ji} = C'_{ji}\rho(v_{ji}, T). \qquad \text{(C.3)}$$

Whereas the upward transition probability per unit time for an electron is equal to C_{ij}, the downward transition probability is given by[1]

$$C_{ji} + P_{ji},$$

where P_{ji} is defined as the spontaneous transition probability per unit particle per unit time.

Equating the two total transition probabilities, we have

$$C'_{ji}\rho(v_{ji}, T) + P_{ji} = \frac{N_i}{N_j} C'_{ij}\rho(v_{ji}, T). \qquad \text{(C.4)}$$

if we assume that $N_i/N_j = e^{(E_j - E_i)/kT}$, in accordance with the Maxwell-Boltzmann statistics given by Eq. (13.1), Eq. (C.4) becomes

$$\rho(v_{ji}, T) = \frac{P_{ji}}{C'_{ij} e^{hv_{ji}/kT} - C'_{ji}}. \qquad \text{(C.5)}$$

---

[1] Here we implicitly assume that $T \to 0$ as $\rho \to 0$ so that we can neglect thermal transitions from state i to j for finite ρ. If $\rho \neq 0$, T is finite and is designated as the *spin temperature*.

If this expression for ρ is to be consistent with that given by (C.1), the following relations between coefficients must be satisfied:

$$C'_{ij} = C'_{ji} \tag{C.6}$$

and

$$P'_{ji} = \frac{8\pi h v_{ji}^3}{c^3} C'_{ji}. \tag{C.7}$$

It further follows from Eqs. (C.2), (C.3), and (C.6) that

$$C_{ij} = C_{ji}. \tag{C.8}$$

We assumed this relationship in Section 3, Chapter 13.

INDEX

Adler, R., 182
Adler tube, 160, 176–178
ammonia, inversion frequency of, 241
ammonia maser (*see* maser, ammonia)
Ashkin, A., 182
atomic clock, 230, 243
atomic system, equation of state for, 1
attenuation constant, 6, 10
Avagadro's number, 261
avalanche breakdown, 202, 203–205
  breakdown condition in, 205
  ionization coefficient, 204
  multiplication factor in, 204–205, 206
avalanche diodes, 5, 84, 205–206
  application of, 206
  breakdown voltage of, 205
  comparison with Zener diodes, 205–206
avalanche $p-n$ junction devices, 210–211
  application of, 211
avalanche-transit-time diodes, 206–208
  operating characteristics of, 210–211

backward wave amplifier, 124
backward wave oscillator, 4, 120, 121–125
  direction of energy flow in, 121
  feedback via electron beam in, 122
  folded line structure used in, 108
  initiation of oscillation in, 133
  interdigital structure used in, 108
  lossy structure in, 124
  oscillation condition for, 122-124
  phase and group velocities in, 113
  transmission line model for, 121–122
  waves in, 122–125
  wide frequency band of, 125
band structure, 1
  of heavily-doped $p-n$ junction, 4
  of semiconductor, 5

Baranowski, J. J., 211
Barber, M. R., 228
Batdorf, R. L., 212
bean coupling, 81, 87, 91
bean loading, 81, 87, 91
Beck, A. H., 128
Bennett, W. R., Jr., 251, 255
Bessel function, 20
binomial theorem, 244
blackbody radiation, 263
Bloembergen, N., 255, 239
Bobroff, D. L., 182
Boltzmann's constant, 135, 193, 230, 261, 263
boundary conditions, electromagnetic, 1, 3, 4, 6, 11–15, 101, 102, 127–128, 219
  at conductor of finite conductivity, 23
  for normal component of **B**, 13
  for normal component of **D**, 13
  for tangential component of **E**, 14
  for tangential component of **H**, 15
  for tangential **E** at perfect conductor, 24, 106, 160, 244, 246
  for tangential **H** at perfect conductor, 24
Bowman-Manifold, 144
Boyd, G. D., 250, 255
Brady, D. P., 227
Branch, G. M., 182
Brand, F. A., 211
Brewster angle, 247, 251, 252
Brillouin field, 147, 152
  for TWT, 149
Brillouin flow, 145–148, 149, 150, 152, 153, 161
bulk semiconductor as microwave amplifiers, 184
Burns, G., 255
Burrus, C. A., 211

cadmium telluride, 221
capacitance, depletion layer, 4
capacitive probe, 214
Carlson, R. O., 255
cathodes,
    emission current density from, 137, 138–139, 141, 142
    slope in microwave tubes, 142, 143, 144
    space-charge-limited operation of, 139–141, 142
    temperature-limited operation of, 139–140
cathode material,
    composite, 138–139
    melting temperature of, 138
    oxide-coated, 138
    pure metal, 138
    work function of, 138, 139
cavities (*see* resonant cavities and laser cavity)
cavity gap,
    convection current in, 83–88
    displacement current in, 83
    equivalent admittance of, 87, 89, 94
    equivalent beam admittance of, 94–95
    equivalent circuit for, 74, 82, 83, 88, 94
    external admittance of, 88
    transit time of electrons in, 86, 98
characteristic impedance,
    ratio between, 32
    of TEM mode, 31
    of waveguides, 31–32, 33, 34, 35, 120–121
charge conservation law, 78–79, 84, 109
Chow, W. F., 211
Chu, L. J., 256
Chynoweth, A. G., 228
circuit equation, 83, 87, 110, 111, 122, 124, 125
classical mechanics, 229
coaxial lines, 29–30
    approximate cutoff wavelength for TE modes in, 30
    approximate cutoff wavelength for TM modes in, 30, 54
    cutoff frequencies of TE modes in, 29–30
    cutoff frequencies of TM modes in, 29–30
    in lighthouse tube, 70

Cohen, B. G., 212
collision frequency, 161
conduction current, 49
conductivity, conductor, 49
confined flow, 148–149
Connes, P., 248, 255
contact potential difference, 186, 196
continuity, equation of, 3, 49, 160
convection current, 49, 109
Cook, J. S., 116
Copeland, J. A., 227, 228
Coulomb repulsion forces between electrons, 134, 144, 154, 155, 156
    in cylindrical beam, 145, 146, 147, 148
    effect on modulation voltage, 160
Coulomb's inverse square, 55–56
Coulton, M., 135, 211, 228
crystalline field, 239, 240, 241
crystals,
    box model of, 135
    cubic, 240
    periodic nature of, 135, 218
Cuccia, C. L., 182
Cuccia electron coupler, 177–178
    noise elimination in, 177, 183
Cummings, R. L., 96
Cutler, C. C., 138
cyclotron frequency, 148, 176, 178
cyclotron waves, 160, 176–178

Debye length, 161
decay time, of charge in conductor, 10
DeGrasse, R. W., 255
DeLoach, B. C., 211, 212
density of states, 136, 137, 196
Denton, R. T., 182
diamagnetic crystal, 240
Dicke, R. H., 255
dielectric breakdown (*see* avalanche breakdown)
dielectric constant, of ionized gas, 161
dielectric loss, 242, 243
dielectric tensor, 2, 6
Dill, F. H., 255
directional coupler, 42, 43
dispersion, medium, 41, 42
displacement current, 16, 109
divergence theorem, 71

Dow, D. G., 228
drift velocity, in conductor, 49
Druesue, M. A., 211
Dumke, W. P., 255

Eastman, 228
Einstein, A., 59
Einstein coefficients, 233, 263–264
  and power absorption, 234
Einstein mass-energy relation, 133
electric and magnetic forces, ratio between, 82
electrodes,
  shapes in microwave tubes, 142
  voltage variation between, 140–141
electrolytic tank, 144
electromagnetic field,
  discontinuity in, 12
  inertia of, 53
  interaction with electron beam, 2
  microscopic origin of, 7–8
  relationship between longitudinal and transverse components, 21–22, 31
  stress on cavity wall, 54
  stress tensor, 52–53
  in terms of longitudinal and transverse components, 31
electron beam,
  electron oscillation in, 155–156
  focusing of, 134
  formation of, 134
  magnetostatic focusing of, 144–154
  microwave devices using, 1, 3
  modulation of, 4
  preacceleration of, 4
electron emission, random effect of, 138
electron-field interaction, in cavity gap, 82–88
electron gun,
  design of, 142–144
  Pierce gun, 134
electron spin resonance, 239
electron velocity, relativistic, 133
electronic motion,
  convection current due to, 75
  displacement current due to, 75
  in electric field, 70
  induced current due to, 71–73, 75

electronics equation, 2, 84, 87, 88, 92, 111, 122, 125
electrostatic field, of charge complex, 57
electrostatic focusing, 145, 154, 161
electrostatic potential, of charge complex, 57
Elliott, R. S., 63
Emde, F., 20
energy absorption, spin in e.m. field, 234
energy band diagram, 4
energy conservation law, 77, 130, 136, 141, 146, 243, 260
  in complex system, 45, 50–52
energy density, 17, 50–52
energy, interaction, between spin and field, 240
energy levels,
  number of possible, 240
  paramagnet, as function of field, 240
energy relation, general, 50
Esaki diode, 184
Esaki, L., 211

Fabry-Perot resonator, 246, 247, 248, 250
Feher, G., 255
Fenner, G. E., 255
Fermi level, 135, 191, 194, 195, 203
Fermi surface, 228
Fermi-Dirac probability, 137, 196
Fermi-Dirac statistics, 135
ferrite circulator, 197–198, 242
ferrite isolation, 243
ferrite losses, 243
ferrites, 2, 6, 180
  use in TWT, 116
ferromagnetic amplifier, 180–181
  traveling-wave, 181
Field, L. M., 108, 144
Fisher, T., 211
Floquet's periodicity theorem, 100–102, 103
Fox, A. G., 250, 255
Fukatsu, Y., 211
Fukui, H., 211

gain—bandwidth product,
  limitation, 4, 66, 74–75, 99
  of pentode amplifier, 75

Galilean transformation, 58, 257, 259
gallium arsenide, 197, 213, 214, 215, 217, 221, 222, 225, 226, 227, 228, 253
gallium phosphide, 221
garnets, 180
Gartner, W. W., 211
gas constant, 262
Gauss's theorem, 12, 13, 50, 145
Gibbens, G., 212
Gilden, M., 211
Gordon, J. P., 241, 255
Gouban, G., 255
group velocity, 17, 40–42, 75
  and energy propagation, 42
Gummell, H. K., 211
Gunn diodes, 1, 5, 161, 213, 228
  amplifier noise figure of, 226
  comparison of maximum power vs. frequency with other oscillators, 225–226
  contacts on, 213, 214, 225
  current waveform in, 218, 224
  dipole layer in, 222, 223
  efficiency of, 224, 225
  electron drift velocities in, 221, 222
  heat flow in, 225
  high field domain in, 5, 214, 217–218, 222, 223, 224, 226, 227
  maximum power vs. frequency for, 225
  microwave amplification using, 226–227
  nucleation centers in, 224
  oscillation frequency of, 214, 224, 225, 226, 227
  power output of, 224, 225
  requirements that band structure must satisfy in, 221
  space charge waves in, 227
  technique of producing, 225
  tuning of, 228
Gunn effect, 213–217, 221
  average electric field within high-field domain, 217, 223
  $\partial V/\partial t$ vs. x in, 214–216
  effect of circuit impedance on, 214, 224
  influence of light irradiation on, 214
  influence of magnetic field on, 214
  influence of surface condition on, 214
  period of oscillation in, 214

Gunn effect *continued*
  threshold electric field in, 213, 214, 215, 217, 221, 222
  velocity of high-field domain in, 217, 224
  $V(t)$ vs. t in, 214–215
Gunn, J. B., 217–228
gyromagnetic ratio, 241

Hahn, N. C., 154
Hakki, B. W., 228
Halkias, C. C., 212
Hall, R. N., 255
Hankel functions, 20
Haus, H. A., 182
Heffner, H., 182
helium dewar, 242, 243
helix, 75, 108, 116–121
  cold, 120
  impedance of, 120–121
  matching into, 120–121
helix, sheath model of, 116–117
  cutoff wave number in, 117, 119–120
  electromagnetic boundary conditions for, 117, 118
  field distribution in, 117–119
  phase velocity vs. frequency for, 120–121
  secular equation for, 119, 120
Helm, R., 144
Hemenway, C. L., 135, 211, 228
Henry, R. W., 135, 211, 228
Herriot, D. R., 251, 255
Higgins, V. J., 211
high-field emission, 134
Hilsum, C., 228
Hines, M. E., 138, 211, 212
Hoeffinger, B., 212
Hogan, C. L., 182
holography, 254
Huygen sources, 250

Ikola, R. J., 212
impact ionization, 205
IMPATT diodes, 1, 5, 211
impedance, input, 35
indium antimonide, 221
indium arsenide, 221
indium phosphide, 213, 214, 221
infrared maser, 229, 247

Jahnke, E., 20
James, R. P., 228
Javan, A., 251, 255
Jepsen, R. L., 182
Johnson, K. M., 212
Johnston, J. B., 212
Josenhans, J. G., 212

Kaminsky, G., 212
Kato, H., 211
KDP crystal, 254
Kennedy, W. K., 228
Kerr shutter, 254
Keyes, R. J., 255
Kingsley, J. D., 255
Kirchkoff's law, 165
Kito, S., 212
Klystron amplifier, 1, 4, 73–74, 76, 78–82, 229
  amplitude modulation of, 96
  antibunch in, 78
  bandwidth of, 82
  collector heating in, 97
  collector voltage in, 97
  conversion efficiency of, 82, 98
  crossing of electron trajectories in, 155
  density modulation in, 74, 76–78, 158
  electron bunches in, 74, 77, 78, 89
  with floating cavity, 82, 98
  frequency modulation in, 98
  frequency translation using, 96
  gain of, 96
  with more than three cavities, 82
  optimum drift tube length in, 81
  output current in, 79, 81, 89, 96–97
  phase modulation of, 96–97, 98
  power output of, 81
  pulse modulation of, 96
  sawtooth modulation for, 82, 98
  shot noise in, 96
  source of power in, 163
  space charge effect in, 81, 82
  transit time between cavities in, 79, 89, 91, 97
  use for voltage or power amplification, 96
  use in linear accelerator, 126–127
  use in microwave transmitters, 96

Klystron amplifier *continued*
  velocity modulation in, 74, 76–77, 98, 158
Klystron oscillator, 76, 88–91
  equivalent circuit for, 90
  optimum feedback for, 90, 98
  phase shift in, 89, 91
  relative tuning of cavities in, 91
  starting current of, 90
Knight, K. L., 227
Knight, S., 228
Kogelnik, 250, 255
Kompfner, R., 116
Kovel, S. R., 212
Krag, W. E., 255
Kroemer, H., 228
Kronig-Kramers relation, 161

Lagrange's method, 261
Lagrangian, 151, 152
Lagrangian equations, 152
Langmuir-Child law, 142, 161
Laplace's equation, 143
laser, 5, 229, 230, 231, 243, 248, 256
  advantages of, 229
  applications of, 230, 253–255
  first successful experimental, 251
  gas, 247, 251, 253
  injection, 252, 253, 255
  liquid, 251
  modulation of, 230, 254
  monochromaticity of, 251, 252
  nature of output of, 252
  neodymium-glass, 251
  population inversion in, 231–234, 254
  ruby, 251
  solid state, 252, 253
  tunability of, 253
laser cavity, 244, 245, 246, 247, 248
  confocal, 247
  electromagnetic field in, 248–250
  losses in, 245, 246, 247, 248, 250
  mode interference in, 244, 245
  nonconfocal, 248
  reflectors in, 249, 251, 252
  stability diagram for, 250, 251
laser radiation, nature of, 230, 252

laser structures, 243–247
Lasher, G. J., 255
Lax, B., 255
lead inductance effects, 4, 66–70, 99
  methods of minimizing, 69–70
Lee, C. A., 212
Lempicki, A., 255
Levi, R., 139
Li, T., 255
lighthouse tube, 70
linear accelerator, 4, 55, 99, 114, 125–129
  Coulomb repulsion between electrons in, 126
  field distribution in, 127–128
  magnetostatic focusing for, 65, 126
  ω–β diagram for, 129
  periodic length in, 126
  relativistic effects in, 55, 64–65, 99, 125–126
  secular equation for, 128–129
  use in high energy physics, 125
  waveguide mode used in, 107, 127
Liu, S. G., 212
longitudinal pumping, 181
Lorentz condition, 46
Lorentz force law, 1, 3, 4, 45, 47, 49, 52, 60, 61, 62, 71, 72, 82, 109, 110, 131
Lorentz, H. A., 7, 60
Lorentz transformation, 58–60, 62, 257–259
Louisell, W. H., 182
LSA, 5, 213, 226
  advantages of, 226
  operating characteristics for, 226

MacLachlan, N. W., 151, 154, 182
magnetic moment, 231
magnetization, 245
  microscopic origin of, 8
magnetostatic focusing, 81
  in TWT, 145
  weight of magnetic structure in, 149
magnetron oscillator, 4, 76, 99, 126, 129–132
  anode structure, 129–132
  back heating in, 132
  compared with TWT and BWO, 129

magnetron oscillator *continued*
  cutoff magnetic field for, 130
  cycloid drift velocity and phase velocity in, 131
  efficiency of, 129
  electron trajectories in, 132
  energy exchange between electrons and r.f. field in, 132
  linear, 131
  mode separation in, 132
  operating frequencies of, 129
  π-mode of, 131, 133
  power output of, 129
  resonant modes in, 130
  source of power in, 163
  tuning of, 132
Maiman, T. H., 251, 255
Marcuwitz, N., 33
Marinaccio, L., 212
masers, 5, 229, 230, 231, 243, 251
  advantages of, 229
  ammonia, 234, 235, 243
  bandwidth of, 230, 243
  emitted power from, 238, 239, 256
  first three level, 239
  first two level, 241
  frequency stability of, 230
  as frequency standard, 243
  gain-bandwidth product of, 242
  gain of, 243
  material, 5
  noise in, 229, 242
  optimum value of N in, 239
  push-pull, 256
  radiation power of, 5
  relation between pump and signal frequencies, 236, 238
  signal transition in, 239
  source of power in, 163
  spontaneous emission in, 229
  three level, 235, 236, 237, 239, 240, 241, 256
  traveling wave, 243
  two level, 234, 237, 241, 256
Mathieu's equation, 154, 165
  solution to, 165–166
Maxwell-Boltzman statistics, 136, 260, 263
Maxwell-Lorentz equations, 7

Maxwellian distribution, 137, 141, 260–262
Maxwell's equations, 1, 2, 3, 4, 6–8, 11, 12, 21, 22, 45, 47, 49, 57, 71, 101, 109, 117, 127, 139, 145, 157, 160, 219, 248
McCumber, D. E., 228
McIntyre, R. S., 212
McWhorter, A. L., 255
Meissner, K. W., 255
Mendel, J. T., 149
Metcalf, G. F., 154
Michelson, A. A., 59
microwave electronics,
  conventional devices in, 1, 4
  iteration solution to problems in, 4
  scope of, 1
  solid state devices in, 1
microwave generation and amplification,
  in bulk semiconductors, 1, 5, 213
  in junction semiconductors, 1, 4, 184, 213
  in Read diode and in p–n junctions, 208
Mihran, T. G., 182
Millman, J., 212
Misawa, T., 212
momentum conservation in composite system, 45, 52–53
Morley, E. W., 59
Mosher, C. H., 228

Napoli, L. S., 212
Nathan, M. I., 255
Newton's second law, 3, 4, 72, 109, 110, 131, 147, 157
noise figure, amplifier chain, 229–230
North, D. O., 73

Ohm's law, 49, 219
optical maser, 229, 247
orbital angular momentum, 239

paramagnet, definition of a, 239
paramagnetic resonance, 240, 241
  power saturation in, 234
paramagnetic specimen, 229, 239, 240, 241, 242, 243, 256
parametric amplifiers, 1
  amplification in, 171–173
  coupling element in, 4, 163, 167, 170, 181–182, 183

parametric amplifiers *continued*
  coupling element nonlinear and time varying, 175, 181
  degenerate, 168
  as down converter, 167, 168, 174
  electron beam as coupling element in, 163, 164, 175–178
  energy conversion in, 163, 168, 176, 181
  ferrite as coupling element in, 163, 164, 180–181
  frequency mixing in, 163, 166, 170, 181
  general equivalent circuit, 167–168
  1/2 power bandwidth, 173, 174
  idler circuits in, 167, 168, 169, 170, 171, 172, 182, 183
  input-output isolation in, 168, 183
  mechanically-pumped, 163–167, 183
  negative resistance in, 171, 173, 184
  noise figure of, 167, 173–174, 176
  oscillation condition in, 171, 182
  power gain in, 173, 174
  proper phase of pump source in, 164, 168, 171, 183
  pump resonant circuit, 166–167, 169
  relationship between pump and source frequency, 164, 168
  semiconductor as coupling element 163, 164, 178–180
  signal amplification in, 168
  simplified equivalent circuit of, 169
  sources of noise in, 174
  summary of operation of 181–182
  as up converter, 167, 168, 174
parametron, 164
partition function, 230
Pauli exclusion principle, 135
pentode amplifier,
  equivalent circuit of, 67, 74
  gain of, 74
  half-power bandwidth of, 74–75
  input admittance of, 68
  input capacitance of, 66
  input resistance of, 69
periodic focusing, 149–154
  beam contours in, 152–153
  beam divergence in, 151–152, 153, 154
  comparison with Brillouin flow, 149–151
  optimum focusing in, 152–153

272 Index

periodic structure, 4, 75, 99
  anisotropic material in, 104
  direction of power flow in, 109, 125, 133
  disk-loaded cylindrical waveguide, 101, 107, 127
  group velocity of spatial harmonies in, 103, 104, 105–106
  higher order modes in, 107
  in presence of electron beam, 99, 109–114
  "nonsynchronous active modes" in, 124–125
  ω–β diagram of, 104–108, 132, 218–219
  pass band in, 99, 106, 107, 219
  passive properties of, 99, 100–108
  phase velocities of spatial harmonics in, 103, 104, 105–106
  reciprocal and periodic nature of, 104–106
  r.f. fringing field at input of, 115
  spatial harmonics in, 99, 100, 102–103
  spatial harmonic amplitudes in, 106–107
  stopbands in, 99, 106, 107, 219
permeability tensor, 2, 6, 15
perveance, electron gun, 142
phase constant, 6, 10, 111
phase velocity, 17, 75, 243
  in waveguides, 40–42, 99, 100
photo emission, 134, 161
photon, energy of, 234, 238
Pierce gun, 143–144, 145
Pierce, J. W., 114, 124, 125, 138, 143, 144, 148, 182
Planck's constant, 4, 218, 263
plane wave, 9–10, 117, 247
plasma,
  in magnetic field, 2, 6
  mode of oscillation in, 157
  in solids, 161
Plasma frequency, 156, 157, 158, 175
plasma oscillations, 134, 156–158, 160
p–n junction, 184, 252, 253
  barrier potential in, 186
  barrier thickness in, 186, 189
  barrier voltage of, 194, 195
  behaviour of, 178–179, 184–186
  heavily doped, 4
  impurity concentration in, 189, 190

p–n junction *continued*
  injection current in, 193, 202
  I–V characteristics of, 202
  punch-through voltage of, 207, 208, 209
  reverse saturation current of, 193, 205
Poisson's equation, 1, 4, 139, 141, 143, 185
polarization, microscopic origin of, 8
population difference, 237
population inversion, 231–234, 236, 237, 238, 241, 251, 253, 256
  by adiabatic passage, 241
  via collision excitation, 252
  methods of, 234, 237
  by preselection, 241
power flow, 17, 50–51
Poynting vector, 32, 42, 51
  average value of, 54
principle of detail balance, 232
Prokhorov, A. M., 247, 255
propagation constant, 6, 10, 17
  of transmission line with electron beam, 110, 112–113
  of waveguide, 18, 37

quantum electronics, 229
quantum mechanics, 229
quantum numbers, 239
Quate, C. F., 149, 182
Quist, T. M., 255

radiation pressure, 254
radio astronomy, 230
Ragan, G. L., 32
Ramo, S., 30, 33
rate equations, 232, 233, 237, 256
Read diodes, 184
  avalanche multiplication in, 208–209, 210
  comparison with p–n junction, 208, 210–211
  critical field in, 209
  doping profile of, 206–207, 208
  electric field distribution in, 207–208, 209, 210–211
  negative resistance in, 208
  phase relations in, 209–210, 212
  power output of, 210
  space charge region in, 208, 210
Read, W. T., 212

Rediker, R. H., 255
reflection coefficient, 17, 34
reflex klystron, 91–96, 145
  beam focusing in, 98
  coarse tuning of, 91, 97
  collecting of electrons in reflected beam of, 98
  electron-beam feedback in, 91
  electron bunch in, 92–93
  electronic tuning of, 91, 96, 97
  frequency deviation as function of voltages of, 95–96
  oscillation frequency of, 91
  output as function of voltages, 95–96
  output current, 93–94
  relationship between frequency and voltages in, 93, 95
  round trip transit time in, 92–93
  use in microwave measurements, 97
  velocity modulation in, 91, 93
relativistic mass, 61, 126
Relativity, special theory of, 58–60, 257
  electric and magnetic field transformation in, 62–63
  force transformation in, 60–61, 126
  fundamental postulates of, 59, 60
  Lorentz contraction in, 65
  time dilation, 65
  transformations in, 57–63
relaxation time, 236, 237, 238
  requirement in a three-level maser, 235, 236, 237
resonant cavities, 4, 17, 35–40, 243, 244
  complex frequency of, 35, 37, 39
  coupled, 107, 127, 129
  dielectric, 242
  dominant mode in, 244, 247
  energy stored in, 39
  external Q of, 38
  field in, 243
  finite wall losses in, 39
  foreshortened coaxial, 43–44
  loaded Q of, 38
  lumped analogon of, 35–36, 37, 39
  non-rectangular types, 39–40, 105
  power loss in, 39
  Q of, 35–36
  rectangular, 36–39, 244

resonant cavities *continued*
  relation between Q and complex frequency for, 36
  resonance frequencies of TE_{mnl} in, 39, 43
  TEM mode in, 44
  total Q of, 38
  unloaded Q of, 38
retarded potential, 45, 47, 54
Rich, J. A., 116
Richardson-Dushman equation, 137
Ridley, B. K., 228
Robinson, N. H., 182
Ruby, 240, 242, 251

Samuelson, H., 255
saturation pumping, 235
scalar potential, 46, 57, 139, 151
Scharfetter, D. L., 211
Schawlow, A. L., 247, 251, 255
Schelkunoff, S. A., 32, 255
Schrodinger's equation, 1, 4, 5, 132, 184, 187, 190, 218
  boundary conditions for, 132, 188, 218
  E–k diagram from, 133, 218
Schulz-DuBois, E. O., 255
Schwering, F., 255
Scovil, H. E. D., 239, 255
secondary emission, 134
Seidel, H., 255, 239
semiconductor,
  conduction band in, 5, 191, 194, 196, 202, 203, 219, 221, 228, 253
  drift velocity in, 221–222, 223
  effective carrier temperature, in, 219
  effective mass in, 195, 219, 220, 222
  E–k diagram of, 218–219
  electron-hole pair creation in, 203, 204, 208
  energy gap in, 191, 194, 195, 203, 253
  holes in, 203, 204, 209, 210, 212, 214
  impurity profile in, 1, 4, 5
  impurity scattering in, 220
  impurity states in, 191
  intrinsic, 191, 208
  I–V characteristic of, 5
  large scale impurity diffusion in, 212
  lattice temperature in, 219
  mobilities in, 219, 220, 221, 222

274 Index

semiconductor *continued*
  noise emitted by, 213
  recombination in, 253
  scattering-limited velocity in, 209, 210
  space charge instability in, 217
  subbands in, 219, 221, 222
  thermal ionization in, 203
  two-valley model of, 217–222
  valence band in, 5, 191, 194, 196, 202, 203, 219, 221
semiconductor devices, bulk,
  comparison with junction types, 213
  conditions for occurrence of negative resistivity in, 220, 221
  negative resistivity in, 213, 217, 219, 222, 223, 226, 227
Shockley, W., 228
shot noise, electron beams, 138, 229
Slater, J. C., 35
Slichter, C. P., 232, 241, 256
slow wave structure, 243
Smith, W. V., 256
Soltys, T. J., 255
Soohoo, R. F., 256
Sorokin, P. P., 256
space charge effects,
  in electron beams, 142, 143
  in vacuum tubes, 73
space charge waves, 134, 156, 159, 162, 163
  amplification of, 175–176
Spangenberg, K. R., 144
specific heat, 262
spin angular momentum, 239
spin states, 241
  distribution of electrons among, 1, 5, 231
  of electron in magnetic field, 137, 231
spin temperature, 263
spin-lattice relaxation time, 233, 234
standing wave ratio, 17, 34
state equation, 5, 235
statistical mechanics, 229, 230
statistical steady state, 230, 233, 234, 235, 237
stimulation emission, 229, 238, 253
Stirling's formula, 260
Stoke's theorem, 14
Stratton, J. A., 41, 42, 52, 256
Sugimoto, S., 212

Suhl, H., 116, 182
superconducting solenoid, 243
superheterodyne receiver, mixing in, 163
Swan, C. B., 212

TE modes, 22
  boundary conditions for, 24, 25, 28, 37
  cutoff frequency in cylindrical guide, 28, 29
  cutoff frequency in rectangular guide, 26, 27
  cutoff wave number in cylindrical guide, 28
  cutoff wave number in rectangular guide, 26
  cutoff wavelength in cylindrical guide, 28
  E_z of, 25
  permissible wave numbers in rectangular guide, 25
  relationship between longitudinal and transverse components for, 22
TEM mode, 26, 29, 43
Terman, F. E., 69
Teubner, B. G., 7
thermal energy, 221
  of emitted electrons, 141
thermal transitions, 232
thermionic emission, 134–139
thermodynamic equilibrium, 230, 232, 234, 235, 261
Thim, H. W., 228
Tien, P. K., 182
TM modes, 22
  boundary conditions for, 24, 25, 28
  cutoff wave number in cylindrical guide, 28
  cutoff wave number in rectangular guide, 26
  cutoff wavelength in rectangular guide, 26, 27
  E_z of, 25
  permissible wave numbers in rectangular guide, 25
  relationship between longitudinal and transverse components for, 22
Townes, C. H., 241, 247, 251, 255
Townsend discharge, 203

Index 275

transistors, 197, 255
transit time devices, 5
transit time effects, 4, 66, 70–74, 99
 grid admittance due to, 73
 grid conductance due to, 73
 grid impedance due to, 72
 minimization of, 73
transition probability
 induced, 233, 234, 235, 237, 239, 263
 natural 233, 235, 256, 263
transmission coefficient, quantum-mechanical, 188–190, 193, 195
transmission line, parallel plate, 16
 radiol, 128
transmission line equations, 110, 114
traveling wave, backward, 113, 114, 122
traveling wave tubes, 1, 4, 75, 99, 111, 113, 120, 229
 attenuator in, 116, 133
 frequency translation using, 96
 gain of, 114–116
 gain parameter of, 112, 113, 114–121
 lossy structure in, 124
 periodic structures for, 108
 source of power in, 163
 transmission line model for, 109, 114, 125
 undesirable oscillation in, 116
 waves in, 114–115
triodes
 input capacitance of, 66, 69
 input resistance of, 69
tunnel diodes, 1, 184, 191–193
 advantages and disadvantages, 197
 amplifier gain of, 201–202
 as amplifier or oscillator, 198–200
 applied voltage range of, 196
 barrier thickness in, 184, 186, 189, 190, 191, 195, 202–203
 characteristics, 4, 191–194, 196–197
 d.c. stability criterion for, 199, 200
 dynamic conductance of, 196, 199, 200
 energy band diagram of, 191–194
 equivalent circuit for, 198
 as high-speed bistable switch, 198–199
 hysteresis effect in, 199
 input-output isolation for, 197–198
 intrinsic stability of, 200
 $I–V$ characteristics of, 192, 198, 201, 212

tunnel diodes *continued*
 noise in, 197
 operating frequency range of, 197
 peak current in, 193, 196, 197
 peak forward voltage of, 196, 197
 semiconductor materials for, 197
 sensitivity to nuclear radiation of, 197
 series-connected, 224
 small signal equivalent circuit for, 196
 stability diagram of, 200
 stability of, 198–200
 transcient response of, 199
 tunneling current in, 193–196
 valley current of, 193, 196, 197
tunneling current, electron supply for, 184
tunneling, quantum-mechanical, 4, 134, 184–191, 202

vacuum tubes,
 density modulation and acceleration in, 72, 77
 high frequency limitation of, 66, 99
 residual gas in, 154, 162
Van Vleck, J. H., 7, 8
Vane, A. B., 228
varactor diodes, 178–180
Vartanian, P. H., 182
vector potential, 46–47, 151
velocity of light, constancy of, 59
Venohara, M., 227–228

Walker, L. R., 138
Wang, S., 212
Watkins, T. B., 228
wave equation, 6, 8–10, 117
 inhomogeneous, 45–47, 54
 and Schrödinger's equation, 5
wave equation, general for plasma, 160
wave equation, solution to, 8–11, 17–21, 22
 in cylindrical coordinates, 19–21
 for good conductor, 11
 for poor conductor, 11
 in rectangular coordinates, 17–19
wave function,
 of an electron, 70, 137, 187, 212
 in a metal, 136, 137
wave impedance, 6, 11, 35
 for TE modes, 22, 31, 245

wave impedance *continued*
    for TM modes, 23, 31
wave mechanics, 49
wave number, cutoff, 18
waveguides, 17
    below cutoff, 18, 44
    cutoff frequency of, 18, 38
    cutoff in, 243
    cutoff wavelength in, 18
    dimension and wavelength, 27
    guide wavelength in, 243
    impedance, 17, 34–35
    matching section between, 43
    obstacles in, 31, 33, 34
    ω–β diagram of, 41–42, 104–105
waveguides, cylindrical, 27–29, 245
    dominant mode in, 29, 245, 246
    maximum operating range of, 29
    mode fields as $f \to \infty$ in, 246–247
waveguide junction, 31, 43
    discontinuity susceptance at, 32–33
waveguide, rectangular, 25–26
    cutoff frequencies vs. guide dimension, 25
    dominant mode in, 26, 31, 36

waveguide, rectangular *continued*
    field distribution in dielectric-loaded, 133
    group velocity in dielectric-loaded, 133
    maximum operating range of, 26–27
    phase velocity in dielectric-loaded, 133
Weber, S. E., 116
Weiss, M. T., 182
Wiegmann, W., 212
WKB approximation, 199
work function, 135

Yocom, W. H., 149

Zeeman energy, 240
Zeeman levels, 5
Zeiger, H. J., 241, 255
Zener breakdown, 202, 203, 211
Zener, C., 212
Zener diodes, 5, 202–203
    application of, 206
    breakdown voltage of, 205
    comparison with avalanche diodes, 206
    depletion layer width, 202–203, 205–206
    doping level in, 202–203
    tunneling in, 203–204

Hollow cyl. w/G $\qquad D/\lambda \geq .586 \qquad TE_{11}$

$\qquad\qquad\qquad D/\lambda \geq .766 \qquad TM$

[rectangle diagram with height b and width a]

$\lambda_c = 2a$

$\dfrac{a}{\lambda} > .5$